# CHEMICAL APPLICATIONS OF
THERMAL NEUTRON SCATTERING

*United Kingdom Atomic Energy Authority Research Group*

# CHEMICAL APPLICATIONS OF THERMAL NEUTRON SCATTERING

Edited by
B. T. M. WILLIS

Harwell Series

OXFORD UNIVERSITY PRESS · 1973

*Oxford University Press, Ely House, London W.1*

Glasgow  New York  Toronto  Melbourne  Wellington
Cape Town  Ibadan  Nairobi  Dar es Salaam  Lusaka  Addis Ababa
Delhi  Bombay  Calcutta  Madras  Karachi  Lahore  Dacca
Kuala Lumpur  Singapore  Hong Kong  Tokyo

© Oxford University Press 1973

Printed in Great Britain by
William Clowes & Sons Limited
London, Colchester and Beccles

# *Preface*

Since the early 1950s, there has been intense activity in the application of neutron-beam methods to the study of the microscopic properties of condensed matter. Most of the earlier work was in the area of solid-state (or liquid-state) physics, but in the past few years there has been a remarkable extension into several branches of chemistry. This volume, which has developed from a series of lectures given at a Harwell Summer School, summarizes the progress to date in using a thermal neutron beam as a tool in chemistry.

Both inelastic and elastic scattering processes are discussed. From inelastic experiments, involving the measurement of the change in energy of neutrons on scattering, 'dynamical' properties, such as molecular energy levels in solids or atomic motion in liquids, can be examined. Elastic scattering experiments, in which no energy change takes place on scattering, give information about 'structural' properties, such as the arrangement of atoms or of atomic magnetic moments in the material.

Chapter 1 covers the basic theory of elastic and inelastic scattering of thermal neutrons; and Chapter 2 gives a brief account of the principal experimental methods. The remaining chapters describe the application of thermal neutron scattering to a wide range of topics, including molecular spectroscopy; structure and dynamics of polymers, glasses, and liquids; chemical bonding; and defects in non-stoichiometric compounds. Each topic is discussed at an introductory level with no specialized knowledge expected of the reader: those interested in pursuing a particular subject further can turn to the General References listed at the end of each chapter. The individual chapters are written by leading scientists with direct experience at Harwell of neutron-beam research; indeed, several authors have pioneered the application of thermal neutron scattering in their chosen branch of chemistry.

Apart from the illustrations supplied by authors, the sources of all diagrams are indicated in the legends, and I gratefully acknowledge the permission of the copyright holders. SI units are used throughout, although there are lapses into those old-fashioned units, the Ångström, the barn and the fermi. It is a pleasure to acknowledge help from Dr. G. C. Stirling and Professor Nico Norman in critically reading the MSS.

B. T. M. WILLIS

A.E.R.E.,
*Harwell.*
October, 1972

# Contents

| | |
|---|---|
| LIST OF IMPORTANT SYMBOLS | xii |
| UNITS AND PHYSICAL CONSTANTS | xv |
| | |
| 1. BASIC THEORY OF THERMAL NEUTRON SCATTERING BY CONDENSED MATTER: *C. G. Windsor* | 1 |
| 1.1. Introduction | 1 |
|    1.1.1. Scope | 1 |
|    1.1.2. Neutrons are marvellous! | 1 |
| 1.2. Scattering theory: elastic scattering | 4 |
|    1.2.1. Scattering from a single nucleus | 6 |
|    1.2.2. Scattering from many nuclei. Coherent and incoherent scattering | 7 |
| 1.3. Nuclear Bragg scattering | 9 |
|    1.3.1. Powder scattering | 12 |
|    1.3.2. Single-crystal scattering | 14 |
| 1.4. Scattering theory: inelastic scattering | 15 |
|    1.4.1. The eigenstate formulation | 15 |
|    1.4.2. The time-dependent formulation | 16 |
|    1.4.3. The space–time representation | 18 |
| 1.5. Neutron scattering and vibrational states of molecular crystals | 21 |
|    1.5.1. Vibrational states of a molecule | 21 |
|    1.5.2. Phonons in molecular crystals | 24 |
|    1.5.3. The coherent scattering from phonons | 26 |
|    1.5.4. Incoherent phonon scattering | 27 |
| 1.6. References | 29 |
| | |
| 2. EXPERIMENTAL TECHNIQUES: *G.C. Stirling* | 31 |
| 2.1. Introduction | 31 |
| 2.2. General features of neutron beam experiments | 31 |
|    2.2.1. Units | 31 |
|    2.2.2. Sources | 33 |
|    2.2.3. Energy (wavelength) selection | 35 |
|    2.2.4. Detectors | 39 |
|    2.2.5. Shielding | 40 |
|    2.2.6. Polarized neutrons | 41 |

## CONTENTS

| | |
|---|---|
| 2.3. Instruments for neutron elastic scattering | 41 |
|     2.3.1. Powder diffractometer | 41 |
|     2.3.2. Single-crystal diffractometer | 41 |
|     2.3.3. Diffuse scattering time-of-flight apparatus | 42 |
| 2.4. Instruments for neutron inelastic scattering | 43 |
|     2.4.1. Triple-axis spectrometer | 43 |
|     2.4.2. Beryllium-filtered-detector spectrometer | 45 |
|     2.4.3. Twin-chopper spectrometer | 45 |
|     2.4.4. Beryllium-filtered-beam spectrometer | 47 |
|     2.4.5. Rotating-crystal spectrometer | 47 |
| 2.5. References | 48 |
| **3. NEUTRON INELASTIC SCATTERING AND MOLECULAR SPECTROSCOPY:** *J. W. White* | 49 |
| 3.1. Introduction | 49 |
| 3.2. Molecular vibrations and the molecular force field | 50 |
| 3.3. Incoherent scattering cross-section for molecules | 51 |
| 3.4. Selection rules and the effects of molecular symmetry | 53 |
| 3.5. Aromatic molecular crystals and the harmonic approximation | 63 |
| 3.6. Isotope substitution method for vibrational assignments | 69 |
| 3.7. Lattice vibrations and the phonon density of states | 72 |
| 3.8. Conclusions | 76 |
| 3.9. References | 77 |
| **4. MODELS FOR CALCULATING THE PROPERTIES OF PHONONS IN MOLECULAR CRYSTALS:** *G. S. Pawley* | 78 |
| 4.1. Introduction | 78 |
| 4.2. Group theoretical assignments | 88 |
| 4.3. Model calculations | 93 |
| 4.4. References | 96 |
| **5. NEUTRON SCATTERING STUDIES OF THE DYNAMICS OF POLYMER CHAINS:** *G. Allen* | 97 |
| 5.1. Background | 97 |
|     5.1.1. Polymer molecules | 97 |
|     5.1.2. The states of polymerized matter | 99 |
| 5.2. The dynamics of the main polymer chains in crystals | 100 |
|     5.2.1. The normal modes of vibration of polymer chains in crystals | 101 |
|     5.2.2. Incoherent inelastic neutron scattering | 104 |
|     5.2.3. Experimental results from incoherent inelastic scattering | 104 |
|     5.2.4. Coherent inelastic neutron scattering | 107 |
| 5.3. Side-group motion in polymers | 109 |
| 5.4. Molecular motion in rubbers | 112 |
| 5.5. References | 117 |

## CONTENTS

6. **ATOMIC AND MOLECULAR MOTION IN LIQUIDS BY THERMAL NEUTRON SCATTERING:** *J. G. Powles* — 118
   - 6.1. Introduction to the liquid state — 118
   - 6.2. 'Incoherent' translational motion — 120
   - 6.3 The velocity auto-correlation function — 128
   - 6.4. 'Coherent' translational motion — 133
   - 6.5. Molecular liquids — 139
   - 6.6. References — 144

7. **STRUCTURE AND ATOMIC MOTION IN GLASSES:** *A. J. Leadbetter* — 146
   - 7.1. The glassy state — 146
   - 7.2. Structure — 148
   - 7.3. Atomic motions — 154
     - 7.3.1. General remarks — 154
     - 7.3.2. The frequency distribution—incoherent scattering and the incoherent approximation — 155
     - 7.3.3. Coherent scattering — 159
   - 7.4. References — 171

8. **THE STRUCTURE OF LIQUIDS BY NEUTRON SCATTERING:** *D. I. Page* — 173
   - 8.1. Introduction — 173
   - 8.2. The structure factor $S(Q)$ — 174
     - 8.2.1. Theory — 174
     - 8.2.2. Application to neutron scattering — 176
     - 8.2.3. Corrections to data — 178
   - 8.3. Monatomic liquids — 181
   - 8.4. Binary liquids — 185
   - 8.5. Molecular liquids — 193
     - 8.5.1. General — 193
     - 8.5.2. Tetrahedrally-bonded molecules — 194
     - 8.5.3. Water — 197
   - 8.6. References — 198

9. **HYDROGEN BONDING, AND SOME RESULTS OF ITS STUDY BY NEUTRON DIFFRACTION:** *J. C. Speakman* — 201
   - 9.1. Introduction — 201
   - 9.2. Crystallographic evidence for hydrogen bonding — 203
   - 9.3. The theory of hydrogen bonding — 205
   - 9.4. The application of neutron diffraction to crystal-structure analysis — 207
   - 9.5. A solution of the phase problem by neutron diffraction — 214
   - 9.6. The location of the hydrogen atom in hydrogen bonds — 216
   - 9.7. Ice — 218

CONTENTS

| | | |
|---|---|---|
| 9.8. | Symmetrical hydrogen bonds and their problems | 220 |
| 9.9. | References | 224 |

10. STRUCTURAL STUDIES ON NON-STOICHIOMETRIC COMPOUNDS BY THE BRAGG SCATTERING OF THERMAL NEUTRONS: *A. K. Cheetham*    225

| | | |
|---|---|---|
| 10.1. | Introduction | 225 |
| 10.2. | Theory | 230 |
| 10.3. | Achievement of required accuracy in measuring Bragg intensities | 233 |
| 10.4. | Applications | 235 |
| | 10.4.1. Anion-deficient compounds with the fluorite structure | 235 |
| | 10.4.2. Anion-excess compounds with the fluorite structure | 237 |
| | 10.4.3. Defective sodium-chloride structures | 242 |
| | 10.4.4. Non-stoichiometric hydrides | 246 |
| 10.5. | References | 247 |

11. DIFFUSE SCATTERING AND THE STUDY OF DEFECT SOLIDS: *B. E. F. Fender*    250

| | | |
|---|---|---|
| 11.1. | Introduction | 250 |
| 11.2. | Theory | 251 |
| | 11.2.1. Random solid solutions | 252 |
| | 11.2.2. Random vacancies and interstitials | 253 |
| | 11.2.3. Short-range ordering | 253 |
| 11.3. | Experimental | 257 |
| 11.4. | Results | 259 |
| | 11.4.1. Interstitial atoms | 259 |
| | 11.4.2. Vacancies | 261 |
| | 11.4.3. Interstitials and vacancies | 265 |
| 11.5. | References | 268 |

12. NEUTRON DIFFRACTION AND COVALENCY: *A. J. Jacobson*    270

| | | |
|---|---|---|
| 12.1. | Introduction | 270 |
| 12.2. | Covalency | 271 |
| | 12.2.1. Molecular orbitals for octahedral complexes | 271 |
| | 12.2.2. Magnetic form factors | 275 |
| 12.3. | Experimental methods | 278 |
| | 12.3.1. Polarized neutron experiments | 278 |
| | 12.3.2. Unpolarized neutron experiments | 285 |
| | 12.3.3. Measurement of paramagnetic scattering | 288 |
| 12.4. | Survey of the experimental results | 289 |
| | 12.4.1. Neutron measurements | 290 |
| | 12.4.2. Comparison with resonance data | 291 |
| 12.5. | References | 294 |

| | |
|---|---|
| **APPENDIX I.** Neutron scattering data for elements and isotopes | 296 |
| **APPENDIX II.** Properties of the $\delta$-function | 300 |
| **APPENDIX III.** Fourier transforms | 301 |
| **AUTHOR INDEX** | 303 |
| **SUBJECT INDEX** | 307 |

# List of Important Symbols

*(Numbers in brackets indicate equations in which symbol first appears)*

| | |
|---|---|
| $\mathbf{a}_1, \mathbf{a}_2, \mathbf{a}_3$ | basic vectors of unit cell (1.12a) |
| $b$ | scattering amplitude or scattering length (1.7) |
| $b_{coh}$ | coherent scattering length = $\langle b \rangle$ (1.9) |
| $b_{incoh}$ | incoherent scattering length = $(\langle b^2 \rangle - \langle b \rangle^2)^{\frac{1}{2}}$ (1.9) |
| $\mathbf{b}_1, \mathbf{b}_2, \mathbf{b}_3$ | basic vectors of reciprocal cell (1.12b) |
| $B_\kappa$ | temperature factor of atom $\kappa = 8\pi^2 \langle u_\kappa^2 \rangle$ (10.10) |
| $\dfrac{d\sigma}{d\Omega}$ | differential scattering cross-section (1.5) |
| $\dfrac{d^2\sigma}{d\Omega dE'}$ | partial differential cross-section = $\dfrac{1}{\hbar} \dfrac{d^2\sigma}{d\Omega\, d\omega}$ (1.18a) |
| $d$ | spacing of Bragg reflecting planes (1.17a) |
| $D$ | diffusion coefficient (6.3) |
| $D_r$ | rotational diffusion coefficient for a molecule (6.45) |
| $D(\mathbf{r})$ | magnetic moment density (12.9) |
| $E$ | initial energy (usually incident neutron energy) |
| $E'$ | final energy (usually scattered neutron energy) |
| $F, F(\mathbf{Q}), F(\mathbf{h}), F_{hkl}$ | structure factor of chemical unit cell (1.16) |
| $F_{hkl}^{mag}$ | structure factor of magnetic unit cell (12.21) |
| $f$ | force constant (3.2) |
| $f(\mathbf{Q})$ | magnetic form factor (12.9) |
| $f_a(\mathbf{Q})$ | contribution of cation to $f(\mathbf{Q})$ (12.16) |
| $f_b(\mathbf{Q})$ | overlap contribution to $f(\mathbf{Q})$ (12.17) |
| $f_c(\mathbf{Q})$ | contribution of ligand to $f(\mathbf{Q})$ (12.18) |
| $G(\mathbf{r}, t)$ | (time-dependent) pair correlation function (1.21a) |
| $G_s(\mathbf{r}, t)$ | self pair correlation function (1.21a) |
| $G_d(\mathbf{r}, t)$ | distinct correlation function = $G(\mathbf{r}, t) - G_s(\mathbf{r}, t)$ (1.27) |
| $g(r)$ | (static) pair distribution function (6.39) |
| $g_e(r)$ | pair distribution function for equilibrium atomic positions (7.4) |
| $h$ | Planck's constant |
| $\hbar$ | Planck's constant $\div 2\pi$ (1.1) |
| $\mathbf{h}$ | |
| $hkl$ | Miller indices (1.14) |
| $h_1 h_2 h_3$ | |
| $\mathscr{H}$ | Hamiltonian (1.19) |
| $I_{hkl}$ | integrated intensity from $hkl$ reflecting plane (1.17b) |
| $I(\mathbf{Q}, t)$ | intermediate scattering function (6.24) |
| $j$ | label for branch of phonon dispersion relation (1.33) |
| $\mathbf{k}$ | wave vector of incident neutron (1.2) |
| $k$ | wave number of incident neutron = $2\pi/\lambda$ (1.3) |
| $\mathbf{k}'$ | wave vector of scattered neutron (1.2) |
| $k'$ | wave number of scattered neutron (1.3) |
| $k_B$ | Boltzmann's constant |

# LIST OF IMPORTANT SYMBOLS

| | |
|---|---|
| $m_n$ | mass of neutron (1.1) |
| $M$ | mass of atom (1.33) |
| $m_\kappa$ | occupancy number of $\kappa$th site in average cell (10.7) |
| $N$ | number of atoms (nuclei) of same type (1.9) |
| $\mathbf{n}$ | position vector of point in direct lattice (1.11) |
| $N_1, N_2, N_3$ | number of unit cells along $\mathbf{a}_1, \mathbf{a}_2, \mathbf{a}_3$ axes (1.14) |
| $N_o$ | total number of unit cells = $N_1 N_2 N_3$ (1.15) |
| $N_c$ | number of unit cells per unit volume (1.17b) |
| $n_s$ | population factor for phonons (1.33a) |
| $n_0$ | number of neutrons per unit volume (2.3) |
| $p$ | magnetic scattering amplitude of atom (12.21) |
| $\mathbf{Q}$ | scattering vector = $\mathbf{k} - \mathbf{k}'$ (1.4b) |
| $Q$ | magnitude of scattering vector = $4\pi \sin\theta/\lambda$ for elastic scattering (1.6) |
| $\mathbf{q}$ | wave vector of phonon (1.32) |
| $q$ | wave number of phonon |
| $q_{max}$ | phonon wave number at boundary of Brillouin zone |
| $q'$ | reduced wave number of phonon = $q/q_{max}$ |
| $q_i$ | mass-weighted Cartesian displacement co-ordinates (3.1) |
| $Q_i$ | normal co-ordinates (5.2) |
| $\mathbf{r}$ | general position vector (1.1) |
| $\mathbf{r}'$ | position vector of a nucleus (1.2) |
| $\mathbf{R}$ | position vector of atom in a crystal (= $\mathbf{n} + \boldsymbol{\rho}$) (1.11) |
| $S(\mathbf{Q}, \omega)$ | coherent scattering law (1.25a) |
| $S_s(\mathbf{Q}, \omega)$ | incoherent scattering law (1.25b) |
| $S_o(\mathbf{Q}, \omega)$ | classical coherent scattering law (1.26b) |
| $\tilde{S}(\mathbf{Q}, \omega)$ | symmetrical scattering law |
| $S(\mathbf{Q})$ | coherent structure factor = $\int S(\mathbf{Q}, \omega) d\omega$ (6.38) |
| $S_{pq}(\mathbf{Q})$ | partial structure factor (8.17) |
| $S$ | spin quantum number of atom (12.21) |
| $t$ | time variable (1.18b) |
| $T$ | absolute temperature (1.17c) |
| $\langle u^2 \rangle$ | mean-square amplitude of vibration of atom (1.35) |
| $\mathbf{u}$ | thermal displacement vector of atom (1.28) |
| $\mathbf{U}(\mathbf{q})$ | polarization vector for phonons (1.33a) |
| $v$ | volume of unit cell (1.15) |
| $v_n$ | speed of neutron (2.1) |
| $v_s$ | speed of sound |
| $W$ | exponent of Debye–Waller factor (1.30) |
| $w(t)$ | width function for diffusion (6.23) |
| $Z(\omega)$ | density of phonon states |
| $\alpha, \alpha(\mathbf{h})$ | phase angle of Bragg reflexion $\mathbf{h}$ (10.13) |
| $\gamma$ | magnetic moment of neutron in nuclear magnetons (12.21) |
| $\delta_{ij}$ | Kronecker delta function, defined by $\delta_{ij} = 1, i = j$; $= 0, i \neq j$ (1.13) |
| $\delta(x)$ | Dirac delta function (1.15) |
| $\eta$ | magnetization vector |
| $2\theta$ | scattering angle (1.6) |

## LIST OF IMPORTANT SYMBOLS

| | |
|---|---|
| $\theta_B, \theta_{hkl}$ | Bragg angle for sample |
| $\theta_M$ | Bragg angle for monochromator |
| $\theta_A$ | Bragg angle for analyser |
| $\Theta(x)$ | amplitude of displacement of molecule |
| $\lambda$ | wavelength (1.6) |
| $\boldsymbol{\lambda}$ | polarization vector for neutron |
| $\mu_N$ | nuclear magneton |
| $\mu_B$ | Bohr magneton |
| $\nu$ | frequency = $\omega/2\pi$ (5.3) |
| $\boldsymbol{\rho}$ | position vector of atom in unit cell (1.11) |
| $\rho$ | number density of atoms (6.40) |
| $\rho(xyz)$ | nuclear density at point in unit cell with fractional co-ordinates $xyz$ (10.12) |
| $\sigma_{tot}$ | total scattering cross-section = $\sigma_{coh} + \sigma_{incoh}$ (1.7) |
| $\sigma_{coh}$ | coherent cross-section (1.10b) |
| $\sigma_{incoh}$ | incoherent cross-section (1.10c) |
| $\sigma$ | one-phonon cross-section (1.31) |
| $\sigma_{abs}$ | absorption cross-section |
| $\boldsymbol{\tau}$ | reciprocal lattice vector (1.15) |
| $\tau$ | time-of-flight of neutron (2.2) |
| $\phi$ | angle between scattering vector (**Q**) and polarization vector (**U**) (7.16) |
| $\omega$ | vibrational (circular) frequency. $\hbar\omega = E - E'$ is the energy transferred to the neutron in a scattering process (1.18a) |
| $\omega_j(\mathbf{q})$ | frequency of phonon of wave vector **q** and dispersion branch $j$ (1.33a) |
| $\Omega$ | solid angle (1.5) |

# Units and Physical Constants

## Units and Physical Constants

| Unit | Symbol or Abbreviation | Value in SI units | Value in CGS units |
|---|---|---|---|
| Ångström | Å | $10^{-1}$ nm | $10^{-8}$ cm |
| Barn | b | $10^{-28}$ m$^2$ | $10^{-24}$ cm$^2$ |
| Fermi | fm | $10^{-15}$ m | $10^{-13}$ cm |
| Electron volt | eV | $1.602 \cdot 10^{-19}$ J | $1.602 \cdot 10^{-12}$ erg |
| Optical wave number | cm$^{-1}$ | $1.985 \cdot 10^{-23}$ J | $1.985 \cdot 10^{-16}$ erg |

10 meV are equivalent to:

| | | | |
|---|---|---|---|
| Neutron speed | | $1.38 \cdot 10^3$ m s$^{-1}$ | $1.38 \cdot 10^5$ cm s$^{-1}$ |
| Neutron time-of-flight | | 724 $\mu$s m$^{-1}$ | 7.24 $\mu$s cm$^{-1}$ |
| Optical wave number | | — | 80.7 cm$^{-1}$ |
| Temperature | | 116 K | — |
| Angular frequency | | $1.52 \cdot 10^{13}$ Hz | — |

| Physical constant | Symbol or Abbreviation | Value in SI Units | Value in CGS Units |
|---|---|---|---|
| Neutron rest mass | $m_n$ | $1.673 \cdot 10^{-27}$ kg | $1.673 \cdot 10^{-24}$ g |
| Electron rest mass | $m$ | $9.110 \cdot 10^{-31}$ kg | $9.110 \cdot 10^{-28}$ g |
| Planck's constant | $h$ | $6.626 \cdot 10^{-34}$ J s | $6.626 \cdot 10^{-27}$ erg s |
| | $\hbar$ | $1.055 \cdot 10^{-34}$ J s | $1.055 \cdot 10^{-27}$ erg s |
| Boltzmann's constant | $k_B$ | $1.381 \cdot 10^{-23}$ K$^{-1}$ | $1.381 \cdot 10^{-16}$ erg K$^{-1}$ |
| Charge on proton | $e$ | $1.602 \cdot 10^{-19}$ C | $4.803 \cdot 10^{-10}$ esu |
| Speed of light in vacuum | $c$ | $2.998 \cdot 10^8$ m s$^{-1}$ | $2.998 \cdot 10^{10}$ cm s$^{-1}$ |
| Bohr magneton | $\mu_B$ | $9.274 \cdot 10^{-24}$ J T$^{-1}$ | $9.274 \cdot 10^{-21}$ erg G$^{-1}$ |
| Nuclear magneton | $\mu_N$ | $5.050 \cdot 10^{-27}$ J T$^{-1}$ | $5.050 \cdot 10^{-24}$ erg G$^{-1}$ |
| Magnetic moment of neutron | — | $-9.65 \cdot 10^{-27}$ J T$^{-1}$ | $-9.65 \cdot 10^{-24}$ erg G$^{-1}$ |

# 1 Basic Theory of Thermal Neutron Scattering by Condensed Matter

*By* C. G. WINDSOR

*Atomic Energy Research Establishment, Harwell*


## 1.1. Introduction

*1.1.1. Scope*

This chapter has been written for the newcomer to the various techniques using the scattering of thermal neutrons. Its aim is to provide him with the theoretical background necessary for following the other chapters, which deal principally with the practical application of neutron scattering to the solution of problems in chemistry.

For a more general and rigorous account of all aspects of the theory of thermal neutron scattering the reader is referred to the excellent book by Marshall and Lovesey (1971).


*1.1.2. Neutrons are marvellous!*

Thermal neutrons are a unique probe for studying both the structure and the dynamics of the condensed state. To examine the structure of matter, we need a scattering process in which the radiation has a wavelength comparable in magnitude with the spacing between atoms, whereas to investigate the dynamics of matter, the scattered radiation must have a frequency comparable with the vibrational frequencies of the atoms in the sample. X-rays meet the first requirement and infra-red radiation the second. Thermal neutrons simultaneously satisfy *both* requirements. Thus thermal neutron scattering by atomic nuclei manages to combine, in the same experiment, the spatial measurements that can be made using X-ray diffraction with the vibrational measurements obtained from infra-red spectroscopy.

There is a second reason for the uniqueness of the interaction between neutrons and nuclei. The amplitude of the neutron wave scattered by an individual nucleus depends on the nature of the nucleus, and so varies between different isotopes of the nucleus and between the two spin states of the neutron–nucleus system. This variation of the scattering amplitude from atom to atom of the same species gives rise to a process known as incoherent scattering, which is characterized by the absence of interference between the waves scattered by different atoms. (The nearest counterpart to incoherent scattering in the X-ray field is the X-ray scattering from a completely disordered alloy containing several chemical species.) The cross-section consists, therefore, of a coherent part, with interference occurring between the scattered waves, and an incoherent part. The magnitude of each part can be altered by changing the isotopic composition of the sample; each can be separately measured and each gives quite different information.

The experimental neutron-scattering cross-section can be subdivided further into an elastic cross-section, involving no transfer of energy from the radiation to the sample, and an inelastic cross-section, with energy transfer. The coherent elastic-scattering cross-section gives, in common with X-ray diffraction, a time-averaged view of the positions of the atoms in an ordered structure. For this type of study, neutrons have an advantage over X-rays in showing clearly the positions of hydrogen atoms, which scatter X-rays only weakly. Inelastic scattering gives us complete information on the vibrational states of atoms or molecules. For materials in single-crystal form, neutrons can excite only those vibrational modes possessing a given frequency and wave vector. The dispersion relations, giving the dependence of frequency on wave vector, can be measured by coherent scattering experiments on such materials: it is then possible, with the aid of a model, to work out the forces between all pairs of atoms when they are displaced from their equilibrium positions.† More restricted information— the simple spectrum of vibrational states—is provided by incoherent neutron scattering experiments and also by infra-red spectroscopy. To obtain the entire spectrum by optical means, it is necessary to consider the overtones and combination bands, where the normal selection rules governing the coupling of the dipole-moment fluctuations with the radiation no longer hold. There are no analogous selection rules for inelastic neutron scattering. In the case of a molecular crystal, we may study by neutron spectroscopy not only the internal vibrational modes of the molecule, but also the vibrational modes of the lattice and the interactions between the lattice and internal modes.

For a liquid, there is an embarrassing richness of information provided by thermal neutron scattering (see Chapter 6). A liquid is without long-range order, but the atomic positions are still highly correlated over short distances and over a time scale between $10^{-14}$ and $10^{-11}$ s, so that both vibrational and diffusive motions can be observed. We owe to Van Hove (1954) the formalism for interpreting the complicated scattering patterns observed from liquids in terms of space-time correlation functions between pairs of atoms. The classical meaning of such functions is the probability that, if there is an atom at position $r_0$ at time $t_0$, there will be another atom, or possibly the same atom, at position $r_0 + r$ at time $t_0 + t$. The coherent scattering gives us the sum of the *self* correlation function (referring to the motion of the same atom) and the *distinct* correlation function, whereas the incoherent scattering gives the self correlation function alone. No other type of scattering experiment gives a knowledge of these functions. Even the simplest of liquids, such as monatomic liquid argon,

† Leigh *et al.* (1971) have shown that the interatomic forces are not *uniquely* determined when the frequencies only of the vibrations are known; a unique solution requires the measurement of the polarization vectors (see p. 26) as well as the dispersion relations from the neutron experiment.

give neutron scattering results that are new and unexpected (see p. 133); molecular liquids present a largely unexplored field. Molecular systems can give rise to two new types of ordered phases between the perfect solid and liquid phases. These are 'liquid crystals' in which the orientations of the molecules in space are fixed while their positions are liquid-like, and 'plastic crystals', where the positions are solid-like but the molecular orientations are disordered. We can expect, in the near future, studies of both types of crystal by neutron scattering.

If the atoms contain unpaired electron spins, there is additional magnetic scattering whose intensity is often comparable with that scattered by the nucleus. Magnetic scattering in itself constitutes a large subject, employing concepts that are analogous to many of those we shall use for nuclear scattering. Thus elastic magnetic scattering gives the spin ordering pattern (the 'magnetic structure') of a magnetic crystal, while inelastic magnetic scattering leads to a determination of the vibrational spin-wave modes and, from these, the interactions between the spin vectors. Covalency in magnetic systems can be studied by observing the spatial distribution of the unpaired spin density in the magnetic unit cell—this is the only aspect of magnetic scattering dealt with in this book, and an outline of the basic theory is presented in Chapter 12.

In this introductory chapter we shall discuss the scattering from atomic nuclei. Our main concern will be with nuclear scattering from solids, and we shall use crystalline naphthalene as a typical chemical system illustrating a number of the points that emerge. The crystal structure of this molecular compound is shown in Fig. 1.1. Only brief mention will be made of the nuclear scattering from liquids, as the basic principles are covered in Chapters 6 and 8.

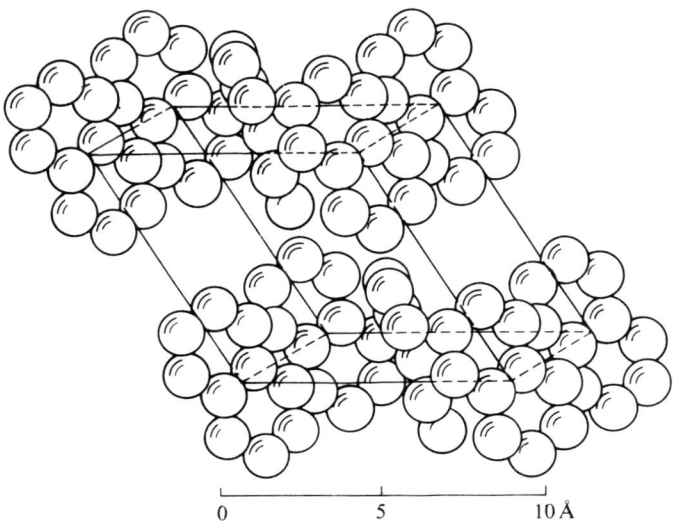

FIG. 1.1. Positions of the molecules in the unit cell of naphthalene. For clarity, carbon atoms only are shown. (After Abrahams, Robertson, and White 1949.)

## 1.2. Scattering theory: elastic scattering

Let us consider the scattering of a neutron beam by a particle *fixed* at $\mathbf{r}'$ (see Fig. 1.2). The incident beam is represented by the plane wave $\exp(i\mathbf{k}\cdot\mathbf{r})$, where $\mathbf{r}$ is the position vector and $\mathbf{k}$ the wave vector. The magnitude of $\mathbf{k}$ is $2\pi/\lambda$, with $\lambda$ the de Broglie wavelength. Since $\mathbf{k}$ is a vector and $\lambda$ a scalar, we shall use the wave vector of the neutron beam in preference to its wavelength.

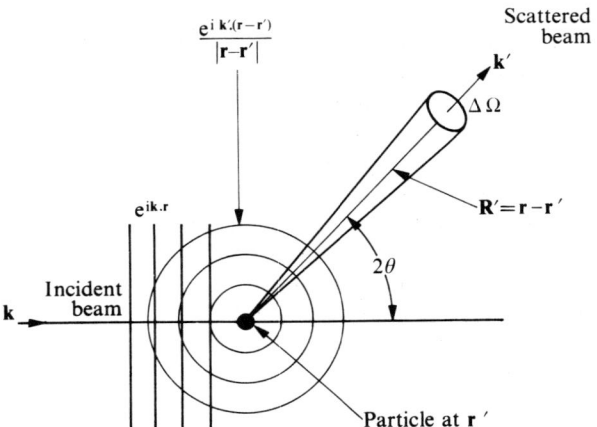

FIG. 1.2. Scattering of plane wave by spherically symmetrical potential field.

$\mathbf{k}$ is related to the neutron velocity $\mathbf{v}_n$ by

$$\hbar\mathbf{k} = m_n\mathbf{v}_n$$

($\hbar$ is $h/2\pi$ and $m_n$ the neutron mass), and to the neutron energy $E$ by

$$E = \tfrac{1}{2}m_n v_n^2 = \hbar^2 k^2/(2m_n).$$

Measuring $k$ in $\text{Å}^{-1}$, $v_n$ in m s$^{-1}$ and $E$ in cm$^{-1}$, these expressions become

$$k = 1588\cdot 3\, v_n,$$

and

$$E = 4\cdot 2163\cdot 10^7 v_n^2 = 1\cdot 6714\cdot 10^5 k^2.$$

Thus for a neutron beam thermalized to a moderator temperature $T_{\text{mod}} = 40°\text{C}$, $E \approx k_B T_{\text{mod}} = 218$ cm$^{-1}$, and $k = 3\cdot 61 \text{Å}^{-1}$.

The wave function for the scattered neutrons can be written as a spherical wave emanating from the point $\mathbf{r}'$. It is shown in standard textbooks (e.g. Schiff 1955) that the solution of the Schrödinger wave equation

$$-\frac{\hbar^2}{2m_n}\nabla^2\psi(\mathbf{r}) + V(\mathbf{r})\psi(\mathbf{r}) = E\psi(\mathbf{r}) \tag{1.1}$$

for a spherically symmetrical scattering potential $V(r)$ and for large distances $|\mathbf{r} - \mathbf{r}'|$ is

$$\psi(\mathbf{r}) = e^{i\mathbf{k}\cdot\mathbf{r}} - \frac{1}{4\pi} \int \frac{e^{i\mathbf{k}'\cdot(\mathbf{r}-\mathbf{r}')}}{|\mathbf{r}-\mathbf{r}'|} \frac{2m_n}{\hbar^2} V(\mathbf{r}')\psi(\mathbf{r}')d\mathbf{r}'. \tag{1.2}$$

The second term in eqn (1.2) represents a superposition of all the spherical waves of wave vector $\mathbf{k}'$ scattered from a source of strength $(2m_n/\hbar^2)V(\mathbf{r}')\psi(\mathbf{r}')$ at $\mathbf{r}'$; $d\mathbf{r}'$ is a volume element in $\mathbf{r}'$ space ($d\mathbf{r}' = d^3 r'$). Putting the wave function $\psi(\mathbf{r}')$ in this term equal to that of the incident beam gives the first Born approximation, when (with $\mathbf{R}' = \mathbf{r} - \mathbf{r}'$) we have

$$\psi(\mathbf{r}) = \exp(i\mathbf{k}\cdot\mathbf{r}) - \frac{1}{4\pi} \int \frac{\exp[i\mathbf{k}'\cdot(\mathbf{R}'+\mathbf{r}')]}{|\mathbf{R}'|} \exp(i(\mathbf{k}-\mathbf{k}')\cdot\mathbf{r}')$$
$$\times \frac{2m_n}{\hbar^2} V(\mathbf{r}')d\mathbf{r}'. \tag{1.3}$$

Now if there are $n_0$ neutrons incident on unit area in unit time, and if $d\Omega$ is an element of solid angle in which the number of neutrons scattered from $\mathbf{r}'$ is counted, then one expects that this number is proportional both to $n_0$ and to $d\Omega$. The factor of proportionality is called the differential scattering cross-section and is denoted $d\sigma/d\Omega$. In our case, the plane wave term $\exp(i\mathbf{k}\cdot\mathbf{r})$ represents a wave of unit density and velocity $\hbar k/m_n$, so that $n_0$ is $\hbar k/m_n$. The spherically-scattered wave, $f\exp(i\mathbf{k}'\cdot\mathbf{R}')/R'$, represents a wave of density $|f|^2/R'^2$ and velocity $\hbar k'/m_n$, where the amplitude $f$, from eqn (1.2), is

$$f = -\frac{1}{4\pi} \int \exp[i(\mathbf{k}-\mathbf{k}')\cdot\mathbf{r}'] \frac{2m_n}{\hbar^2} V(\mathbf{r}')d\mathbf{r}'. \tag{1.4a}$$

The number of scattered neutrons per second crossing the area $R'^2 d\Omega$ in the solid angle $d\Omega$ is

$$\frac{|f|^2}{R'^2} \cdot \frac{\hbar k'}{m_n} \cdot R'^2 d\Omega,$$

and so,

$$\frac{d\sigma}{d\Omega} = \frac{1}{N} \cdot \frac{|f|^2}{R'^2} \cdot \frac{\hbar k'}{m_n} \cdot R'^2$$

$$= \frac{k'}{k} \left| \frac{1}{4\pi} \int \exp[i(\mathbf{k}-\mathbf{k}')\cdot\mathbf{r}'] \frac{2m_n}{\hbar^2} V(\mathbf{r}')d\mathbf{r}' \right|^2.$$

The scattering is elastic for fixed nuclei, i.e. $k = k'$. Writing

$$\mathbf{Q} = \mathbf{k} - \mathbf{k}', \tag{1.4b}$$

the differential cross-section is

$$\frac{d\sigma}{d\Omega} = \left| \frac{1}{4\pi} \int \exp(i\mathbf{Q}\cdot\mathbf{r}') \frac{2m_n}{\hbar^2} V(\mathbf{r}')d\mathbf{r}' \right|^2. \tag{1.5}$$

## BASIC THEORY

The vector **Q** is known as the scattering vector and its magnitude depends on the scattering angle $2\theta$ through the relation (see Fig. 1.3),

$$Q = 2k \sin \theta = 4\pi (\sin \theta)/\lambda. \tag{1.6}$$

The expressions (1.3) or (1.5) show that the amplitude of scattering is proportional to the Fourier transform of the scattering potential $V(\mathbf{r})$, just as

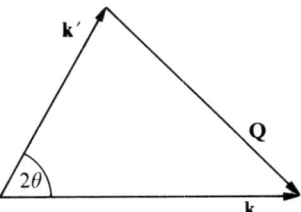

FIG. 1.3. Relation between wave vectors k for the incident neutron and k' for the scattered neutron. $2\theta$ is the scattering angle, and $k' = k = 2\pi/\lambda$ for elastic scattering.

in X-ray diffraction the amplitude scattered coherently by a single atom is proportional to the Fourier transform of the charge density of the atom.

### 1.2.1. Scattering from a single nucleus

$V(\mathbf{r})$ is the interaction potential between the neutron and the atomic nucleus. This is a short-range potential, restricted to nuclear dimensions, which are of the order of $10^{-14}$ m. For thermal neutrons the magnitude of **Q** is a few Å$^{-1}$, so that the $\exp(i\mathbf{Q}\cdot\mathbf{r}')$ term in eqn (1.4a) has hardly begun to deviate from unity before the scattering potential has declined to nothing. The scattering amplitude is therefore

$$f = -\frac{1}{4\pi} \int_{\text{nucleus}} \frac{2m_n}{\hbar^2} V(\mathbf{r})\, d\mathbf{r},$$

which is a constant, independent of **Q**. The quantity $-f$ is called the scattering length of the nucleus, and is denoted by $b$; $b$ is usually positive, although it can be negative, or even complex. It varies from nucleus to nucleus, in a seemingly random way determined by the nuclear structure, and for a given atom it differs from one isotope to another. Since $b$ is independent of **Q**, the differential cross-section $d\sigma/d\Omega$ is isotropic for a single nucleus, and the total cross-section is immediately given by integrating over all solid angles,

$$\sigma_{\text{tot}} = \int \frac{d\sigma}{d\Omega}\, d\Omega = 4\pi b^2. \tag{1.7}$$

Lists of scattering cross-sections for the nuclides are given in *Neutron Cross-sections* (Hughes and Schwartz 1958). Note that in neutron scattering studies samples of macroscopic dimensions are used, so that the relevant cross-sections are usually the bound-atom values. In our discussion above we assumed that the scattering atom was fixed at **r'**, but for the free-atom treatment it is necessary to use centre-of-mass coordinates, replacing $m_n$ in the Schrödinger wave eqn (1.1) by the reduced mass

$$\mu = m_n \frac{A}{A+1},$$

where $A$ is the ratio of the nuclear and neutron masses. The free-atom scattering length $a$ is related to the bound-atom scattering length $b$ by

$$b = \frac{A+1}{A} a.$$

*1.2.2. Scattering from many nuclei. Coherent and incoherent scattering*

The amplitude scattered by a single nucleus at **R** is $b_\mathbf{R}$, where the subscript **R** serves to remind us that the scattering length, $b$, varies irregularly from isotope to isotope, and, indeed, from one nucleus to another of the *same* isotope if there is a non-zero nuclear spin. The phase of the scattered radiation is $\exp(i\mathbf{Q}\cdot\mathbf{R})$, so that the differential scattering cross-section for an assembly of many nuclei, all assumed to be stationary, is

$$\frac{d\sigma}{d\Omega} = \left| \sum_\mathbf{R} b_\mathbf{R} \exp(i\mathbf{Q}\cdot\mathbf{R}) \right|^2 = \sum_{\mathbf{R},\mathbf{R}'} b_\mathbf{R} b_{\mathbf{R}'} \exp[i\mathbf{Q}\cdot(\mathbf{R}-\mathbf{R}')]$$

$$= \sum_\mathbf{R} b_\mathbf{R}^2 + {\sum_{\mathbf{R},\mathbf{R}'}}' b_\mathbf{R} b_{\mathbf{R}'} \exp[i\mathbf{Q}\cdot(\mathbf{R}-\mathbf{R}')], \qquad (1.8)$$

where $\Sigma'$ indicates a summation not including $\mathbf{R} = \mathbf{R}'$. The first term in eqn (1.8) is $N\langle b_\mathbf{R}^2\rangle$, where $N$ is the number of nuclei and the brackets $\langle\ \rangle$ denote the average value. There is no correlation between $b_\mathbf{R}$ and $b_{\mathbf{R}'}$ and so $\langle b_\mathbf{R} b_{\mathbf{R}'}\rangle = \langle b_\mathbf{R}\rangle\langle b_{\mathbf{R}'}\rangle = \langle b_\mathbf{R}\rangle^2$, and the second term can be written

$$N\langle b_\mathbf{R}\rangle^2 {\sum}' \exp[i\mathbf{Q}\cdot(\mathbf{R}-\mathbf{R}')]$$

$$= -N\langle b_\mathbf{R}\rangle^2 + N\langle b_\mathbf{R}\rangle^2 \sum \exp[i\mathbf{Q}\cdot(\mathbf{R}-\mathbf{R}')],$$

where we have restored $\mathbf{R} = \mathbf{R}'$ in the summation $\Sigma$. Thus finally

$$\frac{d\sigma}{d\Omega} = N(\langle b^2\rangle - \langle b\rangle^2) + N\langle b\rangle^2 \left|\sum \exp(i\mathbf{Q}\cdot\mathbf{R})\right|^2. \qquad (1.9)$$

The term $N(\langle b^2\rangle - \langle b\rangle^2)$ in eqn (1.9) can also be written as $N\langle(b-\langle b\rangle)^2\rangle$, showing that it depends on the mean-square deviation of the scattering lengths from their average value. It is known as the incoherent scattering cross-section, and is zero for nuclides of zero spin (e.g. $^{12}C$, $^{16}O$). If we ignore the effect of the Debye–Waller factor, the incoherent elastic scattering is isotropic and gives a

## BASIC THEORY

uniform background to the coherent scattering, represented by the remaining term

$$N\langle b \rangle^2 \left| \sum_{\mathbf{R}} \exp(i\mathbf{Q}.\mathbf{R}) \right|^2$$

in eqn (1.9). Note that the coherent scattering cross-section takes into account interference effects, arising from the relative displacements of the nuclei in the assembly. The incoherent scattering cross-section has no phase term and so gives us no information about relationships between different atoms.

The distinction between coherent and incoherent scattering can be expressed in an equivalent way, as follows. Let $b_\mathbf{R}$, the scattering amplitude for the nucleus at $\mathbf{R}$, be written

$$b_\mathbf{R} = \langle b \rangle + \Delta b_\mathbf{R}, \tag{1.10a}$$

where $\langle b \rangle$ is the amplitude averaged over all $\mathbf{R} \left( = \dfrac{1}{N} \sum_\mathbf{R} b_\mathbf{R} \right)$ and $\Delta b_\mathbf{R}$ is the deviation from the average due to isotopic and spin effects. Then the intensity scattered from the whole assembly of nuclei can be considered as the sum of two parts, the first depending on the square of $\langle b \rangle$ and the second on the mean value of the square of the deviation from the average, $\Delta b_\mathbf{R}$. The first part is the coherent scattering, as it involves the phase difference between the radiation scattered by different nuclei, and the second part is the incoherent scattering.

From eqn (1.7) we have that the total scattering cross-section per atom is

$$\sigma_{tot} = 4\pi \langle b^2 \rangle.$$

The 'coherent scattering cross-section per atom' is

$$\sigma_{coh} = 4\pi \langle b \rangle^2 \tag{1.10b}$$

and the difference between the two is the 'incoherent scattering cross-section per atom'

$$\sigma_{incoh} = 4\pi(\langle b^2 \rangle - \langle b \rangle^2). \tag{1.10c}$$

Table 1.1 lists a few examples of the cross-sections for commonly occurring atoms.† The units are barns, or $10^{-28}$ m$^2$ atom$^{-1}$. Carbon and oxygen are coherent scatterers, as they each contain predominantly one isotope with zero nuclear spin. By contrast, hydrogen has an anomalously large incoherent scattering cross-section of 80 b, because the non-zero nuclear spin ($=\frac{1}{2}$) gives rise to two scattering amplitudes of opposite sign with a weighted average close to zero. Deuteration changes the scattering from primarily incoherent to primarily coherent. Structural investigations of hydrogenous compounds are made easier by substituting deuterium for hydrogen, thus removing the uniform incoherent background. In inelastic scattering from hydrogenous compounds, the large incoherent cross-section may be an advantage as it enables the molecular

† A complete list for all elements of the periodic table is given in Appendix I (p. 296).

vibrations involving the motion of hydrogen atoms to be readily identified in the vibration spectrum (see Chapter 3).

TABLE 1.1

*Neutron cross-sections of light atoms* (b)

|           | $\sigma_{tot}$ | $\sigma_{coh}$ | $\sigma_{incoh}$ |
|-----------|------|------|------|
| Hydrogen  | 81·5 | 1·8  | 79·7 |
| Deuterium | 7·6  | 5·6  | 2·0  |
| Carbon    | 5·6  | 5·6  | 0·0  |
| Nitrogen  | 11·4 | 11·1 | 0·3  |
| Oxygen    | 4·2  | 4·2  | 0·0  |

## 1.3. Nuclear Bragg scattering

We now consider the scattering from a crystal in which the atoms lie in a regular (periodic), three-dimensional array. The atomic position vector **R** is

$$\mathbf{R} = \mathbf{n} + \boldsymbol{\rho},$$

where **n** is a lattice vector referring to a particular unit cell and $\boldsymbol{\rho}$ is the position within the cell (see Fig. 1.4). The coherent scattering amplitude $b_\mathbf{R}^{coh}$ varies from

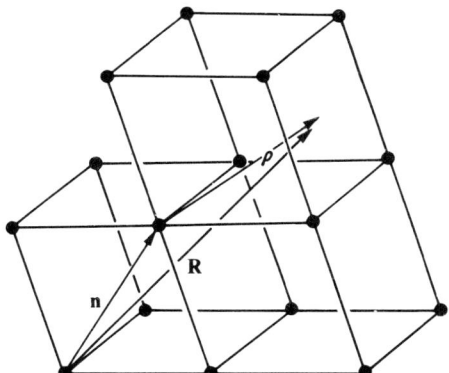

FIG. 1.4. Definition of vectors **R**, **n**, $\rho$.

one atomic species to another in the unit cell, but not between cells. The coherent cross-section is therefore

$$\left(\frac{d\sigma}{d\Omega}\right)_{coh} = \left|\sum_\mathbf{R} b_\mathbf{R}^{coh} e^{i\mathbf{Q}\cdot\mathbf{R}}\right|^2$$

$$= \left|\sum_\mathbf{n} \exp(i\mathbf{Q}\cdot\mathbf{n}) \sum_{\boldsymbol{\rho}} b_{\boldsymbol{\rho}}^{coh} e^{i\mathbf{Q}\cdot\boldsymbol{\rho}}\right|^2$$

$$= \left|\sum_\mathbf{n} \exp(i\mathbf{Q}\cdot\mathbf{n})\right|^2 \times \left|\sum_{\boldsymbol{\rho}} b_{\boldsymbol{\rho}}^{coh} e^{i\mathbf{Q}\cdot\boldsymbol{\rho}}\right|^2. \quad (1.11)$$

BASIC THEORY

If $a_1, a_2, a_3$ are the basic vectors of the unit cell,

$$n = n_1 a_1 + n_2 a_2 + n_3 a_3, \tag{1.12a}$$

where $n_1, n_2, n_3$ are integers. It is convenient to refer $Q$ to a set of axes reciprocal to those used for n:

$$b_1 = \frac{2\pi\, a_2 \times a_3}{a_1 . (a_2 \times a_3)}, b_2 = \frac{2\pi\, a_3 \times a_1}{a_1 . (a_2 \times a_3)}, b_3 = \frac{2\pi\, a_1 \times a_2}{a_1 . (a_2 \times a_3)}. \tag{1.12b}$$

$b_1, b_2, b_3$ are the basic vectors of the reciprocal cell and are related to $a_1, a_2, a_3$ by

$$a_i . b_j = 2\pi \delta_{ij} \tag{1.13}$$

with $\delta_{ij}$ the Kronecker delta.† Fig. 1.5 shows the direct unit cell and the reciprocal cell for crystalline naphthalene.

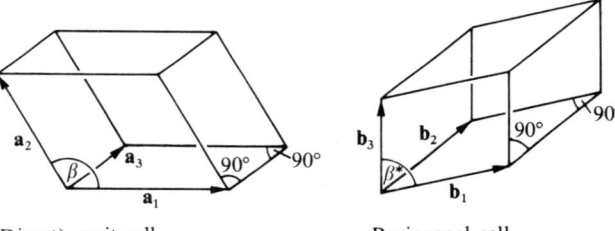

(Direct) unit cell
$a_1 = 8.27$ Å, $a_2 = 5.97$ Å,
$a_3 = 8.67$ Å, $\beta = 123°$

Reciprocal cell
$b_1 = 0.76$ Å$^{-1}$, $b_2 = 1.25$ Å$^{-1}$,
$b_3 = 0.86$ Å$^{-1}$, $\beta^* = 57°$

FIG. 1.5. Unit cell and reciprocal cell of naphthalene.

Writing

$$Q = h_1 b_1 + h_2 b_2 + h_3 b_3,$$

the term

$$\left| \sum_n \exp(iQ . n) \right|^2$$

in eqn (1.11) reduces to

$$\frac{\sin^2 \pi N_1 h_1}{\sin^2 \pi h_1} . \frac{\sin^2 \pi N_2 h_2}{\sin^2 \pi h_2} . \frac{\sin^2 \pi N_3 h_3}{\sin^2 \pi h_3}, \tag{1.14}$$

† Eqn (1.13) is the expression used by most solid-state physicists, whereas crystallographers usually omit the factor $2\pi$. Crystallographers often refer to $|a_1|, |a_2|, |a_3|$ as $a, b, c$, and to $|b_1|/2\pi, |b_2|/2\pi, |b_3|/2\pi$ as $a^*, b^*, c^*$.

where $N_i (i = 1, 2, 3)$ is the number of unit cells along the $\mathbf{a}_i$ axis. This function peaks at integral values of $h_i$, that is, at the points of the reciprocal lattice. Integrating it over the region close to one reciprocal lattice point, defined by $\mathbf{Q} = \boldsymbol{\tau}$, we get $N_0 (2\pi)^3/v$, where $N_0 = N_1 N_2 N_3$ is the total number of unit cells, each of volume $v$. Thus in the limit of large $N_1, N_2, N_3$ eqn (1.14) reduces to a $\delta$-function† of weight $N_0 (2\pi)^3/v$, repeated at each reciprocal lattice point $\boldsymbol{\tau}$, and the cross-section in eqn (1.11) becomes

$$\frac{d\sigma}{d\Omega} = N_0 \frac{(2\pi)^3}{v} \sum_{\boldsymbol{\tau}} \delta(\mathbf{Q} - \boldsymbol{\tau}) \times |F_{\boldsymbol{\tau}}|^2, \quad (1.15)$$

with

$$F_{\boldsymbol{\tau}} = \sum_{\boldsymbol{\rho}} b_{\boldsymbol{\rho}}^{coh} e^{i\boldsymbol{\tau} \cdot \boldsymbol{\rho}}. \quad (1.16)$$

$F_{\boldsymbol{\tau}}$ is known as the structure factor of the unit cell. If the factor $2\pi$ is omitted in eqn (1.13), so that

$$\mathbf{a}_i \cdot \mathbf{b}_j = \delta_{ij},$$

the structure factor equation is then the one used in crystallography:

$$F_{\boldsymbol{\tau}} = \sum_{\boldsymbol{\rho}} b_{\boldsymbol{\rho}} e^{2\pi i \boldsymbol{\tau} \cdot \boldsymbol{\rho}}$$

$$= \sum_{\boldsymbol{\rho}} b_{\boldsymbol{\rho}} e^{2\pi i (h_1 x_1 + h_2 x_2 + h_3 x_3)}.$$

The indices $h_1, h_2, h_3$, often denoted $hkl$, are the 'Miller indices' of the reflecting planes in the crystal, and $x_1, x_2, x_3$ are the fractional atomic coordinates.

The delta function $\delta(\mathbf{Q} - \boldsymbol{\tau})$ in eqn (1.15) ensures that the scattering is limited to values of the scattering vector $\mathbf{Q}$ equal to the set of reciprocal lattice vectors $\boldsymbol{\tau}$. The directions of the vectors $\boldsymbol{\tau}$ lie along the normals to the various Bragg reflecting planes, and the distances $|\boldsymbol{\tau}| = 2\pi/d$ are inversely proportional to the plane spacing $d$. Recalling that

$$Q = 4\pi \sin\theta/\lambda,$$

scattering can occur only at angles given by the Bragg equation

$$1/d = 2 \sin\theta/\lambda. \quad (1.17a)$$

The size and shape of the unit cell may be deduced from the values of $\theta$ at which Bragg reflexions are observed. To deduce the contents of the unit cell, that is, to determine the crystal structure, it is necessary to measure as many as possible of the intensities of the Bragg reflexions and to derive from them the unit cell structure factors $F_{\boldsymbol{\tau}}$.

† See Appendix II (p. 300) for the properties of the $\delta$-function.

## BASIC THEORY

### 1.3.1. Powder scattering

To observe a Bragg reflexion, two conditions must be satisfied. First, the length of the scattering vector $|\mathbf{Q}| = Q$ must satisfy the Bragg equation (1.17). This means that the detector must be set at the correct scattering angle $2\theta$ giving $Q = \tau = 2k \sin \theta$. Second, the direction of the normal to the reflecting planes must be parallel to $\mathbf{Q}$. The simplest way of achieving the second condition is to use a powder, which will always contain a few crystallites of the required orientation. The scattering, at fixed $k = 2\pi/\lambda$, comes out in cones called Debye–Scherrer cones, and these are successively crossed by the detector as the scattering

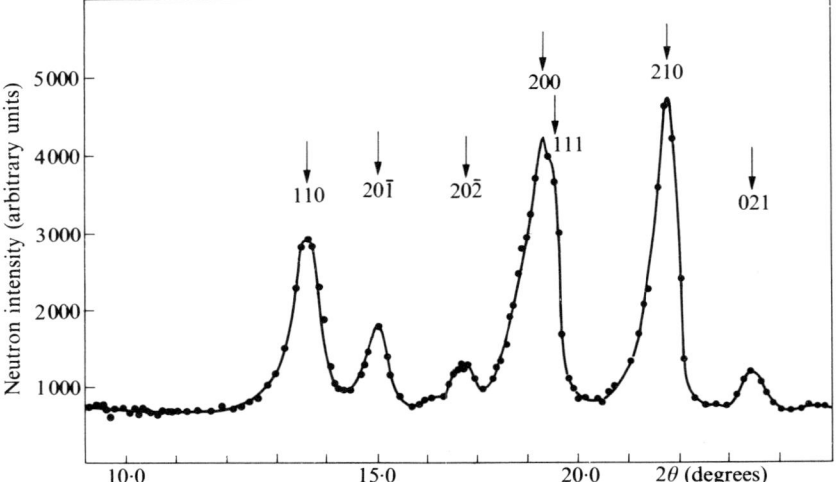

FIG. 1.6. Debye–Scherrer pattern, recorded with neutrons of $\lambda = 1 \cdot 1$ Å, from powdered naphthalene.

angle increases, to give a sequence of Debye–Scherrer lines (Fig. 1.6). The scattering cross-section, integrated over a complete cone, is

$$\sigma(\boldsymbol{\tau}) = \frac{4\pi^3 N_0}{k^2 v} \frac{Z_\tau}{\tau} |F_\tau|^2,$$

where $Z_\tau$ is the number of equivalent reflecting planes associated with $\boldsymbol{\tau}$.

The actual quantity measured in an experiment is the 'integrated intensity' $I_{hkl}$ defined by

$$I_{hkl} = E_0 \omega.$$

$E_0$ is the total number of neutrons scattered into the detector from an incident beam of unit flux, as the detector rotates with uniform angular velocity $\omega$

through the *hkl* cone. The conversion from $\sigma(\tau)$ to $I_{hkl}$ involves certain instrumental and geometrical factors

$$I_{hkl} = \sigma(\tau) \frac{l_s}{2\pi r_0 \sin 2\theta_{hkl}},$$

where $l_s$ is the height of the sample, assumed cylindrical with the cylinder axis normal to the incident radiation, and $r_0$ its radius. The sample is bathed uniformly by the neutron beam, and absorption is neglected.

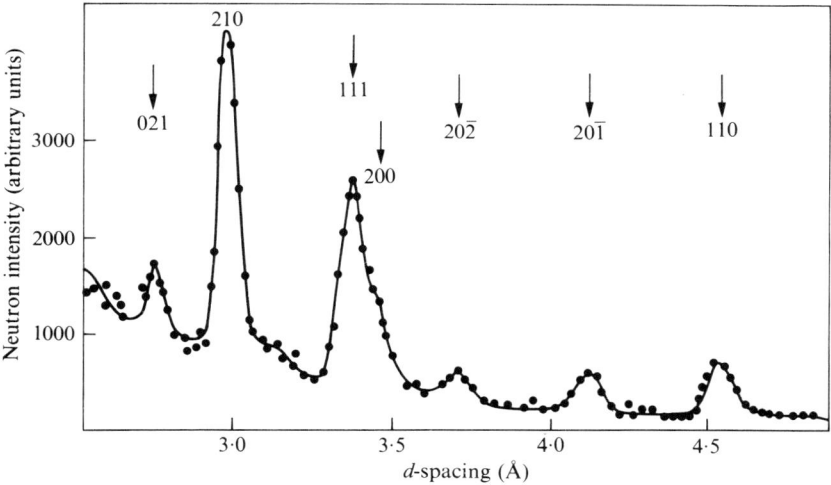

FIG. 1.7. Time-of-flight powder pattern from naphthalene.

An alternative procedure for recording the neutron powder pattern of a sample is to keep $2\theta$ fixed and to satisfy the Bragg equation by varying $k$ (see Chapter 2, §2.2.3.4). The intensity in this case is measured as a function of $k$ or $\lambda$: an example of such a measurement is given in Fig. 1.7. $\lambda$ is readily determined from the flight-time for neutrons passing from the sample to the detector.

The great drawback of the powder method is that for samples of low crystal symmetry, such as naphthalene, $Z_\tau$ is a small number and so higher-order reflexions suffer from severe overlapping. Many symmetry-independent vectors $\tau$ contribute to the observed intensity, and individual determinations of $|F_\tau|^2$ are possible only for the few reflexions that are free from overlapping. However, the experiment itself is very straightforward, and few corrections are required in deriving $|F_\tau|^2$ from $I_{hkl}$ (for example, no extinction corrections are needed). In Chapter 10, work on the crystal structures of non-stoichiometric compounds is described, as deduced from accurate neutron measurements of powdered samples with cubic crystal symmetry.

BASIC THEORY

*1.3.2. Single-crystal scattering*

With a single-crystal sample, there is no difficulty in separating the nuclear Bragg reflexions arising from the various planes in the crystal, as scattering can only occur for a given plane when the sample is correctly aligned to bring $\boldsymbol{\tau}$ parallel to $\mathbf{Q}$. The scattered intensity, integrated over the reflecting region surrounding the reciprocal lattice point, is proportional to $|F_\tau|^2$, eqn (1.15). The measurement can be carried out using a four-circle neutron diffractometer (see Chapter 2). Three circles of the diffractometer, $\phi, \chi, \omega$, are used to orientate $\boldsymbol{\tau}$ along the scattering vector and the fourth to set the detector at the appropriate $2\theta$. The integrated intensity is determined either by scanning through the reflecting position with an $\omega$ motion, in which the detector is either fixed ($\omega$-scan) or coupled to $\omega$ by a 2:1 gearing ($\omega/2\theta$-scan). The formula relating the integrated intensity and $|F_\tau|^2$ is

$$I_{hkl} = \frac{\lambda^3 N_c^2 |F_\tau|^2}{\sin 2\theta} \delta V, \qquad (1.17b)$$

where $N_c$ is the number of unit cells per unit volume and $\delta V$ is the volume of the crystal. It is assumed that the crystal is uniformly bathed in radiation and that there is no absorption from nuclear capture or incoherent scattering processes. (Absorption tends to be much less than in X-ray diffraction, and, in any case, its effect can be readily allowed for in eqn (1.17b).)

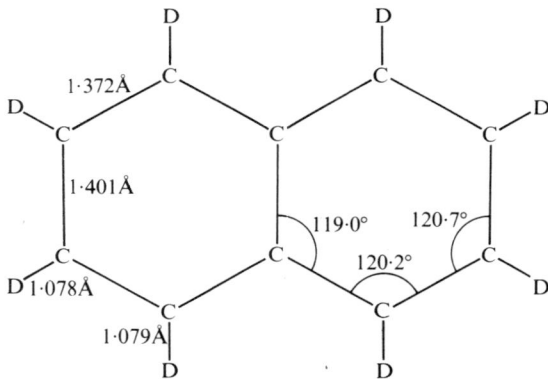

FIG. 1.8. Deuteriated naphthalene: bond lengths and angles determined by Pawley and Yeats (1969).

Pawley and Yeats (1969) have measured the nuclear Bragg intensities of 331 independent reflexions of deuteriated naphthalene, $C_{10}D_8$. The observed values of $|F_\tau|^2$ were compared with the values calculated from the approximate atomic positions given by the X-ray structure determination of $C_{10}H_8$. By adjusting these positions to minimize the differences between observed and calculated $|F_\tau|^2$'s the bond lengths given in Fig. 1.8 were obtained. Pawley and Yeats have shown further that a satisfactory refinement of their data was possible assuming a rigid-body model to calculate the Debye–Waller atomic temperature factors. We shall refer to this model in § 1.5 in considering the vibrational states of molecular crystals.

## 1.4. Scattering theory: inelastic scattering

### 1.4.1. The eigenstate formulation

In our discussion of elastic scattering, we effectively picked out that part of the scattering which does not change the quantum-mechanical state of the system. In fact, the scattering amplitude, given by

$$\sum_{\mathbf{R}} b_{\mathbf{R}} \exp(i\mathbf{Q}\cdot\mathbf{R}),$$

can also excite transitions from initial states $E$ to final states $E'$, provided that the energy conservation law is satisfied (Fig. 1.9). This gives rise to the inelastic scattering cross-section, that is, the cross-section for those scattering processes in

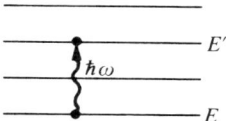

FIG. 1.9. Transition from $E$ to $E'$ accompanied by energy change $\hbar\omega$ of scattered radiation.

which there is a change in energy of the incident radiation. The energy transferred to the radiation is denoted $\hbar\omega$, where $\omega$ has the dimension of rotational frequency.

The nuclear inelastic scattering cross-section is proportional to:

(a) the squared modulus of the matrix elements of the scattering amplitude between initial and final states,

$$\sum_{\mathbf{R},\mathbf{R}'} \langle i | b_{\mathbf{R}'} e^{-i\mathbf{Q}\cdot\mathbf{R}'} | f \rangle \langle f | b_{\mathbf{R}} e^{i\mathbf{Q}\cdot\mathbf{R}} | i \rangle;$$

(b) the probability that the initial state is occupied at temperature $T$,

$$P_i = \frac{e^{-E/k_B T}}{\sum_i e^{-E/k_B T}}; \qquad (1.17c)$$

(c) the $\delta$-function which ensures that the energy-conservation condition is satisfied,

$$\delta(E - E' - \hbar\omega);$$

## BASIC THEORY

(d) the ratio of the final and incident velocities, which is needed in transforming final and initial particle densities into fluxes,

$$\frac{k'}{k}.$$

Collecting these four factors together, we have finally the following formula for the differential cross-section $d^2\sigma/d\Omega dE'$, representing the fraction of neutrons scattered per nucleus per unit area into a solid angle $\Delta\Omega$ and with neutron energy in the range $\Delta E'$:

$$\frac{d^2\sigma}{d\Omega dE'} = \frac{k'}{k} \sum_{i,f} P_i \sum_{\mathbf{R},\mathbf{R}'} \langle i | b_{\mathbf{R}'} e^{-i\mathbf{Q}\cdot\mathbf{R}'} | f \rangle$$
$$\times \langle f | b_{\mathbf{R}} e^{i\mathbf{Q}\cdot\mathbf{R}} | i \rangle \delta(E - E' - \hbar\omega). \tag{1.18a}$$

This expression can be evaluated directly if the eigenstates and energy levels of the scattering system are known. This is not always possible, and so, before proceeding further, it will be useful to consider two other equivalent formulations of the cross-section formula.

### 1.4.2. The time-dependent formulation

The eigenstate formulation expresses the scattering in terms of the variables $\mathbf{Q}$ and $\omega$ where $\hbar\mathbf{Q}$ is the momentum transferred from the neutron to the scattering system and $\hbar\omega$ is the energy transfer. The real world is in $\mathbf{r}$ and $t$ which are their Fourier-transformed variables.

The time variable is introduced through the integral representation of the $\delta$-function (see Appendix II, p. 300):

$$\delta(x) = \frac{1}{2\pi} \int_{-\infty}^{\infty} e^{ixt} \, dt. \tag{1.18b}$$

How this expression works can be seen qualitatively by putting cosines in the right-hand side. All the different frequencies have random phases and tend to cancel, except at the pole of the $\delta$-function (Fig. 1.10). Recalling that

$$\delta(\hbar x) = \frac{1}{\hbar}\delta(x),$$

we have

$$\delta(E - E' - \hbar\omega) = \frac{1}{2\pi\hbar} \int_{-\infty}^{\infty} e^{i(t/\hbar)\cdot(E - E' - \hbar\omega)} \, dt.$$

The next step is to apply the time-dependent Schrödinger equation

$$i\hbar \frac{\partial \psi(t)}{\partial t} = \mathcal{H}\psi(t)$$

giving

$$\psi(t) = e^{-i\mathcal{H}t/\hbar} \psi(0),$$

where $\psi(t)$ is the wave function at time $t$ and $\psi(0)$ the wave function at time zero. When $\psi$ is the eigenstate of state $i$ with energy $E_i$,

$$\mathcal{H}\psi_i = E_i\psi_i$$

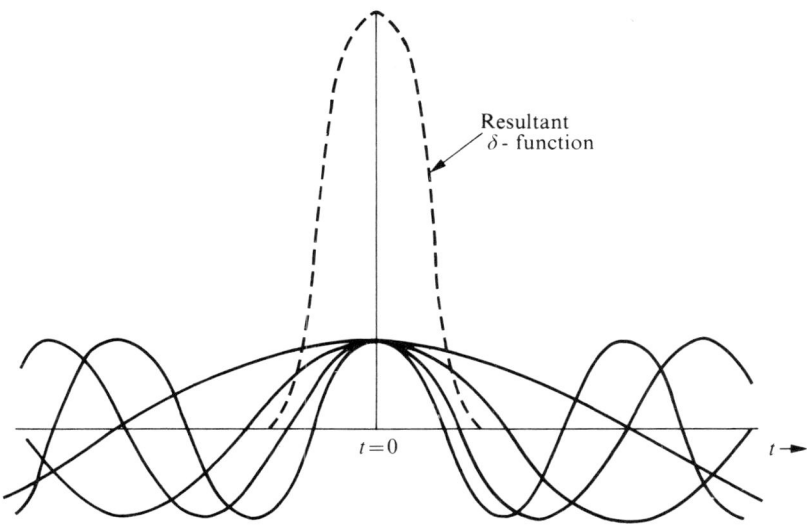

FIG. 1.10. Integral representation of delta function.

and

$$\psi_i(t) = e^{-iE_i t/\hbar} \psi_i(0).$$

We have then that

$$\langle f | b_\mathbf{R}\, e^{i\mathbf{Q}\cdot\mathbf{R}} | i \rangle \delta(E - E' - \hbar\omega)$$

$$= \frac{1}{2\pi\hbar} \int_{-\infty}^{\infty} e^{-i\omega t} \langle f e^{-itE'/\hbar} | b_\mathbf{R}\, e^{i\mathbf{Q}\cdot\mathbf{R}} | e^{itE/\hbar} | i \rangle \, dt$$

$$= \frac{1}{2\pi\hbar} \int_{-\infty}^{\infty} e^{-i\omega t} \langle f | e^{-it\mathcal{H}/\hbar} b_\mathbf{R}\, e^{i\mathbf{Q}\cdot\mathbf{R}} | e^{it\mathcal{H}/\hbar} | i \rangle \, dt$$

$$= \frac{1}{2\pi\hbar} \int_{-\infty}^{\infty} e^{-i\omega t} \langle f | b_\mathbf{R}\, e^{i\mathbf{Q}\cdot\mathbf{R}(t)} | i \rangle \, dt, \qquad (1.19)$$

remembering that $\exp(itE/\hbar)$ is a number and so can be moved freely to the left or right in the operator expressions. Eqn (1.19) gives the scattering amplitude

## BASIC THEORY

from atom **R** written in terms of a time-dependent operator **R**(*t*). All the states are now at time zero, and we may immediately sum over initial states, with probability $P_i$, by taking the thermal expectation value $\langle \ \rangle_T$.

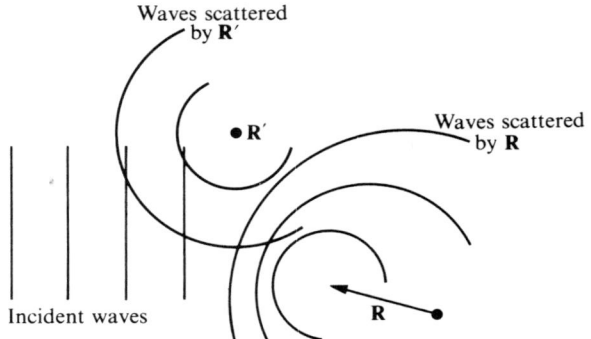

FIG. 1.11. Interference between stationary centre at **R**′ and moving centre at **R**.

We arrive then at a second basic equation for the differential scattering cross-section:

$$\frac{d^2\sigma}{d\Omega\, dE'} = \frac{k'}{k}\frac{1}{2\pi\hbar}\int_{-\infty}^{\infty} dt\, e^{-i\omega t} \sum_{\mathbf{R},\mathbf{R'}} \langle b_{\mathbf{R'}}\, e^{-i\mathbf{Q}\cdot\mathbf{R'}(0)}\, b_{\mathbf{R}}\, e^{i\mathbf{Q}\cdot\mathbf{R}(t)} \rangle_T. \tag{1.20}$$

In this formulation the scattering can be seen to arise from the interference between a wave scattered from a stationary centre at **R**′ and a retarded wave scattered from the moving centre at **R** (Fig. 1.11).

### 1.4.3. The space–time representation

Finally, let us perform the Fourier transformation over space as well as time. We do this by writing eqn (1.20) as

$$\begin{aligned}
\frac{d^2\sigma}{d\Omega\, dE'} &= \frac{k'}{k}\frac{1}{2\pi\hbar}\int_{-\infty}^{\infty} dt\, e^{-i\omega t} \sum_{\mathbf{R},\mathbf{R'}} \langle b_{\mathbf{R}} b_{\mathbf{R'}} \rangle \langle e^{-i\mathbf{Q}\cdot\mathbf{R'}(0)}\, e^{i\mathbf{Q}\cdot\mathbf{R}(t)} \rangle_T \\
&= \frac{k'}{k}\frac{1}{2\pi\hbar}\int_{-\infty}^{\infty} dt\, e^{-\omega t}\Big\{\langle b_{\mathbf{R}} \rangle^2 \sum_{\mathbf{R},\mathbf{R'}} \langle e^{-i\mathbf{Q}\cdot\mathbf{R'}(0)}\, e^{i\mathbf{Q}\cdot\mathbf{R}(t)} \rangle_T \\
&\quad + (\langle b_{\mathbf{R}}^2 \rangle - \langle b_{\mathbf{R}} \rangle^2) \sum_{\mathbf{R}} \langle e^{-i\mathbf{Q}\cdot\mathbf{R}(0)}\, e^{i\mathbf{Q}\cdot\mathbf{R}(t)} \rangle_T\Big\} \\
&\equiv N\frac{k'}{k}\frac{1}{2\pi\hbar}\int_{-\infty}^{\infty} dt\, e^{-i\omega t} \sum_{\mathbf{r}} e^{i\mathbf{Q}\cdot\mathbf{r}}\Big\{\langle b_{\mathbf{R}} \rangle^2 G(\mathbf{r},t) \\
&\quad + (\langle b_{\mathbf{R}}^2 \rangle - \langle b_{\mathbf{R}} \rangle^2)\, G_s(\mathbf{r},t)\Big\},
\end{aligned} \tag{1.21a}$$

where we have, by definition,

$$N \sum_{\mathbf{r}} G(\mathbf{r}, t) e^{i\mathbf{Q}\cdot\mathbf{r}} = \sum_{\mathbf{R},\mathbf{R}'} \langle e^{-i\mathbf{Q}\cdot\mathbf{R}'(0)} e^{i\mathbf{Q}\cdot\mathbf{R}(t)} \rangle_T \quad (1.21b)$$

and

$$N \sum_{\mathbf{r}} G_s(\mathbf{r}, t) e^{i\mathbf{Q}\cdot\mathbf{r}} = \sum_{\mathbf{R}} \langle e^{-i\mathbf{Q}\cdot\mathbf{R}(0)} e^{i\mathbf{Q}\cdot\mathbf{R}(t)} \rangle_T, \quad (1.21c)$$

with $N$ the number of scattering nuclei. $G(\mathbf{r}, t)$ and $G_s(\mathbf{r}, t)$ are the famous Van Hove correlation functions (Van Hove 1954). Note that we have separated the cross-section into two parts: the first part involves the square of the average scattering length $\langle b \rangle^2$ of the nucleus, and the second part the mean-square deviation, $\langle (b - \langle b \rangle)^2 \rangle$, from the average scattering length. In analogy with the terminology introduced in § 1.2 for elastic scattering, the two contributions are called the coherent and incoherent scattering cross-sections:

$$\frac{d^2\sigma_{\text{coh}}}{d\Omega\, dE'} = N \frac{k'}{k} \frac{1}{2\pi\hbar} \int_{-\infty}^{\infty} dt\, e^{-i\omega t} \sum_{\mathbf{r}} e^{i\mathbf{Q}\cdot\mathbf{r}} \langle b \rangle^2 G(\mathbf{r}, t) \quad (1.22)$$

and

$$\frac{d^2\sigma_{\text{incoh}}}{d\Omega\, dE'} = N \frac{k'}{k} \frac{1}{2\pi\hbar} \int dt\, e^{-i\omega t} \sum_{\mathbf{r}} e^{i\mathbf{Q}\cdot\mathbf{r}} (\langle b^2 \rangle - \langle b \rangle^2) G_s(\mathbf{r}, t). \quad (1.23)$$

The formulae (1.21b) and (1.21c) for the Van Hove correlation functions are quantum-mechanical expressions, and so we expect the correlation functions to be complex quantities. However, in the classical limit the functions are real and can be given a simple physical interpretation. The operators $\mathbf{R}'(0)$ and $\mathbf{R}(t)$ then commute and we have

$$G(\mathbf{r}, t) = \frac{1}{N} \sum_{\mathbf{R},\mathbf{R}'} \delta(\mathbf{r} + \mathbf{R}'(0) - \mathbf{R}(t))$$

and

$$G_s(\mathbf{r}, t) = \frac{1}{N} \sum_{\mathbf{R}} \delta(\mathbf{r} + \mathbf{R}(0) - \mathbf{R}(t)).$$

$G(\mathbf{r}, t)$ gives the probability that if there is an atom at $\mathbf{R}'(0)$ at time zero, there will be an atom at $\mathbf{R}(t)$ at time $t$. The self correlation function $G_s(\mathbf{r}, t)$ similarly gives the probablity that if a given atom is at $\mathbf{R}(0)$ at time zero, the same atom will be displaced to $\mathbf{R}(t)$ at time $t$.

## BASIC THEORY

Eqns (1.22) and (1.23) can also be written

$$\frac{d^2\sigma_{coh}}{d\Omega\, dE'} = N\frac{k'}{k}\frac{\langle b\rangle^2}{2\pi\hbar}\iint dr\, dt\, e^{i(\mathbf{Q}\cdot\mathbf{r}-\omega t)} G(\mathbf{r}, t) \qquad (1.24a)$$

and

$$\frac{d^2\sigma_{incoh}}{d\Omega\, dE'} = N\frac{k'}{k}\frac{(\langle b^2\rangle - \langle b\rangle^2)}{2\pi\hbar}\iint dr\, dt\, e^{i(\mathbf{Q}\cdot\mathbf{r}-\omega t)} G_s(\mathbf{r}, t). \qquad (1.24b)$$

The integrals in eqns (1.24a) and (1.24b) represent four-dimensional $\mathbf{Q}$-$\omega$ Fourier transforms of the Van Hove correlation functions. It is customary to use the following notation to represent these integrals:

$$S(\mathbf{Q}, \omega) = \frac{1}{2\pi}\iint dr\, dt\, e^{i(\mathbf{Q}\cdot\mathbf{r}-\omega t)} G(\mathbf{r}, t) \qquad (1.25a)$$

$$S_s(\mathbf{Q}, \omega) = \frac{1}{2\pi}\iint dr\, dt\, e^{i(\mathbf{Q}\cdot\mathbf{r}-\omega t)} G_s(\mathbf{r}, t), \qquad (1.25b)$$

where the scattering functions $S(\mathbf{Q}, \omega)$ and $S_s(\mathbf{Q}, \omega)$ are known as the coherent and incoherent 'scattering laws'. Thus the relations between the cross-sections and the scattering laws are

$$\frac{1}{\hbar}\frac{d^2\sigma_{coh}}{d\Omega\, d\omega} \equiv \frac{d^2\sigma_{coh}}{d\Omega\, dE'} = N\frac{k'}{\hbar k}\langle b\rangle^2 S(\mathbf{Q}, \omega) \qquad (1.25c)$$

and

$$\frac{1}{\hbar}\frac{d^2\sigma_{incoh}}{d\Omega\, d\omega} \equiv \frac{d^2\sigma_{incoh}}{d\Omega\, dE'} = N\frac{k'}{\hbar k}(\langle b^2\rangle - \langle b\rangle^2) S_s(\mathbf{Q}, \omega). \qquad (1.25d)$$

The space-time correlation functions represent basic dynamic properties of a scattering system, being independent of the properties of the scattered neutron, and other scattering techniques can provide information about these functions. X-ray scattering can tell us about $G(\mathbf{r}, t)$ and infra-red spectroscopy about $G_s(\mathbf{r}, t)$. However, although the same function is measured in experiments with different radiations, different momentum and frequency ranges are covered and the information obtained is not the same for each experiment.

We have noted already that the Van Hove correlation function is a complex function of time, so that when it is Fourier transformed with respect to time it becomes an unsymmetrical function of $\omega$. The eigenfunction formulation, eqn (1.18a), tells us that the scattering laws for energy-loss processes ($\omega > 0$) and for energy gain ($\omega < 0$) are related by

$$S(\mathbf{Q}, \omega) = S(-\mathbf{Q}, -\omega)\exp(\hbar\omega/k_B T), \qquad (1.26a)$$

a property of $S(\mathbf{Q}, \omega)$ which is frequently called the 'detailed balance condition'. In the classical approximation, we put $\hbar \to 0$, so that $S(\mathbf{Q}, \omega)$ is then a symmetrical function in $\omega$ and $G(\mathbf{r}, t)$ is a real quantity. It can be shown that $S_0(\mathbf{Q}, \omega)$, the scattering function calculated on a classical model, is given by

$$S_0(\mathbf{Q}, \omega) = \tfrac{1}{2}[S(\mathbf{Q}, \omega) + S(\mathbf{Q}, -\omega)].$$

Using eqn (1.26a) we may substitute for $S(\mathbf{Q}, -\omega)$ to give

$$S_0(\mathbf{Q}, \omega) = \tfrac{1}{2}S(\mathbf{Q}, \omega)[1 + \exp(-\hbar\omega/k_B T)]$$

or

$$S(\mathbf{Q}, \omega) = \frac{2S_0(\mathbf{Q}, \omega)}{1 + \exp(-\hbar\omega/k_B T)}. \tag{1.26b}$$

Thus if the scattering function is known from classical considerations, a detailed balance-corrected function may be easily calculated.

Finally, we remark that in the analysis of the neutron inelastic scattering from liquids (see Chapters 6 and 8), the correlation function $G_d(\mathbf{r}, t)$ is used. It is defined by

$$G(\mathbf{r}, t) = G_s(\mathbf{r}, t) + G_d(\mathbf{r}, t) \tag{1.27}$$

and is called the 'distinct' correlation function because its classical interpretation is the probability that there is a particle at $\mathbf{r} + \mathbf{r}'$, $t + t'$, when there is a *different* particle at $\mathbf{r}'$, $t'$.

## 1.5. Neutron scattering and vibrational states of molecular crystals

In this section we shall apply the scattering theory outlined in §1.4 to the analysis of the spectrum of neutrons inelastically scattered by a molecular crystal.

### 1.5.1. Vibrational states of a molecule

Many molecular crystals, such as naphthalene, are composed of covalently bonded molecules, with high internal vibrational frequencies, coupled together by much weaker inter-molecular forces. As a first approximation, therefore, we may classify the vibrational states in terms of internal modes, calculated assuming the molecules to be independent, and external modes of lower frequency, calculated assuming the molecules to vibrate as rigid units. Naphthalene, $C_{10}H_8$, with 18 atoms in the molecule has $3 \times 18 = 54$ degrees of freedom, so that there are 54 internal modes of the molecule, each with its characteristic vibration frequency. These frequencies have been calculated by Pawley and Cyvin (1970) from the two-atom force constants in the molecule, and they range from 176 to 3065 cm$^{-1}$ (Fig. 1.12). (See p. xv for relations between energy units.)

The $\delta$-function in the eigenstate formulation, eqn (1.18a), limits the neutron scattering to transitions from an occupied level (or the groupd state) to another level, up or down in energy, since the sum over the initial and final states $i, f$,

## BASIC THEORY

which normally refer to the whole system, may be restricted here to the single molecule. This is exactly the same situation as in infra-red or Raman scattering, but in the optical case there are also stringent selection rules, since the radiation couples to the molecule through its electric dipole moment, and only certain levels are optically active. The scattering conditions are less stringent in the neutron case and may be readily understood by referring to the time-dependent formulation of the scattering cross-section.

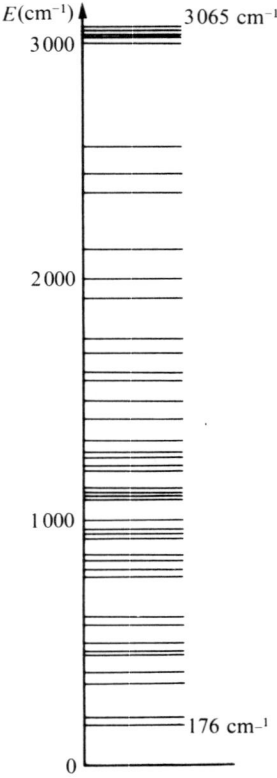

FIG. 1.12. Energy level diagram for internal modes of naphthalene. (After Pawley and Cyvin 1970).

Let us suppose that the displacement $\mathbf{R}(t)$ of the $n$th atom for a given vibrational mode of the molecule is

$$\mathbf{R}(t) = \mathbf{n} + \mathbf{u}_n(t)$$

where $\mathbf{n}$ is the vector from the origin to the mean position of the atom and

$$\mathbf{u}(t) = \mathbf{u}_0 \cos \omega_0 t = \mathbf{u}_0 \frac{\exp(i\omega_0 t) + \exp(-i\omega_0 t)}{2}.$$

$u_0$ is the amplitude of the vibration and $\omega_0$ its characteristic frequency. Substituting in eqn (1.20), we have

$$\frac{d^2\sigma}{d\Omega\, dE'} = \frac{k'}{k} \frac{1}{2\pi\hbar} \int_{-\infty}^{\infty} dt\, e^{-i\omega t} \sum_{n,m} \langle b_n b_m \rangle e^{i\mathbf{Q}\cdot(\mathbf{n}-\mathbf{m})}$$
$$\times \langle e^{i\mathbf{Q}\cdot(\mathbf{u_n}(t)-\mathbf{u_m}(0))} \rangle_T. \qquad (1.28)$$

We now assume that $u_0$ is small compared with $\lambda/2\pi$, where $\lambda$ is the wavelength of the incident neutron beam. The thermal expectation term in eqn (1.28) may then be expanded as a power series in $\mathbf{u}$:

$$\langle e^{i\mathbf{Q}\cdot(\mathbf{u_n}(t)-\mathbf{u_m}(0))} \rangle_T$$

$$= \langle 1 + i\mathbf{Q}\cdot(\mathbf{u_n}(t) - \mathbf{u_m}(0)) + \frac{i^2}{2}[\mathbf{Q}\cdot(\mathbf{u_n}(t) - \mathbf{u_m}(0))]^2$$
$$+ \ldots \rangle_T$$

$$= \langle 1 + i\mathbf{Q}\cdot(\mathbf{u_n}(t) - \mathbf{u_m}(0))$$
$$- \tfrac{1}{2}[(\mathbf{Q}\cdot\mathbf{u_n}(t))^2 + (\mathbf{Q}\cdot\mathbf{u_m}(0))^2 - 2\mathbf{Q}\cdot\mathbf{u_n}(t)\mathbf{Q}\cdot\mathbf{u_m}(0)] + \ldots \rangle_T$$

$$= \langle 1 - \tfrac{1}{2}((\mathbf{Q}\cdot\mathbf{u_n})^2 + (\mathbf{Q}\cdot\mathbf{u_m})^2) + (\mathbf{Q}\cdot\mathbf{u_0})^2 \frac{\exp(i\omega_0 t) + \exp(-i\omega_0 t)}{2}$$
$$+ \ldots + \text{imaginary terms} \rangle_T.$$

Ignoring the imaginary terms, the time integration in eqn (1.28) can be done immediately to give

$$\frac{d^2\sigma}{d\Omega\, dE'} = \frac{k'}{k} \sum_{n,m} \langle b_n b_m \rangle e^{i\mathbf{Q}\cdot(\mathbf{n}-\mathbf{m})} \Big[\big\{1 - \tfrac{1}{2}\langle(\mathbf{Q}\cdot\mathbf{u_n})^2\rangle$$
$$- \tfrac{1}{2}\langle(\mathbf{Q}\cdot\mathbf{u_m})^2\rangle\big\}\delta(\hbar\omega) + \tfrac{1}{2}(\mathbf{Q}\cdot\mathbf{u_0})^2 \delta(\hbar\omega - \hbar\omega_0)$$
$$+ \tfrac{1}{2}(\mathbf{Q}\cdot\mathbf{u_0})^2 \delta(\hbar\omega + \hbar\omega_0) + \ldots\Big]. \qquad (1.29)$$

By taking only the real terms, we have obtained an expression (1.29) which is valid only in the classical limit, but it does indicate a number of important points:

(a) The cross-section contains an elastic term, with $\omega = 0$, and inelastic terms, with $\omega = \omega_0$ and $\omega = -\omega_0$. (The cross-sections for neutron energy loss ($\omega = +\omega_0$) and neutron energy gain ($\omega = -\omega_0$) processes are the same, as expected for the classical approximation which violates the detailed balance condition (1.26a).)
(b) The inelastic terms contain a polarization factor $(\mathbf{Q}\cdot\mathbf{u_0})^2$, showing that the inelastic scattering is a maximum for the scattering vector $\mathbf{Q}$ parallel to the polarization direction $\mathbf{u_0}$ and is zero for $\mathbf{Q}$ at right-angles to $\mathbf{u_0}$.
(c) The vibrational intensities, for a fixed angle between $\mathbf{Q}$ and $\mathbf{u_0}$, are proportional to $Q^2 = 16\pi^2 \sin^2\theta/\lambda^2$, where $2\theta$ is the scattering angle.

BASIC THEORY

(d) Both elastic and inelastic contributions to the cross-section are attenuated by the Debye-Waller factor, which arises because interference effects between atoms on different sites are smoothed out by thermal vibrations. The contents of the curly bracket in eqn (1.29) represent the leading terms of the exponential expression for the Debye-Waller factor

$$e^{-2W} = e^{-(W_n + W_m)},$$

where

$$W_\mathbf{n} = \tfrac{1}{2} \langle (\mathbf{Q} \cdot \mathbf{u_n})^2 \rangle. \tag{1.30}$$

(e) The inelastic cross-section contains both coherent and incoherent scattering components, although the components have different intensities. This can be seen by considering the one-phonon terms in eqn (1.29). Thus

$$\begin{aligned}\frac{d^2 \sigma^1}{d\Omega\, dE'} &= \frac{k'}{k} \sum_{n,m} \langle b_n b_m \rangle e^{i\mathbf{Q}\cdot(\mathbf{n}-\mathbf{m})} \\ &\quad \times \tfrac{1}{2}(\mathbf{Q}\cdot\mathbf{u_0})^2 (\delta(\hbar\omega - \hbar\omega_0) + \delta(\hbar\omega + \hbar\omega_0)) \\ &= \frac{k'}{k} \sum_{n,m} \langle b \rangle^2 e^{i\mathbf{Q}\cdot(\mathbf{n}-\mathbf{m})} \\ &\quad \times \tfrac{1}{2}(\mathbf{Q}\cdot\mathbf{u_0})^2 (\delta(\hbar\omega - \hbar\omega_0) + \delta(\hbar\omega + \hbar\omega_0)) \\ &\quad + \frac{k'}{k} (\langle b^2 \rangle - \langle b \rangle^2) N \tfrac{1}{2}(\mathbf{Q}\cdot\mathbf{u_0})^2 (\delta(\hbar\omega - \hbar\omega_0) \\ &\quad + \delta(\hbar\omega + \hbar\omega_0)), \end{aligned} \tag{1.31}$$

where the superscript in $\sigma^1$ indicates the one-phonon contribution to the cross-section. The first term on the right-hand side of eqn (1.31) is the one-phonon coherent scattering cross-section and the second term the one-phonon incoherent scattering cross-section. The coherent cross-section contains a phase factor $\exp[i\mathbf{Q}\cdot(\mathbf{n}-\mathbf{m})]$ which imposes a condition on the momentum $\hbar\mathbf{Q}$ transferred during scattering: there is no such condition applying to the incoherent cross-section.

## 1.5.2. Phonons in molecular crystals

The normal modes of a molecular crystal must have the same periodicity as the lattice. By Bloch's theorem, the wave functions for a periodic potential are of the form

$$\psi_\mathbf{q}(\mathbf{r}) = \exp(i\mathbf{q}\cdot\mathbf{r}) u_\mathbf{q}(\mathbf{r}) \tag{1.32}$$

where $\mathbf{q}$ is the wave-vector index of the wave function and $u(\mathbf{r})$ is periodic in the lattice. It is easy to show from eqn (1.32) that it is always possible to select $\mathbf{q}$ so that it lies within the first Brillouin zone (of volume $v_a = (2\pi)^3/v$ with $v$ the volume of the unit cell). Thus each normal mode of the crystal becomes a dispersive mode and can be described by a wave-vector **q** within the first Brillouin zone. In naphthalene, there are two molecules per unit cell (see Fig. 1.1), and so we expect 2 x 3 x 18 = 108 modes, all functions of **q**, and these give a complete

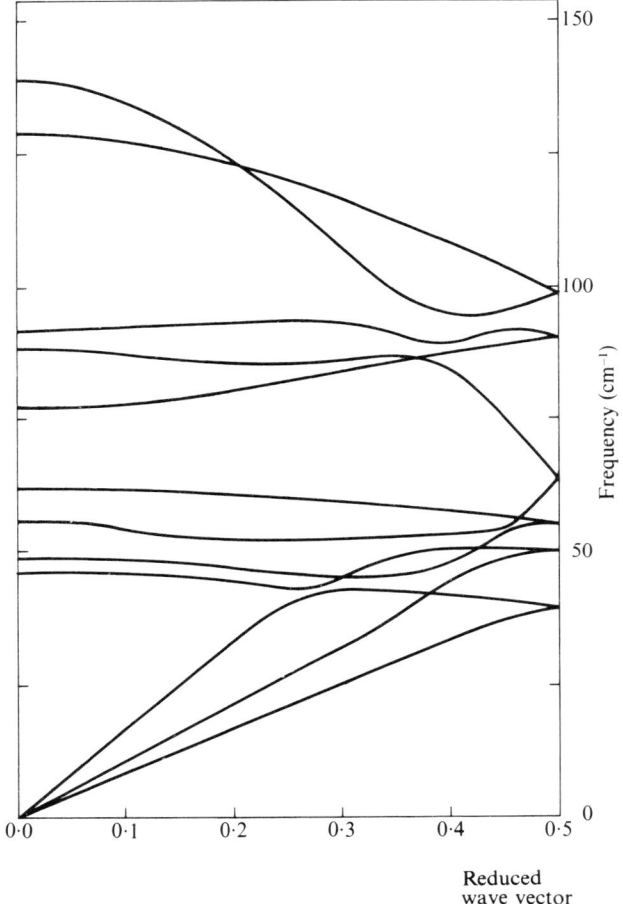

FIG. 1.13. Calculated dispersion curves for naphthalene crystal for wave vectors along the two-fold crystal axis, [010]. (After Pawley 1967).

description of the vibrations of the solid. However, if we make the approximation that the molecule vibrates as a rigid unit, the number of degrees of freedom in the unit cell is reduced to 2 x 3 + 2 x 3 = 12, three translational modes per molecule and three rotational modes. In this approximation, therefore, there are 12 external dispersive modes, and 54 internal non-dispersive modes of higher frequency.

Fig. 1.13 shows the dispersion curves for the twelve phonon modes of

## BASIC THEORY

naphthalene, calculated by Pawley (1967) assuming rigid-body vibrations and a potential function between atoms $i, j$ in the molecule of the form

$$V(r_{ij}) = -\frac{A}{r_{ij}^6} + B \exp(-\alpha r_{ij}).$$

$r$ is the separation of the atoms and $A, B, \alpha$ are constants for a given type of atom pair (C..C, C..H or H..H). There are three acoustic modes, which become the two transverse and one longitudinal sound waves as $q \to 0$, and nine optic modes, which can be measured, in principle, by Raman spectroscopy as $q \to 0$. The phonon modes in Fig. 1.13 reach a frequency of 140 cm$^{-1}$ and so are not far below the lowest molecular vibrational levels, Fig. 1.12. Measurements by neutron coherent inelastic scattering of the phonon modes and of the possible dispersion of the low-lying molecular modes would give important information on the interactions between the two kinds of vibration.

The shifts in the phonon levels caused by relaxing the rigid-body constraint have been calculated by Pawley and Cyvin (1970).

### 1.5.3 The coherent scattering from phonons

The general theory of the scattering of neutrons by phonons is not simple and we shall only quote the results. The scattering is now a sum over the wave vectors $\mathbf{q}$, the various modes $j$, and the atoms within the unit cell. Corresponding to the displacement vector $\mathbf{u}_0$ of the simple harmonic oscillator, §1.5.1, we must now use a vector ('polarization vector') $\mathbf{U}_\rho^j(\mathbf{q})$. The coherent one-phonon cross-section is

$$\frac{d^2\sigma_{\text{coh}}^1}{d\Omega\, dE'} = \frac{(2\pi)^3}{v} \sum_{\mathbf{q},j} \frac{k'}{k} \delta(\hbar\omega - \hbar\omega_j(\mathbf{q})) \times \sum_{\boldsymbol{\tau}} \delta(\mathbf{Q} \mp \mathbf{q} - \boldsymbol{\tau})$$
$$\times \frac{\hbar(n_s + \tfrac{1}{2} \pm \tfrac{1}{2})}{2\omega_j(\mathbf{q})}$$
$$\times \left| \sum_\rho \frac{\langle b \rangle_\rho}{M_\rho} e^{i\mathbf{Q}\cdot\boldsymbol{\rho}}\, \mathbf{Q}\cdot\mathbf{U}_\rho^j(\mathbf{q})\, e^{-W_\rho} \right|^2. \quad (1.33a)$$

Here $\omega_j(\mathbf{q})$ is the characteristic frequency of the mode $j$, $\mathbf{q}$ and $\boldsymbol{\tau}$ is the reciprocal lattice vector. $\langle b \rangle_\rho$ is the coherent scattering length of the atom at $\rho$ in the unit cell, $M_\rho$ is its mass and $e^{-W_\rho}$ its Debye–Waller factor. The upper signs in the middle terms refer to neutron energy loss or phonon creation and the lower signs to neutron energy gain or phonon annihilation. All the phonon modes are harmonic oscillators, and so they have Bose–Einstein population factors $n_s + 1$ for energy loss and $n_s$ for energy gain, where

$$n_s = [\exp(\hbar\omega_j(\mathbf{q})/k_B T) - 1]^{-1}.$$

Most of the important features in eqn (1.33a) have been anticipated already by the scattering formula (1.31), for a simple harmonic oscillator.

We note that there are two δ-functions in eqn (1.33a), one representing conservation of energy and the other conservation of momentum. The momentum δ-function occurs generally in periodic structures since the Bloch wave functions have the form $\exp^{(i\mathbf{q}\cdot\mathbf{n})}$. The matrix element in the cross-section is then

$$\left|\sum_{\mathbf{n}}\langle i|e^{i\mathbf{q}\cdot\mathbf{n}}|f\rangle\right|^2 = \left|\sum_{\mathbf{n}} e^{-i\mathbf{k}\cdot\mathbf{n}} e^{i\mathbf{q}\cdot\mathbf{n}} e^{i\mathbf{k}'\cdot\mathbf{n}}\right|^2$$

$$= N_0 \frac{(2\pi)^3}{v} \sum_{\boldsymbol{\tau}} \delta(\mathbf{Q}-\mathbf{q}-\boldsymbol{\tau}),$$

where we have used the theorem (proved in §1.3):

$$\left|\sum_{\mathbf{n}} e^{i\mathbf{Q}\cdot\mathbf{n}}\right|^2 = N_0 \frac{(2\pi)^3}{v} \sum_{\boldsymbol{\tau}} \delta(\mathbf{Q}-\boldsymbol{\tau}).$$

The existence of the double δ-function means that scattering occurs under special conditions, which are only satisfied occasionally throughout the course of an experiment on a single crystal. (The situation is analogous to the experimental study of nuclear Bragg scattering, discussed in §1.3.2.) One obtains phonon peaks in an experiment, not phonon distributions. The complicated 'dynamic structure factor'

$$\sum_{\rho} \frac{\langle b \rangle_\rho}{M_\rho} e^{i\mathbf{Q}\cdot\boldsymbol{\rho}} \, \mathbf{Q}\cdot\mathbf{U}_\rho^j(\mathbf{q}) \, e^{-W_\rho}, \qquad (1.33b)$$

appearing in the last factor of eqn (1.33a), is not of great importance, as it only affects the intensity of the phonon peaks and not their energy. However, it can provide a means of distinguishing the various modes in a complex phonon spectrum, since the polarization term $\mathbf{Q}\cdot\mathbf{U}_\rho(\mathbf{q})$ is multiplied by the phase term $\exp^{(i\mathbf{Q}\cdot\mathbf{n})}$ and causes different modes to be prominent at different values of $\mathbf{Q}$. The most striking application of these ideas is given by Waeber's work on gallium (Waeber 1969).

*1.5.4 Incoherent phonon scattering*

The incoherent one-phonon scattering yields less information than the coherent one-phonon scattering. The absence of the momentum δ-function means that the scattering consists of a broad distribution in energy. It also means that the relative orientation of the scattering vector $\mathbf{Q}$ and of the reciprocal lattice vector $\boldsymbol{\tau}$ is of little importance, and so it is customary to carry out experimental measurements on powdered samples.

The differential scattering cross-section for the single-phonon process is given by

$$\frac{d^2\sigma_{\text{incoh}}^1}{d\Omega\, dE'} = \sum_{\mathbf{q},j} \frac{k'}{k} \delta(\hbar\omega \mp \hbar\omega_j(\mathbf{q})) \frac{\hbar(n_s + \tfrac{1}{2} \pm \tfrac{1}{2})}{2\omega_j(\mathbf{q})}$$

$$\times \sum_\rho \frac{(\langle b^2\rangle_\rho - \langle b\rangle_\rho^2)}{M_\rho} |\mathbf{Q}\cdot\mathbf{U}_\rho^j(\mathbf{q})|^2 \, e^{-2W_\rho}. \qquad (1.34)$$

## BASIC THEORY

As before, the upper signs refer to neutron energy loss and the lower signs to energy gain. This formula resembles the formula (1.33a) for the coherent one-phonon cross-section, with the important difference that the mode polarization term $|\mathbf{Q}.\mathbf{U}|^2$ has now lost its structure factor coefficient so that the intensity of a mode is proportional to the square of the vibrational amplitude of each atom in that mode and to the corresponding incoherent atomic cross-section, $4\pi(\langle b^2 \rangle_\rho - \langle b \rangle_\rho^2)$.

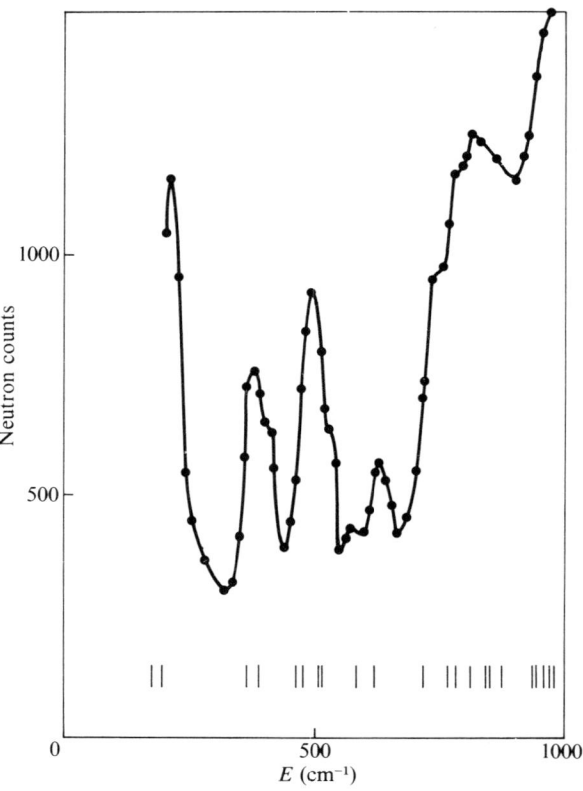

FIG. 1.14. Excitation spectrum (giving phonon density of states) from naphthalene, as measured with a beryllium-filter-detector spectrometer. The stick diagram indicates the energy levels derived from Fig. 1.12.

In the special case of a crystal with cubic symmetry, containing one atom in the primitive unit cell, we can write $\langle (\mathbf{Q}.\mathbf{U})^2 \rangle = \tfrac{1}{3} Q^2 \langle u^2 \rangle$ where $\langle u^2 \rangle$ is the mean-square amplitude of vibration of the atom. Substituting in eqn (1.34):

28

$$\frac{d^2\sigma^1_{incoh}}{d\Omega\,dE'} = \frac{\hbar k'}{k}(\langle b^2\rangle - \langle b\rangle^2)\frac{Q^2\langle u^2\rangle}{6M}(n_s + \tfrac{1}{2} \pm \tfrac{1}{2})\,e^{-2W}$$
$$\sum_{q,j}\frac{\delta(\hbar\omega \mp \hbar\omega_j(q))}{\omega_j(q)}.$$

Because of the $\delta$-function, the summation over $q$ can be transformed to an integration over the density of phonon states $Z(\omega)$:

$$\sum_{q,j}\frac{\delta(\hbar\omega \mp \hbar\omega_j(q))}{\omega_j(q)} = \frac{1}{\hbar}\sum_{q,j}\frac{\delta(\omega \mp \omega_j(q))}{\omega_j(q)} = \frac{3N}{\hbar}\frac{Z(\omega)}{\omega}.$$

The factor $3N$ is the total number of phonon states and $Z(\omega)$ is the normalized density: $\int_0^\infty Z(\omega)d\omega = 1$. The cross-section formula reduces, therefore, to

$$\frac{d^2\sigma^1_{incoh}}{d\Omega\,dE'} = \frac{k'}{k}(\langle b^2\rangle - \langle b\rangle^2)\frac{Q^2\langle u^2\rangle}{2M}(n_s + \tfrac{1}{2} \pm \tfrac{1}{2})\,e^{-2W}N\frac{Z(\omega)}{\omega}.$$

(1.35)

Thus a simple experimental measurement on a powdered sample of the energy distribution of the neutrons scattered in an arbitrary direction will give directly the range in energy of the phonon spectrum.

In the case of hydrogenous materials, such as ordinary undeuteriated naphthalene, the incoherent hydrogen scattering cross-section of 80 b dominates all other scattering processes, and so the observed inelastic scattering is proportional to the density of states of each mode multiplied by the mean-square vibrational amplitude of the hydrogen atoms in that mode. With the use of a beryllium-filter-detector spectrometer (see Chapter 2), the experiment is particularly easy and presents a method of determining the vibrational amplitude of each hydrogen atom by the successive replacement of hydrogen atoms by deuterium atoms. Fig. 1.14 illustrates the results obtained from such an experiment on naphthalene, and further examples are discussed in Chapters 3 and 5. The density of states spectrum, $Z(\omega)$ versus $\omega$, derived from Fig. 1.14 is given in Fig. 3.14.

## 1.6 References

ABRAHAMS, S. C., ROBERTSON, J. M. *and* WHITE, J. G. (1949). *Acta crystallogr.* **2**, 238.
HUGHES, D. J. *and* SCHWARTZ, R. B. (1958). *Neutron cross-sections*, BNL-325. U.S. Govt. Printing Office, Washington D.C. (and later supplements).
LEIGH, R. S., SZIGETI, B. *and* TIWARI (1971). *Proc. R. Soc.* **A320**, 505.
PAWLEY, G. S. (1967). *Phys. Stat. Sol.* **20**, 347.
PAWLEY, G. S. *and* YEATS, E. (1969). *Acta crystallogr.* **B25**, 2009.
PAWLEY, G. S. *and* CYVIN, S. J. (1970). *J. chem. Phys.* **52**, 4073.

SCHIFF, L. I. (1955). *Quantum mechanics.* McGraw-Hill, New York.
VAN HOVE, L. (1954). *Phys. Rev.* **95**, 249.
WAEBER, W. B. (1969). *J. Phys.* C **2**, 882, 903.

*General*

LOMER, W. M. *and* LOW G. G. (1965). Chapter 1 of *Thermal neutron scattering* [*editor: P. A. Egelstaff*] Academic Press, London.

This chapter is a highly condensed treatment of the theory necessary for deriving the cross-section formulae for the various types of neutron scattering processes that can occur in solids and liquids.

MARSHALL, W. C. *and* LOVESEY, S. W. (1971). *Theory of thermal neutron scattering.* Clarendon Press, Oxford.

This is the first, comprehensive, theoretical treatment of all aspects of neutron scattering which are appropriate to the study of the condensed state of matter.

# 2 *Experimental Techniques*

*By* G. C. STIRLING

*Rutherford Laboratory, Chilton*

## 2.1. Introduction

The progress of neutron scattering to its present status, where it can be regarded as a routine technique in chemistry, has followed a pattern familiar with other procedures borrowed from the realm of physics. Neutron scattering processes were first investigated because of their inherent importance in nuclear reactor systems, but it quickly became apparent that thermal neutron scattering could provide unique information in many branches of the physics of condensed matter. Early experiments were devoted to diffraction studies, complementing and extending X-ray work. Subsequently, with the increase in quality of neutron sources. inelastic scattering experiments became practicable, and neutron energy changes were measured as well as diffraction effects. The rapid growth in the number of these measurements, particularly in the field of solid state physics, has been paralleled by the development of experimental techniques to a high level of efficiency. In principle, neutron scattering measurements differ little from those employed in other radiation scattering techniques familiar in chemistry. In neutron elastic scattering, where momentum changes only are measured, the scattered intensity is measured as a function of the direction of scattering with respect to the incident beam and the orientation of the sample. In inelastic scattering, energy changes are determined, and this requires a knowledge of neutron energies before and after scattering. Momentum change is precisely the quantity determined in X-ray diffraction and energy change in Raman spectroscopy.

In the account that follows an outline will be given of the more important methods used in neutron scattering. Brief descriptions will be given of the various instruments used for the investigations covered in later chapters, and so they will be biased towards instruments currently in use at the U.K. research reactors: DIDO and PLUTO at Harwell, and HERALD at Aldermaston. More detailed accounts of particular aspects of instrumentation, and of specific instruments, can be found in sources quoted in the general references (page 48).

## 2.2. General features of neutron beam experiments

### 2.2.1. *Units*

We have seen in the preceding chapter how radiation scattered from condensed matter can be described in terms of the variables $\mathbf{Q}$ and $\omega$, where $\hbar\mathbf{Q}$ is the momentum transferred in the scattering process, and $\hbar\omega$ is the energy transfer.

## EXPERIMENTAL TECHNIQUES

In essence, therefore, a scattering experiment is concerned with the determination of these quantities, and it is fortunate that for neutrons both are readily accessible. We recall that the momentum change is given by

$$\hbar Q = \hbar(k - k')$$

and the energy transfer by

$$\hbar\omega = \frac{\hbar^2}{2m_n}(k^2 - k'^2) = E - E',$$

where $k$ and $k'$ are the incident and scattered wave-vectors and $E$ and $E'$ the corresponding energies. The magnitude of $k$ is $2\pi/\lambda$ where $\lambda$ is the neutron wavelength

$$\lambda = \frac{h}{m_n v_n}, \tag{2.1}$$

and $v_n$ is the neutron velocity. For thermal neutron beams, both the wave properties and the particulate properties (related by eqn 2.1) can be exploited to measure neutron energies. Because of these different approaches, a variety of units has arisen in the description of the quantities measured in neutron scattering experiments.

Table 2.1 shows typical energy and wavelength ranges covered by thermal neutron beams. It is the correspondence of these values with the energies and distances encountered in the solid and liquid state that marks the uniqueness of thermal neutron scattering in the study of condensed matter.

TABLE 2.1

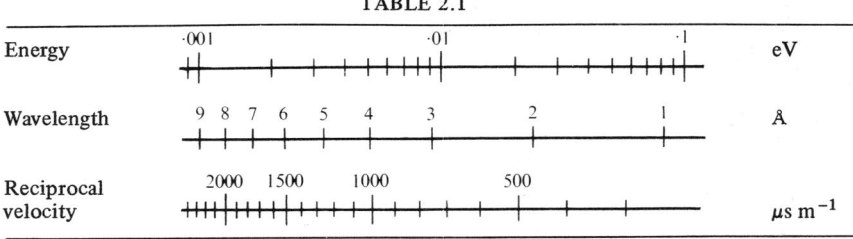

The values in Table 2.1 are related by the useful conversion expression

$$\lambda = 3{\cdot}96 \times 10^{-3}\,\tau = \frac{0{\cdot}286}{\sqrt{E}}, \tag{2.2}$$

where $\lambda$ is the wavelength in Å, $\tau(= 1/v_n)$ is the reciprocal velocity (sometimes called the 'time-of-flight') in $\mu\text{s m}^{-1}$, and $E$ is the energy in eV. In fact the terms wavelength, velocity, and energy are frequently used interchangeably, with the numerical values related according to eqn (2.2).

## 2.2.2. Sources

Neutron scattering research as described in this volume has been made possible by the development of the nuclear fission reactor as a neutron source, and more and more sophisticated experiments have become possible as the available neutron fluxes have been increased. The highest flux reactors in operation today yield maximum thermal fluxes of the order $1\text{-}2 \times 10^{15}$ neutrons/$cm^2$/s, although the majority of work to date has probably been carried out on medium flux reactors such as DIDO and PLUTO at Harwell, with maximum thermal fluxes in the region of $10^{14}$ neutrons/$cm^2$/s.

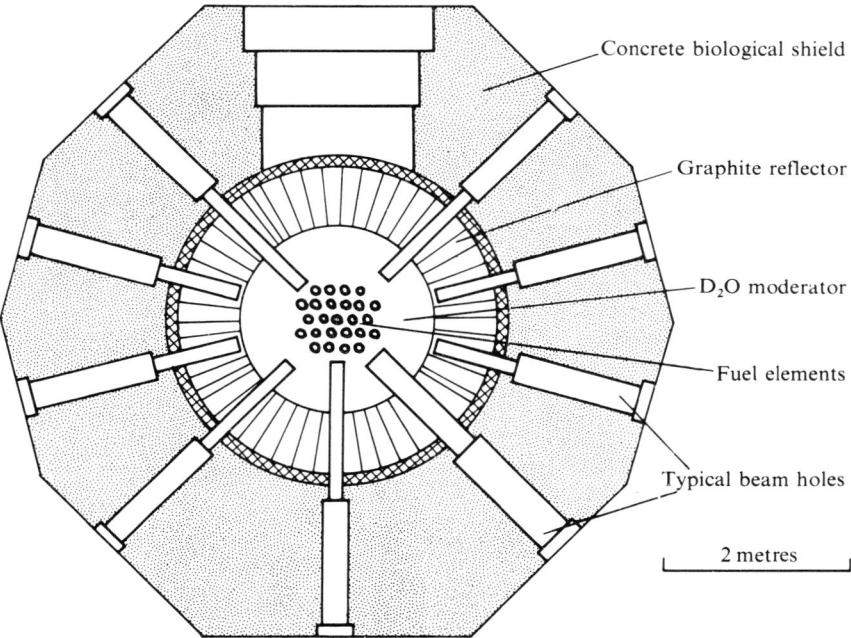

FIG. 2.1. Typical beam hole arrangement in a $D_2O$-moderated reactor (based on the reactor DIDO at Harwell).

Fig. 2.1 shows diagrammatically a typical reactor used for neutron beam research. Fission neutrons are produced in the enriched uranium core and escape into the $D_2O$ where they are moderated to thermal energies. The maximum thermal flux occurs just outside the core, and it is into this region that the majority of beam tubes penetrate. Although the flux in this region (the effective source) may be of the order of $10^{14}\text{-}10^{15}$ neutrons/$cm^2$/s, only a small fraction of the neutrons are moving in the direction of the beam tube and can travel to the experimental position outside the biological shielding. The peak flux in the beam tube is thus reduced by a factor of about $10^5$, determined by the solid angle subtended by the source area. Monochromation of the collimated beam reduces the useful flux still further, yielding a neutron flux at the sample position of about $10^5$-$10^6$ neutrons/cm$^2$/s. This is a relatively low intensity compared with, for example, an X-ray beam from a commercial sealed-off tube, and to obtain adequate counting rates, relatively large specimens and beam areas are required, sometimes of the order of 5 x 5 cm$^2$.

The thermal neutrons have a spectrum of energies that approximates to a Maxwellian distribution at the temperature of the reactor moderator. The neutron flux is $v_n$ times the velocity distribution, and in terms of $\lambda$ the neutron flux spectrum, giving the number of neutrons with a wavelength in the range $\lambda$ to $\lambda + d\lambda$, is

$$n(\lambda)\,d\lambda = \frac{4n_0 v_T}{\pi^{\frac{1}{2}}} \frac{\lambda_T^4\,d\lambda}{\lambda^5} \exp(-\lambda_T^2/\lambda^2). \tag{2.3}$$

Here $n_0$ is the total number of neutrons per unit volume in the source,

$$v_T = (2k_B T/m_n)^{\frac{1}{2}},$$

and

$$\lambda_T = h/m_n v_T.$$

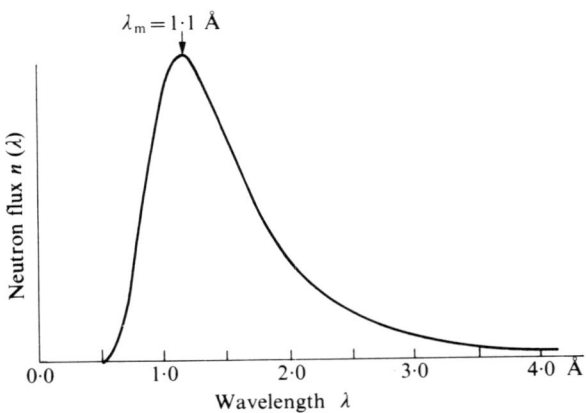

FIG. 2.2. Wavelength spectrum of neutrons for a typical moderator temperature.

The spectrum given by eqn (2.3) has a maximum at the wavelength

$$\lambda_m = h/(5k_B T m_n)^{\frac{1}{2}}.$$

The calculated spectrum of neutrons corresponding to a moderator temperature of 40°C, for which $\lambda_m$ is 1·1 Å, is shown in Fig. 2.2. It should be noted that there is, in practice, a tail of incompletely thermalized neutrons extending to high energies in the MeV region.

Eqn. (2.3) shows how the flux distribution as a function of wavelength can be modified by suitable selection of the moderator temperature $T$. This is achieved, in practice, by inserting in the reactor special 'cold' sources and 'hot' sources, which shift the Maxwellian distribution of the emerging neutrons to longer or shorter wavelengths respectively. For example, a small volume of liquid hydrogen at 20K situated at the inner end of the beam tube can give considerable flux gains for long-wavelength neutrons ($\lambda > 10$ Å), while BeO and graphite at 1000–2000°C give increased fluxes for wavelengths $< 1$ Å. Although the great majority of neutron scattering work has been carried out at continuous-flux reactors, it should be noted that alternative sources are in operation, and promise to be of increasing importance in the future as steady-state reactors approach their technological limit. In pulsed fast reactors, high-intensity neutron pulses are produced by mechanical variation of the reactivity. In sources based on electron linear accelerators (linacs), electrons are stopped by a heavy metal target to produce $\gamma$-rays which in turn yield neutrons by $(\gamma, n)$ or $(\gamma, f)$ reactions. In both pulsed reactors and linac sources, relatively high-energy bursts of neutrons are produced, and are moderated outside the source. Experiments are carried out by time-of-flight techniques.

*2.2.3. Energy (wavelength) selection*

Neutrons emerging from a reactor have a broad, continuous range of energies, and in most experiments it is necessary first to select a narrow band of energies from this distribution. This contrasts strongly with X-ray sources, which produce a characteristic line spectrum superimposed on a white background, so that monochromation essentially involves elimination of the unwanted non-characteristic radiation. Neutron-energy selection can be achieved by Bragg scattering, using single crystals to give well-defined wavelengths or polycrystals to remove a range of wavelengths, by mechanical velocity selectors, or by time-of-flight methods. The method chosen will depend on experimental requirements, and on the desired wavelength $\lambda$ and wavelength spread $\Delta\lambda$.

*2.2.3.1. Crystal diffraction.* Neutrons from the reactor impinging on a single crystal will be diffracted to give specific wavelengths at angles of scattering $2\theta_M$ according to the Bragg relation

$$\lambda = 2d \sin \theta_M$$

where $d$ is the interplanar spacing of the diffracting planes (Fig. 2.3). Perfect single crystals prove unsatisfactory, in practice, because of their low reflected intensity, and mosaic crystals are preferred, with an angular spread ideally matching that of the beam divergence, usually of the order of one degree of arc. Under these conditions, beams with a wavelength spread of a few per cent of the selected wavelength are obtained. Suitable crystals for monochromators are selected on the basis of their reflectivity, mosaic spread, and on the neutron wavelength required. Order contamination (simultaneous reflexion of neutrons

EXPERIMENTAL TECHNIQUES

with wavelengths $\lambda, \lambda/2, \lambda/3$ etc.) can present difficulties for certain types of experiment and so can influence the choice of crystal. Many different crystals have been used; typical examples include aluminium, germanium (mechanically deformed to increase its mosaic spread), and pyrolytic graphite. Notice also that

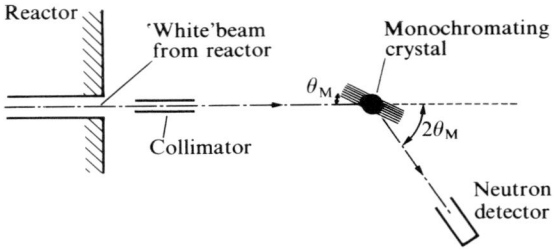

FIG. 2.3. Neutron beam monochromation by single-crystal diffraction

apart from acting as monochromators in selecting a particular band of wavelengths from a white beam, single crystals are also used in analysing the wavelength distribution of scattered beams, for instance in the triple-axis spectrometer (§2.4.1.).

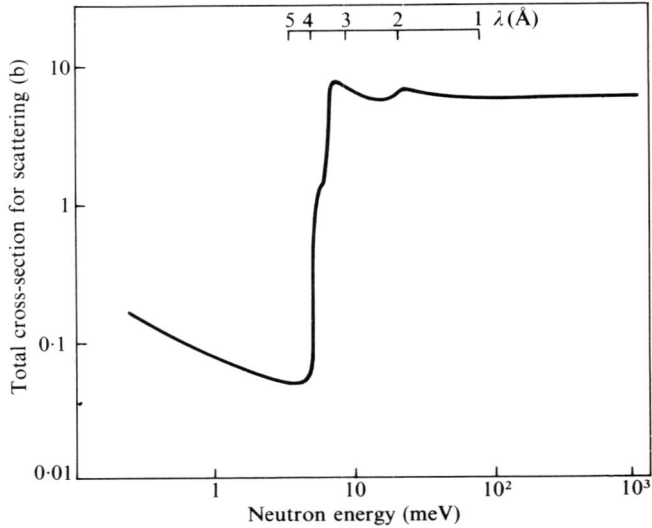

FIG. 2.4. The cross-section of polycrystalline beryllium at 77K, showing the sharp cut-off at $\lambda = 3\cdot96$ Å.

*2.2.3.2. Polycrystalline filters.* In a polycrystalline material there is a maximum spacing between planes ($d_{max}$) corresponding to the lowest-index reflexion. This leads to a maximum wavelength that can be reflected, $\lambda = 2d_{max}$, and Bragg scattering is absent at longer wavelengths. A sharp cut-off in total scattering cross-section results, shown for the case of polycrystalline beryllium in Fig. 2.4.

One way in which this cut-off at 3·96 Å is exploited is in the production of a beam of cold neutrons. A suitable length of polycrystalline beryllium, say 20-40 cm, in a beam emerging from a reactor will remove all neutrons with energies greater than 5 meV, leaving a beam with the relatively narrow energy spread of 0-5 meV.

*2.2.3.3. Mechanical velocity selectors.* A variety of mechanical methods is used for monochromating neutron beams. These usually involve some form of rotating slit device. A simple arrangement for selecting neutrons within a defined velocity range is shown in Fig. 2.5. Two disks opaque to neutrons

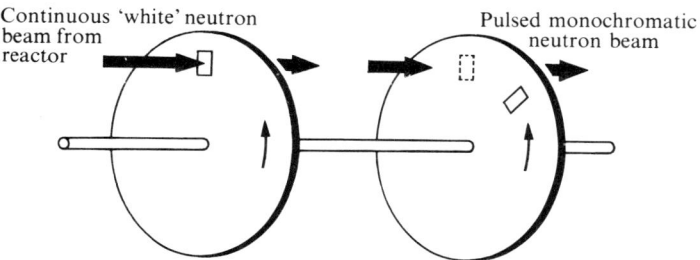

FIG. 2.5. Sketch showing the principle of a mechanical velocity selector.

rotate on the same axis parallel to the neutron beam. A slit in the first disk periodically chops the neutron beam, yielding bursts of neutrons covering the full velocity spectrum. Neutrons within the pulse separate according to their relative velocities, and only those neutrons falling within an appropriate velocity range are passed by the second slit. The mean velocity and the velocity spread are dependent on the time-of-flight between the disks, the relative phase between the slits, and the open times of the slits. In practice, a series of three or four chopping disks is often used, providing greater flexibility in the choice of wavelength and resolution. The method is especially useful at longer wavelengths, say > 3 Å, where troublesome effects can arise with crystal monochromators through order contamination.

Another form of velocity selector is shown in Fig. 2.6, illustrating a neutron chopper of typical design in routine use at Harwell. The chopper is constructed of neutron-absorbing Mg–Cd alloy, with a series of curved slits to transmit neutrons. When spun rapidly about its vertical axis, only those neutrons within a narrow velocity band are transmitted, depending on the speed of rotation and the shape of the slits. All other neutrons collide with the slit walls and are removed from the beam. Further flexibility in selection of $\lambda$ and $\Delta\lambda$ is afforded if a pair of choppers is used in tandem in a similar manner to the device in Fig. 2.5 (see §2.4.3.).

# EXPERIMENTAL TECHNIQUES

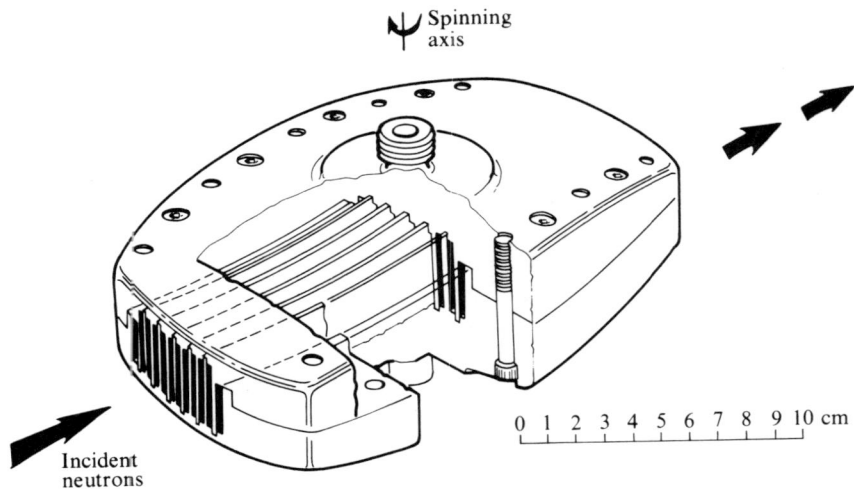

FIG. 2.6. The Harwell chopper monochromator used on the DIDO time-of-flight spectrometer shown in Fig. 2.14. (After Bedford, Dyer, Hall, and Russell 1968).

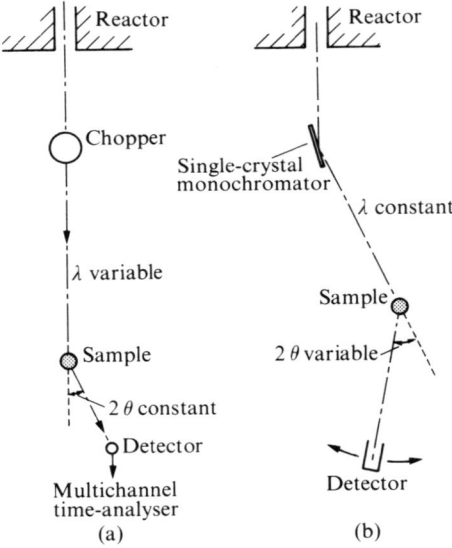

FIG. 2.7. (a) Sketch showing the arrangement for a time-of-flight diffraction experiment. Note that the chopper would not be required with a pulsed neutron source. (b) shows for comparison the conventional monochromatic-beam method. Examples of measurements obtained with the two methods are shown in Figs. 1.6 and 1.7.

*2.2.3.4. Time-of-flight techniques.* The fact that thermal neutron velocities fall in a readily accessible range with flight times of the order of $10^{-5}$–$10^{-3}$ $\mu$s/m provides a convenient method for wavelength analysis of scattered neutrons. The wavelength is given by

$$\lambda = h/m_n v_n$$
$$= (h/m_n)t/l$$

and so can be determined by measuring the time of arrival $t$ of neutrons after traversing the distance $l$. A typical application is the white-beam method of structure analysis referred to in Chapter 1. The technique is illustrated in Fig. 2.7, which also shows, for comparison, the more conventional monochromatic-beam method. The monochromatic method follows X-ray practice, where the intensity of the scattered radiation is measured as a function of the angle of scattering $2\theta$ at constant $\lambda$. Diffraction peaks are observed when the Bragg condition $\lambda = 2d \sin \theta$ is satisfied. In the time-of-flight method, the wavelength spectrum is obtained by measuring the flight times of scattered neutrons at constant $\theta$. The method is particularly good for observing diffraction peaks at large $d$'s, since both the resolution and scattered intensity are enhanced at longer wavelengths. With continuous neutron sources the data collection efficiencies of the two methods are roughly comparable, but time-of-flight methods may prove more important in the future with the advent of high-intensity pulsed neutron sources, where monochromation by diffraction would be extremely wasteful of the available neutrons.

## 2.2.4. Detectors

Because neutrons are non-ionizing in their passage through matter, they are invariably detected by observing the results of secondary nuclear reactions, for example, by the formation of energetic charged particles in (n, p) or (n, $\alpha$) reactions:

$$^{10}B\ (n, \alpha)^{7}Li$$
$$^{3}He\ (n, p)^{3}H$$
$$^{6}Li\ (n, \alpha)^{3}H.$$

The recoil particles are detected either by gas ionization or solid-state scintillation techniques. It is necessary that detectors be as insensitive as possible to unwanted fast neutrons and $\gamma$-rays while retaining a high detection efficiency for thermal neutrons. In addition it is necessary, especially in time-of-flight methods, that the time and position uncertainty of detection be kept low.

In the widely-used boron trifluoride proportional counter shown in Fig. 2.8, detection efficiencies approaching 100 per cent are achieved by enrichment of the $^{10}B$ isotope to approximately 96 per cent with gas pressures of 1-2 atmospheres. The counters are usually of all-metal construction with the outer casing acting as one of the electrodes. Typical dimensions are diameters of 2-5 cm, with lengths up to 50 cm. For maximum efficiency, and when the spatial orientation

## EXPERIMENTAL TECHNIQUES

of the scattered neutrons is desired, the detector is used end-on, whereas in time-of-flight systems, when the location of the detection event in the neutron path is important, the counter is used side-on to achieve adequate time resolution. Helium-3 filled proportional counters are similar in principle and construction to the $BF_3$ counter, with the advantage of a larger nuclear cross-section, leading

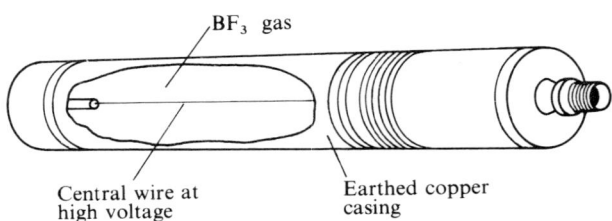

FIG. 2.8. Drawing of a $BF_3$ proportional counter, with cut-away section showing interior.

to relatively greater efficiencies. But its general robustness and lower cost have, to the present, maintained the $BF_3$ counter as the most generally used neutron detector.

In scintillation counters, neutron absorbing nuclei are combined in a solid matrix with a suitable phosphor material. Following neutron absorption the products of the nuclear reaction are detected by a photomultiplier tube. Solid-state detectors have the advantage that the location in space of detected neutrons can be better defined than in the bulkier gas-filled counters. However, the superior discrimination against $\gamma$-rays provided by gas counters, and their generally higher efficiency, have led to the predominant use of the $BF_3$ and $^3$He counters.

### 2.2.5. Shielding

One of the dominating aspects in the design of neutron-beam equipment arises from the great penetrating power of neutrons, particularly those of higher energies, which leads to high levels of stray thermal neutrons and fast neutrons in experimental areas. In addition, there is always an associated $\gamma$-ray component of some magnitude. The resulting effect is a requirement for substantial shielding, both to reduce health hazards to personnel and to reduce spurious background counts at detectors. Slow neutrons are easily absorbed by boron- or cadmium-containing materials, these nuclei having extremely high absorption cross-sections to thermal neutrons (see Appendix I). Fast neutrons are much more difficult to attenuate; usually bulky hydrogenous material such as water, resin, or polythene is used to moderate the neutrons to thermal velocities so that they can then be absorbed by boron or cadmium, resulting in massive shielding that can restrict considerably the manoeuvrability of neutron apparatus. This is seen in the photograph of a spectrometer (p. 42). Stray $\gamma$-

radiation can also be difficult to eliminate, but fortunately this can be discriminated against electronically, particularly with gas-filled proportional counters.

### 2.2.6. Polarized neutrons

Polarized neutrons have so far been little used in chemical applications of neutron scattering,† although an increased interest seems likely in the future, particularly as neutron sources and techniques improve in quality. Polarized beams are most frequently produced by Bragg reflexion from magnetized single crystals. For example, polarization efficiencies greater than 99 per cent can be attained for reflexion from the (111) plane in crystals of cobalt-iron alloy. The polarization can be switched by applying a radio-frequency field along the neutron flight path, which also helps to avoid any depolarization effects. It should be noted that polarized beams produced by Bragg reflexion are also monochromatic in wavelength. When energy selection is not required magnetized mirrors may be used, but these have found little use to date in routine solid-state research.

## 2.3 Instruments for neutron-beam research: Elastic scattering

### 2.3.1. Powder diffractometer

Fig. 2.9 is a layout of a conventional diffractometer used for the examination of polycrystalline materials. Neutrons from the reactor moderator pass through a collimator in the reactor shield. Perfect collimation can only be achieved at the expense of neutron intensity, and in practice a beam divergence of about $\pm 0.3°$ is selected. The beam is monochromated by Bragg reflexion from a single crystal as described in §2.2.3.1. Beam channels at various take-off angles may be provided, to be used as required for different neutron wavelengths. There is extensive fast neutron and $\gamma$-ray shielding, particularly in the vicinity of the primary beam, to reduce background counting levels at the detector. The scattered neutrons are detected by a proportional counter used end-on. Note the large scale of the diffractometer, as indicated in Fig. 2.9. These large dimensions are required because a large cylindrical sample, perhaps 1 cm diameter and 2 cm long is used; the large size enhances the neutron count rate, but to obtain satisfactory resolution of the Debye-Scherrer peaks the apparatus must be scaled up accordingly. A typical diffraction pattern, for polycrystalline naphthalene, is shown in Fig. 1.6. Diffractometers of this type are also used for structure work on liquids, as described in Chapter 8.

### 2.3.2. Single-crystal diffractometer

The four-circle diffractometer is a well-established instrument for crystal-structure analysis by X-ray diffraction, and has been readily adapted for corre-

---

† Their application to the study of covalency in magnetic salts is described in Chapter 12.

## EXPERIMENTAL TECHNIQUES

sponding neutron studies as outlined in Chapter 1. A typical instrument, connected 'on line' to a computer, is shown in Fig. 2.10. In neutron equipment the heavy weight of the detector shielding favours the use of a normal-beam equatorial geometry with the detector moving in a single horizontal plane. Specimen sizes are mainly influenced by the need to achieve adequate counting rates without too much trouble from systematic errors such as extinction; a minimum size of crystal is $c.$ 1 mm. Since data collection continues over days

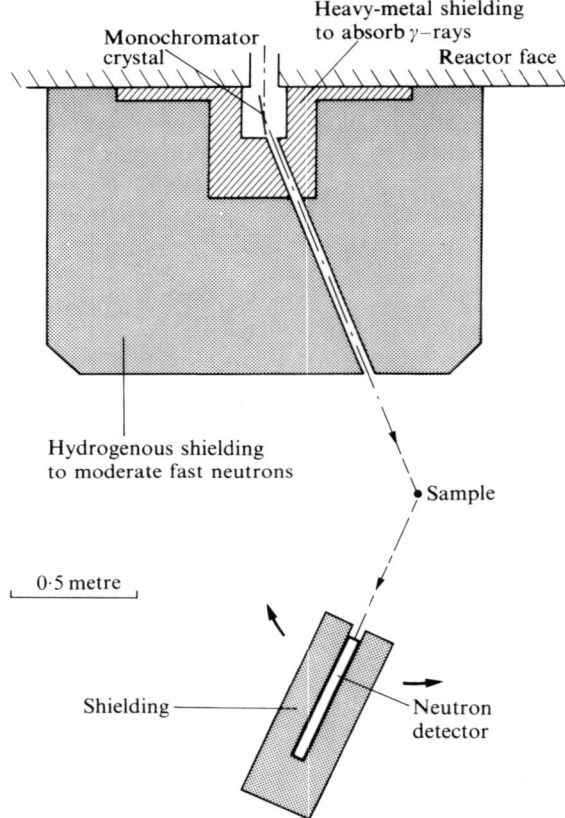

FIG. 2.9. Arrangement of rotating neutron detector for recording diffraction pattern of polycrystalline materials.

or weeks of running, instruments are pre-programmed to operate automatically. Single-crystal results obtained with this kind of instrument are discussed in Chapter 9.

### 2.3.3. Diffuse-scattering time-of-flight apparatus

A time-of-flight instrument (see Chapter 11 and Fig. 11.3) can be used for measuring the weak elastic diffuse scattering associated with the presence of lattice defects in solids. A simple chopper pulses the beam, and the scattered

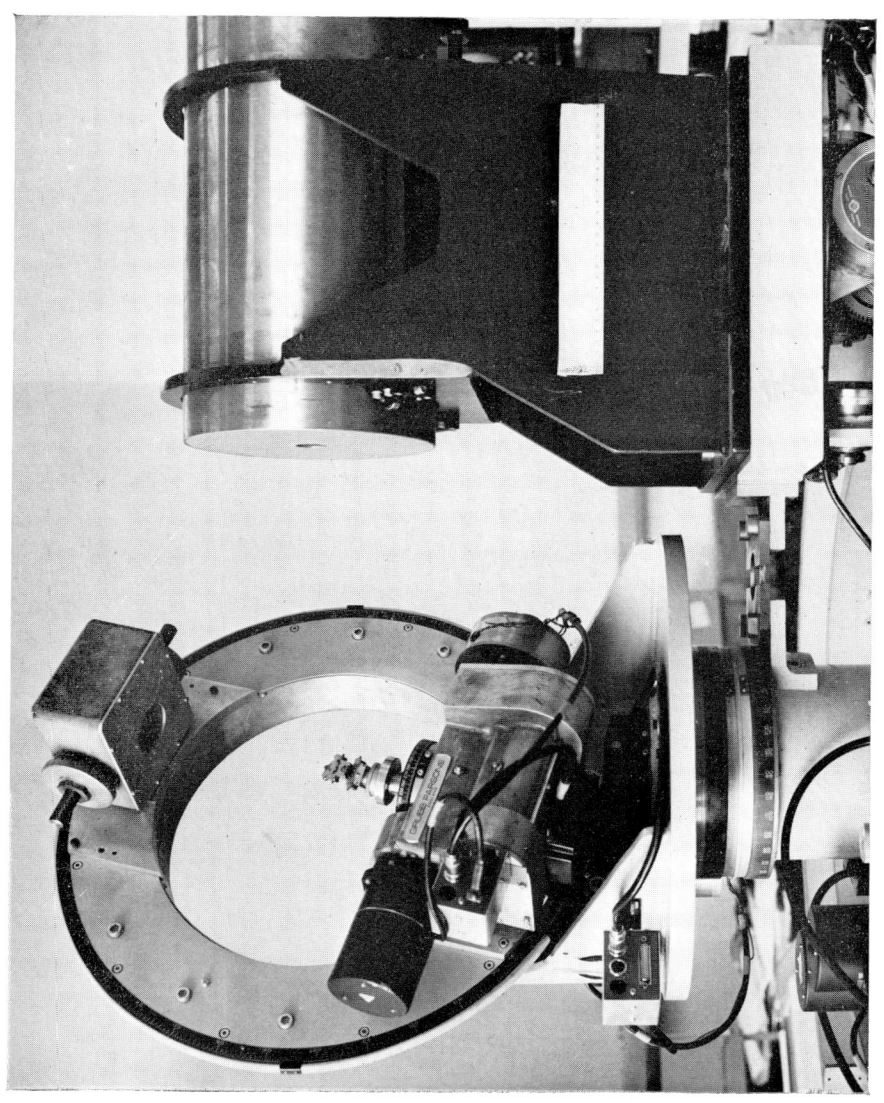

FIG. 2.10. Photograph of the Harwell 'Mark-VI' four-circle diffractometer.

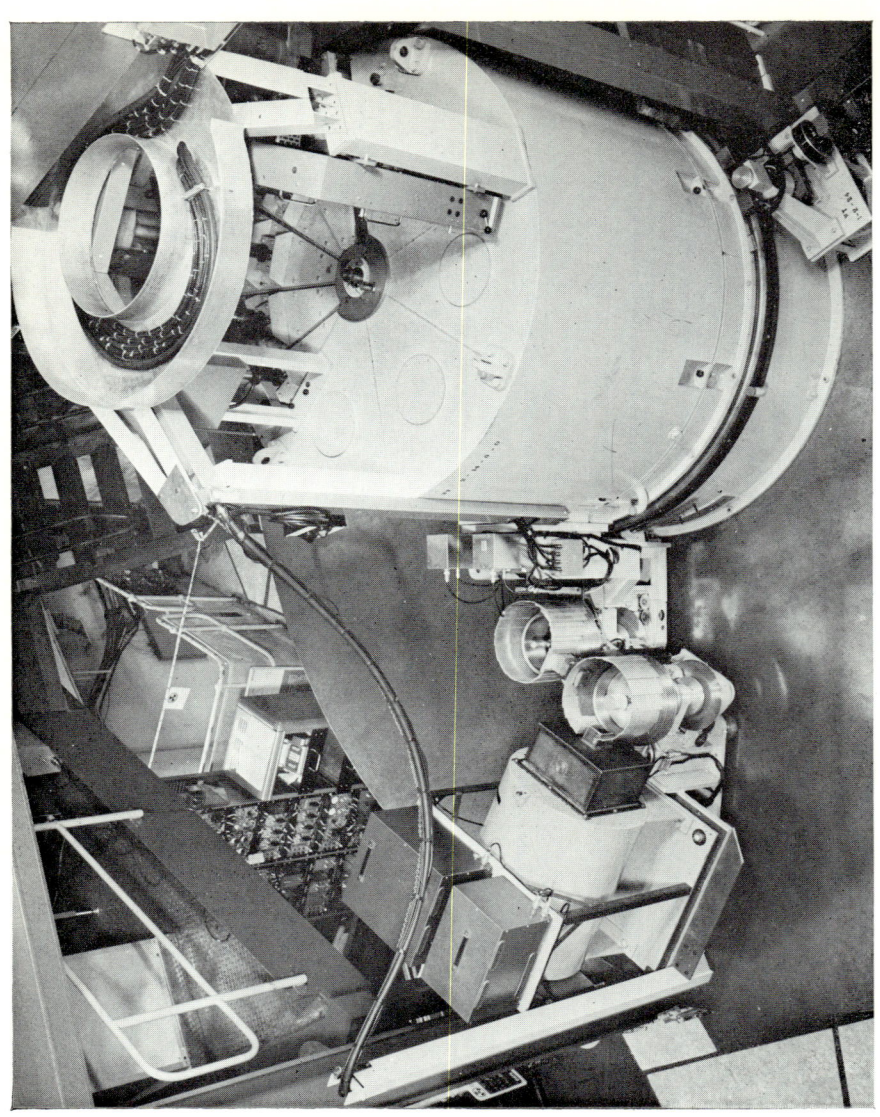

FIG. 2.12. Photograph of the PLUTO three-axis spectrometer. The large shielding drum on the right houses the monochromator; it is 1·8 m in diameter and weighs 7000 kg. (After Chumbley et al. 1968).

neutrons are detected at a number of angles of scattering after traversing a known flight path. The counters are electronically gated in synchronism with the pulses so that the neutrons are counted for only a set period after the burst, corresponding to a particular wavelength range. The wavelength resolution ($\Delta\lambda/\lambda$) of such an instrument may be made as poor as 25 per cent so as to enhance the count rate from the specimen.

It may be noted that if pulses from the detectors in Fig. 11.3 are fed to a multichannel analyser for full time-sorting, the apparatus corresponds to the white-beam diffractometer referred to in Chapter 1. (See also Fig. 2.7.) Under these circumstances the beryllium filter would be removed to give the full white beam from the reactor.

## 2.4 Instruments for neutron beam research: Inelastic scattering

### 2.4.1. Triple-axis spectrometer

In the triple-axis spectrometer both monochromation of the incident neutrons and analysis of the scattered neutrons is carried out by crystal diffraction, see Fig. 2.11 (a). A photograph of the triple-axis spectrometer on the reactor PLUTO at Harwell is shown in Fig. 2.12. Neutrons of energy $E$ are selected from the reactor beam by the single-crystal monochromator. The large shielding drum enables a continuous range of energies to be selected through variation of the monochromator take-off angle $2\theta_M$. Neutrons are scattered from the sample through the angle $2\theta_S$, and final energies are determined from the orientation of the analyser crystal and the detector angle $2\theta_A$. In this arrangement, therefore, both the energy, and momentum, transfer are determined simultaneously. The triple-axis spectrometer is widely used in coherent scattering studies of single-crystal lattice vibrations as described in Chapters 4, 5 and 7. A major advantage is the great flexibility that can be achieved through appropriate adjustment of the angles $2\theta_M, 2\theta_S, 2\theta_A$ and the sample orientation, allowing any point in reciprocal space to be reached for any particular energy transfer.

A multi-angle reflecting-crystal spectrometer, which measures scattering along a straight track in momentum-energy space instead of just at one point, has been developed at Risø in Denmark by Kjems (1970). This utilizes a large analyser crystal in conjunction with a position-sensitive detector. For a point source the position of detection along the counter uniquely defines the momentum and energy of the scattered neutron (see Fig. 2.13). The instrument gains in data-collection rate through the large solid angle in which the neutrons are detected, and is comparable with a monochromatic-beam time-of-flight instrument. At the same time, the instrument preserves the selectivity available with the triple-axis spectrometer in choosing the most suitable range of momentum–energy space to be covered in the experiment. This flexibility is not possible with a time-of-flight instrument, such as the twin-chopper spectrometer (§ 2.4.3.), in which the measurements are recorded over a more or less predetermined range of momentum and energy transfers.

# EXPERIMENTAL TECHNIQUES

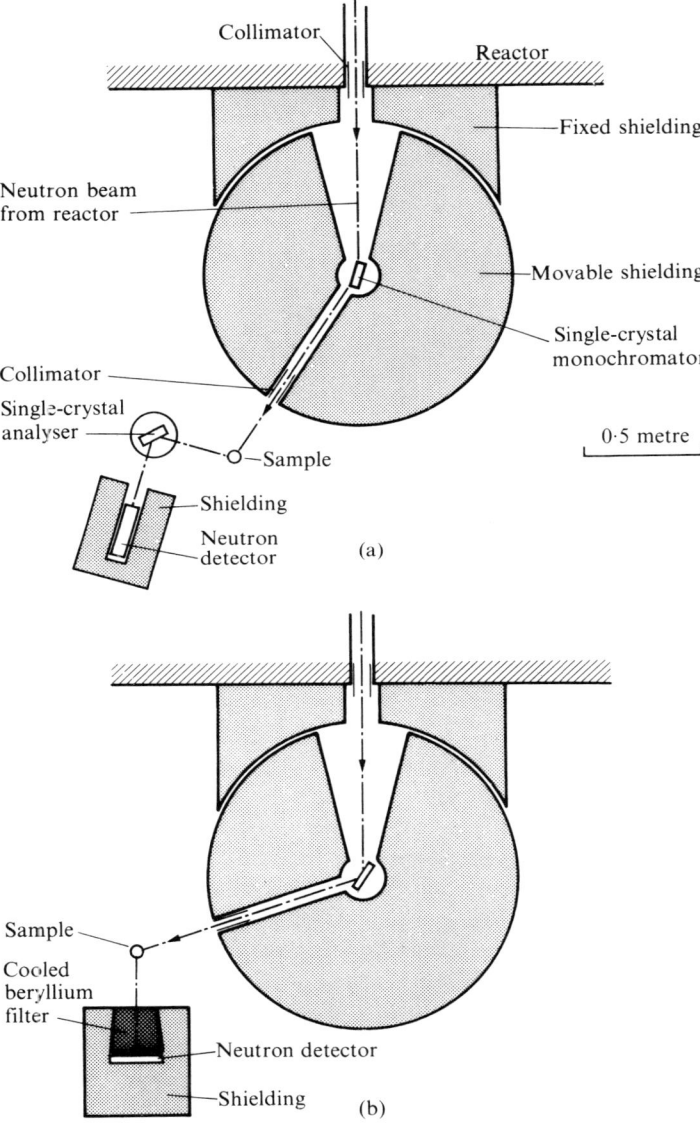

FIG. 2.11. (a) Diagram of a three-axis spectrometer. (b) shows the instrument modified for use as a beryllium-filter-detector spectrometer.

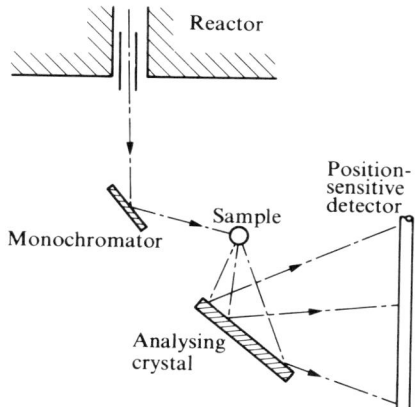

FIG. 2.13. Principle of a multiple-angle crystal spectrometer.

### 2.4.2. Beryllium-filter-detector spectrometer

A variation of the triple-axis spectrometer is provided by the beryllium-filter-detector spectrometer, in which the single-crystal analyser is replaced with a counter preceded by a polycrystalline beryllium filter. This is illustrated in Fig. 2.11(b). In this situation, only those neutrons whose energies after scattering lie in the narrow range 0–5 meV are detected. The arrangement has the advantage that large solid angles of the scattered beam can be utilized, giving faster counting rates. The energy-transfer spectrum is obtained by scanning with a variable initial energy $E$ and subtracting the relatively small final energy. The method is particularly useful for studying molecular energy levels in the range 50–250 meV: examples of experimental data are given in Chapters 1 and 3.

### 2.4.3. Twin-chopper spectrometer

Multichopper spectrometers employ the mechanical and time-of-flight techniques described in §2.2. An example of a twin curved-slot chopper instrument is shown in Fig. 2.14. The neutron beam from the reactor is pulsed and roughly monochromated by the first chopper. The second chopper opens after a preset delay time and acts still further to improve the monochromation. Appropriate adjustment of chopper properties such as slot size and speed of rotation provides considerable flexibility in the choice of beam characteristics. Typically wavelengths in the range 4–10 Å are selected, with resolutions of the order 1–5 per cent in $\Delta\lambda/\lambda$. Energy analysis of the scattered neutrons is by time-of-flight, and data are collected at many angles simultaneously. Spectrometers of this type are especially valuable in inelastic scattering studies applied to liquid dynamics (Chapter 6).

Although crystal spectrometers and monochromatic incident-beam time-of-flight spectrometers carry out essentially the same function in that they both

## EXPERIMENTAL TECHNIQUES

measure neutron energy and momentum changes, it will be noted that they are usually applied to different kinds of problem. The choice of spectrometer is usually obvious. With a triple-axis spectrometer the wide range of available

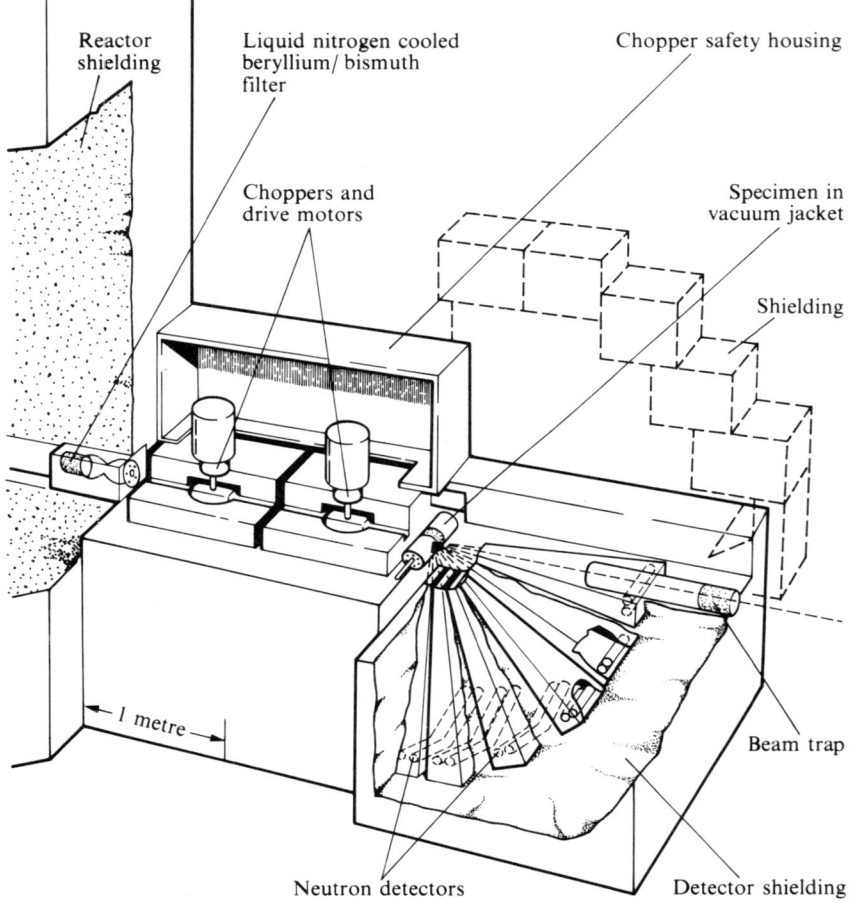

FIG. 2.14. Cut-away drawing of the DIDO time-of-flight spectrometer. A cold neutron beam ($\lambda > 4$ Å) is obtained with a cooled beryllium filter. Monochromation is further improved by a twin-chopper velocity selector. Scattered neutrons are detected at a number of angles $2\theta$, between $0°$ and $90°$ in the vertical plane, simultaneously.

spectrometer settings provides considerable selectivity in choosing any desired point in reciprocal space. The method is most valuable therefore in single-crystal work where, for example, particular phonons can be studied in considerable detail. When a wide range of $S(\mathbf{Q}, \omega)$ is of interest, which is often the case for amorphous and liquid materials, the fact that time-of-flight spectrometers can measure the complete range of energy transfer at a number of angles simultaneously provides a decisive advantage.

## 2.4.4. Beryllium-filtered-beam spectrometer

This spectrometer, which has found widespread use for investigating energy levels in molecules, especially low-frequency vibrational modes in hydrogenous compounds, utilizes the band-pass characteristic of polycrystalline beryllium to produce a roughly monochromatic incident beam. The instrument is shown in its simplest form in Fig. 2.15.

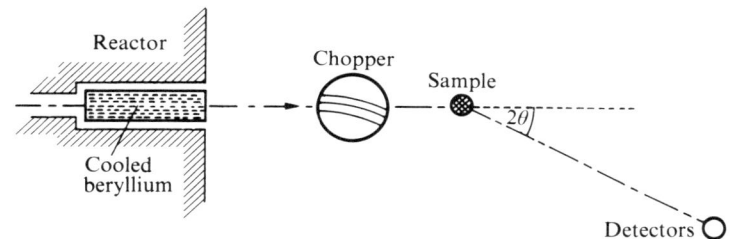

FIG. 2.15. Diagram of a low-resolution beryllium-filtered-beam spectrometer.

The beryllium filter produces a neutron beam with initial energy in the range $0 < E < 5$ meV. After scattering, neutrons are analysed by time-of-flight with a simple chopper pulsing the beam. Because the instrument is necessarily a neutron energy-gain spectrometer, the energy range that can be investigated is limited by the small population factors of upper molecular energy levels; the upper limit is usually about 80 meV for specimens at ambient temperatures. A typical spectrum obtained by this method is shown in Fig. 3.8 (b): the improved resolution which can be obtained with the twin-chopper spectrometer (§ 2.4.3.) is illustrated in Fig. 3.8 (a).

## 2.4.5. Rotating crystal spectrometer

As an example of an instrument that simultaneously utilizes diffraction and mechanical methods for neutron monochromation and analysis, we mention the rotating crystal spectrometer shown in Fig. 2.16. A monochromatic beam is provided by crystal diffraction in the usual way. However, rotation of the monochromator sweeps the diffracting plane through the Bragg reflecting condition for the particular take-off angle $2\theta_M$ of the instrument. The result is a pulsed monochromatic beam which is analysed after scattering by time-of-flight. The instrument is analogous in operation to the multichopper system described above, the output data having the same form.

# EXPERIMENTAL TECHNIQUES

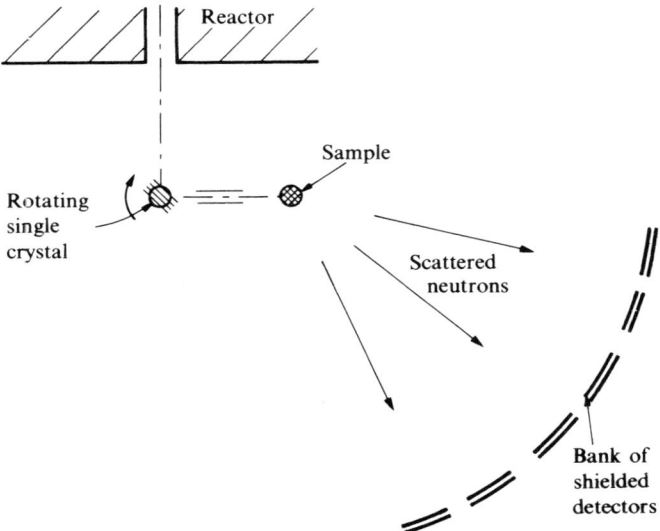

FIG. 2.16. Principle of a rotating crystal spectrometer.

## 2.5 References

BEDFORD, L. A. W., DYER, R. F., HALL, J. W. *and* RUSSELL, M. C. B. (1968). *Rep. U.K. Atom. Energy Auth. (London)* R 5662.

CHUMBLEY, L. C., DYER, R. F. *and* WALLIS, D. E. (1968). *J. Phys.* E Series 2, 1, p. 528.

KJEMS, J. (1970). Risø Research Establishment (Denmark), Report number 229.

*General*
Techniques for elastic scattering are described in the books:

ARNDT U. W. *and* WILLIS B. T. M. (1966). *Single-crystal diffractometry.* Cambridge University Press.

BACON G. E. (1962). *Neutron diffraction (2nd Ed.).* Clarendon Press, Oxford.

WILLIS B. T. M. (*editor*)(1970). *Thermal neutron diffraction.* Oxford University Press.

Inelastic scattering methods are covered in the volume:

EGELSTAFF P. A. (*editor*)(1965). *Thermal neutron scattering.* Academic Press, London,

and in the proceedings of the IAEA Symposia on *Neutron inelastic scattering* (published in several volumes, 1961, 1963, 1965, 1968, by the IAEA, Vienna). A recent publication devoted entirely to inelastic techniques is *Instrumentation for neutron inelastic scattering* (IAEA, Vienna, 1970).

# 3 Neutron Inelastic Scattering and Molecular Spectroscopy

*By* J. W. WHITE

*Physical Chemistry Laboratory, Oxford University*


## 3.1. Introduction

Molecular spectroscopy is one of the most valuable tools in chemistry, both for analysis, and for understanding the nature of the forces holding atoms together to give molecular structures. In studying the ground electronic state of molecules, infra-red and Raman spectroscopy have traditionally been used. Molecular spectra by these methods serve as finger prints for the molecules, and when analysed yield the inter-atomic force constants for the various normal modes of vibration of the molecular structure. For highly symmetrical molecules neither technique alone is sufficient to reveal all of the vibrational frequencies because of optical selection rules, and the full impact is only gained by the use of the two techniques in conjunction. There are even some normal modes (e.g. methyl torsions), where the electric dipole-moment change and polarizibility change of the molecule are both so small that the mode is inaccessible with either method.

Neutron spectroscopy is not limited by selection rules in this way (see §1.4) and all initial and final states may be connected by the scattering operator. Modes such as torsions appear strongly in the neutron spectrum and so the technique can complement both infra-red and Raman spectroscopy. But this is not the only function of neutron spectroscopy in relation to molecular dynamics and molecular force fields. In studies of molecular spectra, a pattern has emerged which shows that the neutron scattering spectrum is itself limited by a type of selection rule which is rather different from those which occur in optical spectroscopy. This arises because the scattering intensity, from a particular moving atom in the molecular vibration, is weighted by the atomic scattering cross-section of that atom, and also by its amplitude of motion. Thus those molecular modes, which lead to large amplitudes of atomic motion for atoms of large scattering cross-section, appear strongly in the incoherent inelastic neutron scattering spectrum. Other modes which either leave the atoms stationary or displace atoms of low scattering cross-section are much less easily seen.

Neutron and optical spectroscopy may therefore be combined in a valuable way to make unambiguous assignments of molecular vibration frequencies. Since the neutron spectra yield both eigen-frequencies and eigen-vectors, the vibrational-amplitude dependent part of optical absorption and scattering intensities can be analysed to further our understanding of molecular electronic properties (Longster and White 1969).

# NEUTRON INELASTIC SCATTERING AND MOLECULAR SPECTROSCOPY

A second unique contribution of incoherent neutron scattering spectroscopy arises from the fact that appreciable momentum is transferred in neutron scattering events. In scattering from molecular solids (and to some extent molecular liquids) this allows the whole density of states for the internal and inter-molecular (external) modes of vibration to be measured, in contrast to the situation in optical spectroscopy where only those modes near the Brillouin zone centre are accessible in single-photon experiments. The use of this information to estimate the dispersion of intra-molecular modes and of external modes is growing. Since the data are easy to collect compared to the laborious procedure for obtaining phonon dispersion curves from molecular crystals, and since the density of states can be connected to macroscopic properties, there is something to be said for using this method in preference to coherent scattering, at least to obtain a broad view of the inter-molecular binding in molecular solids.

This review develops both of these unique aspects of incoherent inelastic neutron scattering spectroscopy for molecular solids. Nothing will be said of phase transitions or other kinetic phenomena in solids, which are now beginning to be studied, nor anything about the dynamics of molecular liquids (see Chapter 6).

## 3.2. Molecular vibrations and the molecular force field

Classical mechanics affords a simple description of vibrations in poly-atomic molecules. The potential energy $2V$ can be expressed as a power series in the atomic displacements $q_i$ as follows:

$$2V = 2V_0 + 2 \sum_{i=1}^{3N} \left(\frac{\partial V}{\partial q_i}\right) q_i$$

$$+ \sum_{i,j=1}^{3N} \left(\frac{\partial^2 V}{\partial q_i \partial q_j}\right) q_i q_j + \cdots \quad (3.1)$$

$$= 2V_0 + 2 \sum_{i=1}^{3N} f_i q_i + \sum_{i,j=1}^{3N} f_{ij} q_i q_j + \cdots, \quad (3.2)$$

where $q_i$ are the mass-weighted Cartesian displacement co-ordinates (Wilson, Decius and Cross 1955) defined by

$$q_i = m_i^{\frac{1}{2}} x_i, \quad (3.3)$$

with $m_i$ the mass and $x_i$ the component of the Cartesian vector displacement. $N$ is the total number of atoms in the molecule. By choosing the zero of energy as that of the equilibrium configuration, all the $q_i$'s are zero for this configuration, and the first and second terms in eqn (3.2) are zero. The potential energy is thus given to the first order of approximation by the sum of quadratic terms such as

$$2V = \sum_{i,j=1}^{3N} f_{ij} q_i q_j. \tag{3.4}$$

This is the harmonic approximation in which the force constants, $f_{ij}$, specify the molecular force field in the limit of small vibrational displacements.

To simplify the problem of computing the vibrational frequencies by taking into account the molecular symmetry, it is convenient to introduce a new set of co-ordinates (normal co-ordinates) which are linear combinations of the mass-weighted co-ordinates $q_i$:

$$Q_k = \sum_{i=1}^{3N} l_{ki} q_i, \quad k = 1, 2 \ldots 3N. \tag{3.5}$$

These normal co-ordinates naturally involve the displacements of many atoms at the same time in a particular normal mode, and the coefficients $l_{ki}$ are chosen so that the vibrational Hamiltonian of the isolated molecule is diagonal. The normal co-ordinates are classified by the covering operations of the undistorted molecular point group.

In the harmonic approximation, the simplest form for the molecular force field is obtained by assuming force constants between nearest neighbour atoms only. This is called the valence force field, and is surprisingly accurate for predicting frequencies of vibration in symmetrical molecules. Improvements to this force field invoke force constants between non-bonded atoms. This is the Urey-Bradley force field. Although the number of force constants chosen for only one molecule may be greater than the number of observable frequencies, a useful property of this potential is that by studying a series of molecules of closely related chemical structure, a single field which is transferable to all molecules and hence can predict a very much larger number of frequencies than there are constants used, can be developed. This is the case for the halogen-substituted benzenes discussed below (Sherrer 1968, Reynolds and White 1969).

## 3.3. Incoherent scattering cross-section for molecules

Incoherent inelastic neutron scattering spectroscopy most closely resembles Raman scattering of light. In the experiment, an incident, monoenergetic beam of neutrons is scattered from a sample through some angle $2\theta$. A detector placed at this angle records the energy spectrum of the scattered neutrons in much the same way as does the analysing spectrograph for inelastic light scattering. Experimental apparatus for doing this is discussed in Chapter 2.

In most of the spectra to be presented in this chapter, the energy analysis is performed by time-of-flight measurement and so the abscissa is proportional to the inverse square-root of the neutron energy instead of the energy transfer itself. Also, most of the experiments are performed with cold neutron up-scattering techniques, where an incident beam of neutrons moderated at liquid hydrogen temperature suffers energy gain in the inelastic scattering from the sample. Because the incident neutrons have such low energies they are unable to excite molecular motions; consequently, the time-of-flight spectra look asymmetrical, with the elastically scattered (zero energy change) neutrons on the right-hand side and with peaks in the inelastic-scattering spectrum at shorter times-of-flight, corresponding to energy gain by the neutron on scattering.

For monoenergetic incident neutrons, the resolution function of the instrument is a delta function and the energy spectrum of the scattered neutrons is given by a modified version of eqn (1.34)

$$\frac{d^2 \sigma_{incoh}}{d\Omega dE'} = \sum_{q,s} \frac{k'}{k} \delta(\hbar\omega \mp \hbar\omega_s(\mathbf{q})) \frac{\hbar(n_s + \tfrac{1}{2} \pm \tfrac{1}{2})}{2\omega_s(\mathbf{q})}$$

$$\times \sum_{\rho} \frac{b_{incoh}^2}{M_\rho} |\mathbf{Q} \cdot \mathbf{U}_\rho(\mathbf{q})|^2 \exp(-2W_\rho), \qquad (3.6)$$

where the sum (over wave vectors $\mathbf{q}$, and modes $s$) is over all internal and external modes of the molecule and the sum over atoms of mass $M_\rho$ at positions $\rho$ in the unit cell is as before. The polarization vectors of these modes are described by $\mathbf{U}_\rho(\mathbf{q})$, and the momentum transfer $\hbar\mathbf{Q}$ is defined in terms of the incident and outgoing wave vectors $\mathbf{k}, \mathbf{k}'$ by

$$\mathbf{Q} = \mathbf{k} - \mathbf{k}'. \qquad (3.7)$$

The Debye-Waller factor $\exp(-2W_\rho)$ includes contributions from both the intra-molecular modes and from the external vibrations. Since, to a good degree of approximation, these modes are well separated in frequency, $W_\rho$ can be represented approximately as the sum of contributions from the two types of motion. When the force constants for the internal and external modes are very different, as in some molecular crystals, it is to be expected that the scattering spectrum will show two distinct regions of scattering at low and high frequency, corresponding to the density of states for the optic and acoustical phonons and the density of states for the intra-molecular modes respectively. It can be seen, by reference to eqn (3.6) and eqn (1.35), that the intensity of scattering is determined by three main factors summarized in eqn (3.8):

$$\frac{d^2 \sigma_{incoh}}{d\Omega dE'} = \frac{k'}{k} \sum_{\nu} (b_{incoh}^\nu)^2 \frac{Q^2 \langle u_\nu^2 \rangle}{2M_\nu} (n_s + \tfrac{1}{2} \pm \tfrac{1}{2})$$

$$\times \exp(-2W) N \frac{Z(\omega)}{\omega}. \qquad (3.8)$$

(The upper sign in $\pm\frac{1}{2}$ refers to neutron energy loss—'down-scattering'—and the lower sign to neutron energy gain—'up-scattering'.) One factor is the density of states spectrum $Z(\omega)$: the scattered intensity will be large for those frequencies $\omega$ for which $Z(\omega)$ is intense. However, at high frequencies, the influence of $Z(\omega)$ is diminished for up-scattering by the Boltzmann factor ($n_s \sim \exp(-\hbar\omega/k_BT)$), by the $1/\omega$ term, and by the Debye-Waller factor whose exponent ($2W = \frac{1}{6}\langle u^2\rangle Q^2$) is large since $Q$ is large at high energy transfers.

The scattering spectrum is modified by two further factors. The density of states is weighted strongly in favour of those atoms, $\nu$, with large incoherent scattering amplitudes $b^\nu_{incoh}$ and with large mean-square amplitudes of vibration $\langle u^2_\nu \rangle$. For hydrogen $b^2$ is about twenty times that for carbon, oxygen, etc. which are the other constituents of organic molecules. Also because of its small mass, the amplitude of vibration is very great. Thus for hydrogenous compounds, the incoherent scattering cross-section is largely dominated by the hydrogen motions.

Reference to the cross-section formula (eqn (3.8)) shows that, in principle, all modes of vibration for the molecule, and the scattering from all atoms in it, should appear in a neutron scattering spectrum. It can easily be anticipated that this wealth of information could well be embarrassing, especially when resolution of the spectrometer is poor. In the early days of neutron spectroscopy there was some despair because of this. A further complication was that for overlapping peaks at poor resolution the combined effect of the Debye-Waller factor and the amplitude term (both depending on $Q^2$) is to cause peaks in the incoherent neutron scattering spectrum to shift as the spectrum is observed at different angles. It was thus very difficult to decide which spectrum to compare with optical spectroscopic measurements. At higher resolution, where the peak overlap is much less pronounced, the variation in the intensities of peaks with momentum transfer can be used to test theoretical predictions of spectra.

To resolve the question of what a neutron spectrum is likely to tell us in relation to optical spectrum measurements, we discuss in the next section the scattering from a number of different molecules. These were carefully chosen to reveal the effects of symmetry in molecular vibrations and to investigate how neutron spectroscopy and optical spectroscopy could be usefully combined to obtain quite new information.

## 3.4. Selection rules and the effects of molecular symmetry

The simplest molecular material from which neutron scattering can be observed is hydrogen and Fig. 3.1 shows the scattering spectra for the solid at 12 K and for the liquid at 15 K. Both the spectra are recorded at three angles of scattering, 20°, 45°, and 90° to the incident neutron beam. The spectra illustrate some of the unique features of neutron spectroscopy to be discussed below.

For solid hydrogen it can be seen that there is an intense elastic scattering peak at a time-of-flight corresponding to the incident neutron energy. Most neutrons are scattered elastically. The intensity of the incoherent elastic scattering does not decrease very greatly as the angle of scattering is increased from 20° to 90°: in other words, the Debye–Waller factor is nearly unity at these temperatures. The other feature in the spectrum is the sharp peak at much shorter times of flight. This corresponds to an inelastic scattering event, where

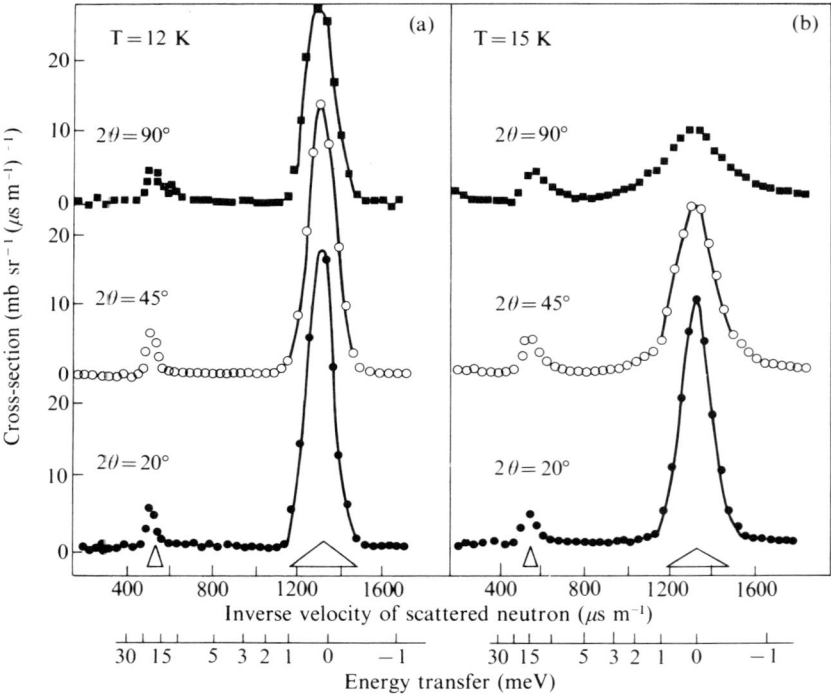

FIG. 3.1. Incoherent scattering spectra from (a) solid, (b) liquid hydrogen. The instrumental resolution is indicated at two points for each set of spectra. (After Egelstaff, Haywood, and Webb 1967).

the neutron has gained energy and the solid hydrogen lost a quantum of rotational energy. The energy difference between the elastic peak and this inelastic feature amounts to 120 cm$^{-1}$ and is equal to the energy difference between the rotational state of the molecule with the rotational angular momentum quantum number $J = 1$ and the rotational state $J = 0$. The $J = 1$ state is appreciably populated, even at 12 K, although all higher energy levels, both rotational and vibrational, have almost negligible population.

That this transition can be observed by neutron spectroscopy immediately demonstrates the freedom of the technique from optical selection rules. Such a transition ($\Delta J = \pm 1$) for a homonuclear diatomic molecule is forbidden in infrared spectroscopy as there is no molecular dipole-moment change in the transition. The transition is also forbidden for Raman spectroscopy where the selection rule $\Delta J = \pm 2$ operates as a consequence of the need for a polarizability change to be associated with Raman scattering transitions. Finally, it is remarkable that the transition can be seen at all, for it indicates that the molecular hydrogen is rotating much as in the gas phase even though hydrogen is a solid at this low temperature.

The spectrum of liquid hydrogen at 15 K shown in Fig. 3.1($b$) reveals one further unique feature of the neutron scattering technique. The inelastic peak corresponding to the $J = 1$ rotational transition is again visible for the liquid, so that rotations are not quenched appreciably by the inter-molecular collisions. These collisions and the rapid random motion of the liquid molecules are manifested in the spectrum, however. The elastic scattering peak changes markedly in intensity and in width as the scattering angle is increased from 20° through 45° to 90°. This behaviour is a consequence of the fact that, when a neutron is scattered by a molecule in the liquid state, Doppler shifts of the neutron's energy may occur, since relative motion between the neutron and the molecule can take place during the time scale of the neutron–molecule interaction. These Doppler shifts amount to an energy broadening and the peak is commonly referred to as the quasi-elastic peak, since the scattering is not strictly elastic any more. The diffusion coefficient for molecular motion can be determined from the energy broadening of this peak (see Chapter 6 for more details of this method). The example of molecular hydrogen shows plainly how all rotational transitions for a molecule may be expected to contribute to the incoherent neutron-scattering spectrum, although selection rules do arise for coherent scatterers. This is confirmed by detailed calculations (Sears 1965, 1967) on poly-atomic molecules as well.

In vibrational spectroscopy, three types of behaviour are possible. Firstly, the neutron and optical spectra may show the same number of vibrational features (although the intensities, in general, are quite different). Secondly, as might be expected from the above, the neutron spectrum shows more vibrational features than the optical spectrum (because of the limitations imposed by optical selection rules). Thirdly, because of the weighting factors apparent in eqn (3.8) the neutron spectrum may contain fewer vibrational features than the optical spectrum. From the point of view of a spectroscopist, this balance of possibilities is rather elegant, and an example from each class will be discussed in the following paragraphs.

Experimental difficulties mentioned above had cast doubt on the reliability of neutron spectroscopy for revealing molecular vibrational frequencies. Therefore, it was essential at the outset to have an example of molecular scattering where the neutron and optical spectra gave the same information. The molecule chosen for this study (Longster and White 1969) was the halogen-substituted ethane,

## NEUTRON INELASTIC SCATTERING AND MOLECULAR SPECTROSCOPY

$CH_2Br\text{-}CHFBr$. It was a molecule whose optical spectrum had not previously been measured but which was otherwise desirable for the comparison, since its low symmetry ensured that there would be no optical selection rule operating to forbid any transition in the infra-red spectrum. Also large dipole moments in the molecule ensured that the infra-red spectrum was intense, and finally its heavy mass minimizes the diffusion occurring in the liquid state and so allowed slightly sharper neutron spectra to be measured. Scattering from the molecule was studied in both the solid and liquid phases with a neutron spectrometer of rather poorer resolution than is now available. The time-of-flight spectrum of the liquid at 293 K is shown in Fig. 3.2.

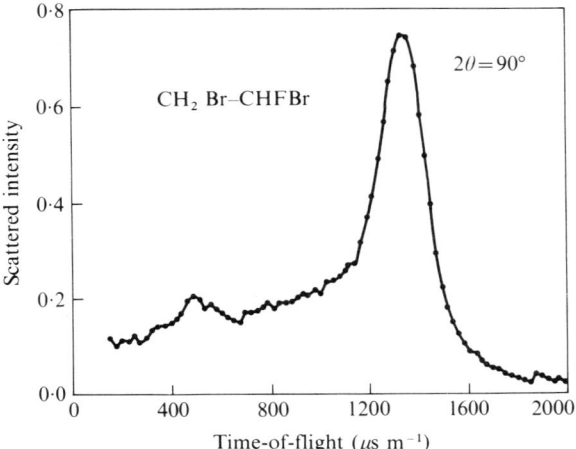

FIG. 3.2. Time-of-flight spectrum of liquid $CH_2Br\text{-}CHFBr$ at 293 K for incident 5·2 Å neutrons and 90° scattering angle. (After Longster and White 1969).

A time-of-flight spectrum measures the intensity of scattered neutrons both as a function of the energy transfer and of the momentum transfer in a scattering event, and so it is not directly suitable for comparison with optical spectroscopic measurements. The neutron spectrum represents a cut through energy-transfer and momentum-transfer space determined by the dispersion curve of the instrument. A suitable function to compare with optical spectroscopic measurements (which all occur at nearly zero momentum transfer) is the density of states spectrum $Z(\omega)$. This was extracted from the experimental curve by the method of Egelstaff (1961) using the expression (6.31) given in Chapter 6. This procedure has the virtue that the data from many scattering angles are combined and the statistics thereby improved. The technique consists in extrapolating the scattering spectra for many different angles of scattering to zero momentum transfer.

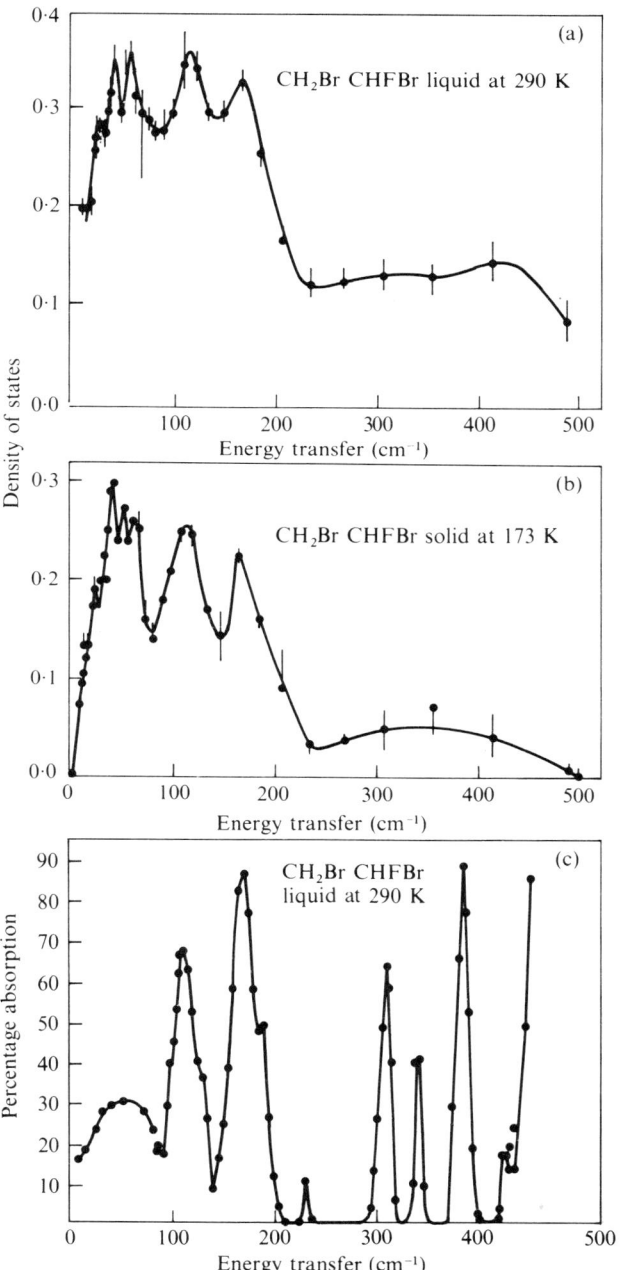

FIG. 3.3. Comparison of hydrogen-amplitude weighted $Z(\omega)$ for (a) liquid, (b) solid $CH_2Br$–CHFBr, compared with (c) its infra-red spectrum. (After Longster and White 1969).

Fig. 3.3 shows the extrapolated hydrogen-amplitude-weighted density-of-states spectrum for $CH_2Br$–$CHFBr$ at room temperature and 173 K, compared with the subsequently measured far-infra-red spectrum of this molecule. It can

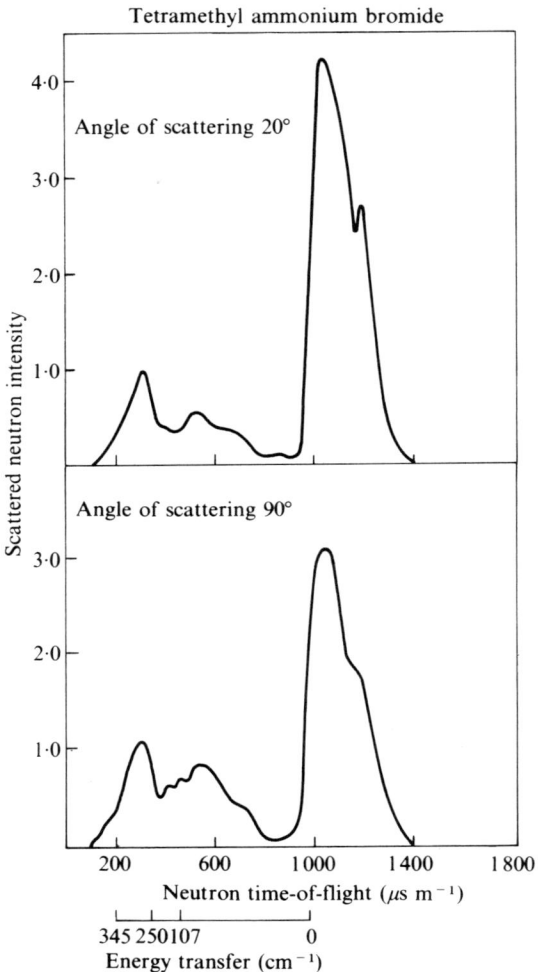

FIG. 3.4. Time-of-flight spectra for $N(CH_3)_4Br$ at 296 K, and at scattering angles 20° and 90° to the incident beam.

be seen that at energy transfers above 200 $cm^{-1}$, the resolution of the spectrometer was insufficient to resolve the features seen in the infra-red spectrum. In the low-frequency region, however, the features seen in the neutron spectrum are reproduced in the infra-red spectrum (although with different intensities). The agreement in frequency is well within the experimental inaccuracies due to resolution, and the different intensities can be expected because of the different factors weighting the density of states in the optical and neutron scattering cases. The lowest-frequency peak in the spectrum, whilst weak in the infra-red, is strongly neutron active, presumably because of the large hydrogen amplitude of vibration involved in it. Conversely, the peak at 165 cm$^{-1}$ is strong in the infra-red by comparison, whilst the neutron scattering is rather weaker. We can be sure that this higher frequency mode is associated with a large electric dipole moment change in the molecule and with rather smaller amplitude of motion, and the two spectra taken together allow tentative assignments of the lowest frequency mode to the torsions and the higher frequency mode to a skeletal vibration.

In the next case there are more features in the neutron scattering spectrum than in the infra-red spectrum (Brown 1970, Jones 1971). Fig. 3.4 shows the time-of-flight spectra for tetramethyl ammonium bromide, taken with a moderate resolution spectrometer at the Aldermaston Reactor, HERALD. At 90° of scattering angle it can be seen that there are two prominent peaks in the inelastic scattering spectrum, with a shoulder on the lowest-frequency edge and a weak peak between the two sharp maxima. Even from the time-of-flight spectrum, the positions can be read off for comparison with optical spectroscopic measurements. Again, however, a more revealing method is to determine the extrapolated density-of-states spectrum, which is shown in Fig. 3.5 directly compared with the far infra-red spectrum of tetramethyl ammonium bromide solid. The internal vibrations of the tetramethyl ammonium ion (which is tetrahedral) are quite analogous to those of uncharged molecules.

By combination of the two techniques of infra-red and neutron spectroscopy, all features in the spectrum may be assigned. Thus the lowest-frequency feature is very intense in the infra-red but only weakly appearing in the neutron spectrum. This is the Reststrahl frequency where tetramethyl and bromide ions move relative to one another, thus producing a very large electric dipole moment change in the vibration. This motion, however, has relatively low amplitude and so the hydrogen displacements are small and the contribution to the hydrogen amplitude-weighted density of states correspondingly weak. The next feature is the shoulder on the strong infra-red band at about 110 cm$^{-1}$ which by constrast is the most intense feature in the neutron scattering. This feature undoubtedly corresponds to the torsional vibrations of the whole tetramethyl ammonium ion. These produce large displacements of the hydrogen atom but are inefficient at generating infra-red intensity. All features at higher frequencies correspond to molecular modes rather than to lattice vibrations. The first of these is the weak band that appears in the neutron scattering but not at all in the infra-red at about 265 cm$^{-1}$. This is the torsional frequency of the $CH_3$ groups of the tetramethyl ammonium ion. At higher frequencies still can be seen the bending vibrations of the tetrahedron which have large hydrogen amplitudes but only rather weak contributions to the infra-red spectrum compared to the intense lattice vibration. Experiments with the beryllium-filter-detector spectrometer (see §2.4) clearly resolve these molecular vibrations (White 1972).

Finally, the spectrum of the molecule Co(CO)$_4$H is a case where the neutron spectrum actually contains fewer vibrational bands than the infra-red spectrum (White and Wright 1970). Since the most interesting molecular vibrational frequencies for this molecule occur in the energy transfer region above 200 cm$^{-1}$, a beryllium-filter-detector spectrometer was used for the measurements. This has rather better

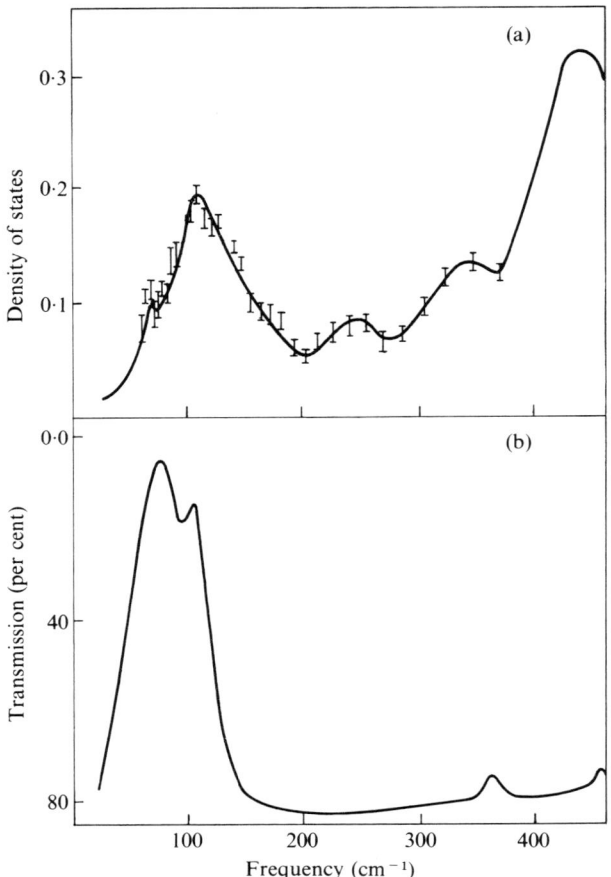

FIG. 3.5. Density-of-states spectrum (a) for N(CH$_3$)$_4$Br compared with its infra-red spectrum (b).

energy resolution than the time-of-flight spectrometer in this energy range. For reasons connected with the spectrometer resolution function, the $Q^2$ dependence is effectively extracted from this spectrum and it can be directly compared with the far infra-red spectrum of the molecular vapour (see Fig. 3.6) (White and Wright 1972).

The distinctive feature here is that the infra-red intensities are proportional to the transition moment, whilst neutron scattering is only strong where there is a large amplitude of motion of the hydrogen. The scattering cross-sections for carbon, oxygen and cobalt are sufficiently small, and their amplitudes of motion are sufficiently diminished by the mass effect, that the hydrogen modes only contribute to the neutron spectrum. The range of energy transfers is too low to observe the hydrogen stretching vibration (with respect to the cobalt), but the

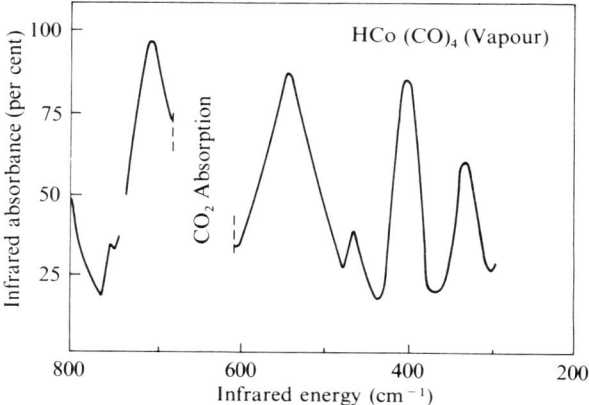

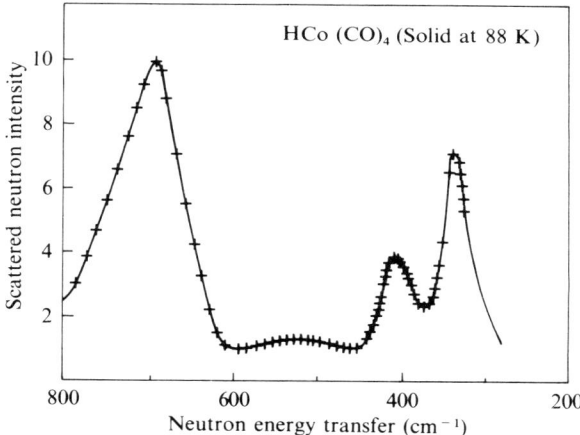

FIG. 3.6. Comparison of neutron and infra-red spectra of HCo(CO)$_4$. (After White and Wright 1970).

hydrogen bending vibration, of $E$ symmetry in the molecular point group $T_d$ or $C_{3v}$, can be seen. This is the intense feature at about 700 cm$^{-1}$. It lies directly underneath the carbonyl bending vibrations which are so strong in the infra-red that the hydrogen mode itself could never be observed by that technique. This $E$-symmetry mode mixes only with other modes of the same symmetry, and so it can be concluded that the bands at 330 cm$^{-1}$ and 410 cm$^{-1}$ are $E$-symmetry deformations of the cobalt carbonyl hydride which have stolen intensity from the $E$-symmetry hydrogen-bending mode. By contrast, the strong mode in the infra-red at about 550 cm$^{-1}$ steals no intensity and therefore belongs to an

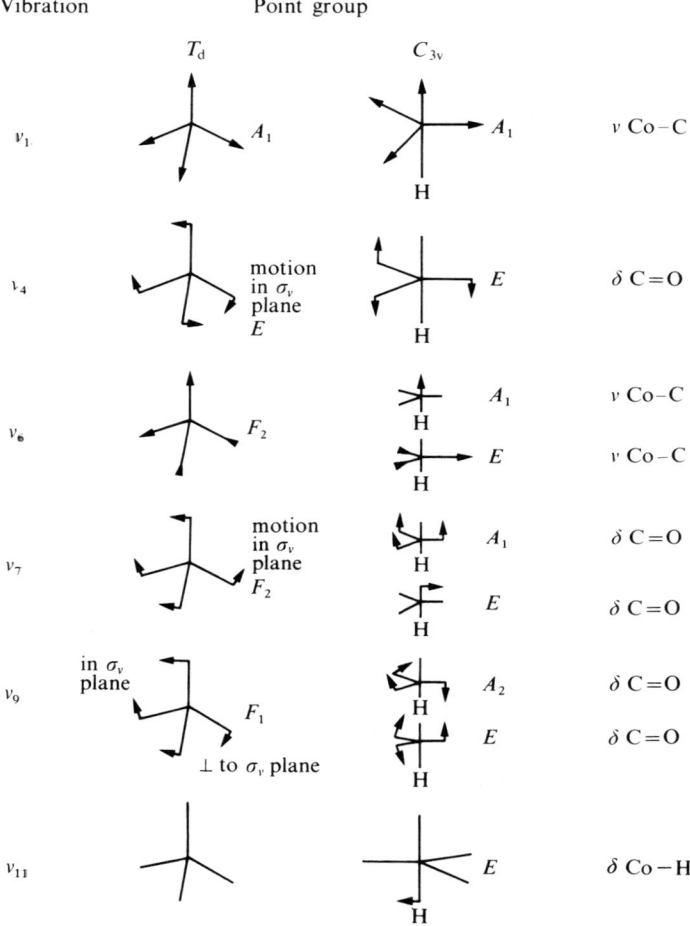

FIG. 3.7. Diagram illustrating normal modes of vibration of HCo(CO$_4$), considered as a tetrahedron ($T_d$) or distorted tetrahedron ($C_{3v}$).

orthogonal class in the molecular point group, $A$. Fig. 3.7 shows diagrams of the normal modes of this molecule considered as a tetrahedron and as a distorted tetrahedron ($C_{3v}$).

A general rule emerges from these comparative spectra. For hydrogenous molecules those modes which are seen strongly are the ones which produce the greatest amplitudes of vibration for the hydrogen atoms. Furthermore, hydrogen amplitude can be stolen by the interaction of modes of the same symmetry. This rule amounts to a type of selection rule in neutron spectroscopy, and can be generalized to include the motions of other light atoms in a heavy-atom environment. It has very important consequences in simplifying what otherwise might be very complicated spectra. An example of this is the interpretation of the spectra of aromatic molecules and molecular crystals discussed below.

## 3.5 Aromatic molecular crystals and the harmonic approximation

In discussions of molecular force fields, aromatic molecules have received a great deal of attention. This is partly because they have highly symmetrical molecular structures and also because of intrinsic interest in the types of force field likely to arise from mixed $\sigma$- and $\pi$-type atomic bonding. In these molecules the molecular framework is laid down jointly by interatomic (e.g. carbon–carbon) electron-pair $\sigma$-bonds and by delocalized electron $\pi$-bonds between the atoms. Much effort has been devoted to obtaining force constants transferable from one molecule to a closely related structure. A particularly good example of this work is that of Sherrer (1968) for the halo-benzene molecules. He has shown how a valence-force field of 49 parameters gives 520 fundamental mode frequencies for 28 different chlorinated-benzene molecules. A Urey–Bradley force field of 38 parameters gives 344 in-plane fundamentals for these molecules, predicting vibrational frequencies to within $\pm 3 \cdot 5$ cm$^{-1}$ of the observed frequencies.

The existence of such a good force field for a molecule makes possible an experimental test of the scattering cross-section formula (e.g. eqn (3.6)) for molecular vibrations, and at the same time the validity of the harmonic approximation used in the molecular force field can be checked. This test has been performed for para-dichloro benzene and symmetrical 1-2-4-5 tetra-chloro benzene by Reynolds and White (1969), and is an important exercise to ensure that the absolute intensities of scattering from molecules can be quantitatively understood. Also the studies on these molecules reflect the importance of the hydrogen-amplitude-weighting selection rule for neutron scattering, mentioned above.

Quantitative studies on the intensities of incoherent scattering from molecules have only been worth-while since the development of higher resolution time-of-flight techniques. This improved resolution has resulted from great care in setting up the elements of the time-of-flight spectrometers, and from the use of higher-resolution choppers (Bunce, Harris, and Stirling 1970). Fig. 3.8 shows the neutron scattering spectra from para-dichloro benzene at room temperature taken with a degraded-resolution spectrometer and with a high-resolution instrument. It can be seen that the overlapping of peaks is greatly reduced at the higher resolution, and sharp features well distinguished from one another

NEUTRON INELASTIC SCATTERING AND MOLECULAR SPECTROSCOPY

can be seen in the spectrum. This removed any doubts that spectra from molecular crystals could indeed be very sharp and devoid of any excessive multi-phonon obscuration.

Fig. 3.9 shows the incoherent scattering spectrum from para-dichloro benzene (1-4 dichloro benzene) at two angles of scattering. The elastic peak has not been shown and a smooth line has been drawn through the spectrum. In the low-frequency region there are many small features on the side of the peaks which have been neglected in drawing this curve. These show up as significant singularities

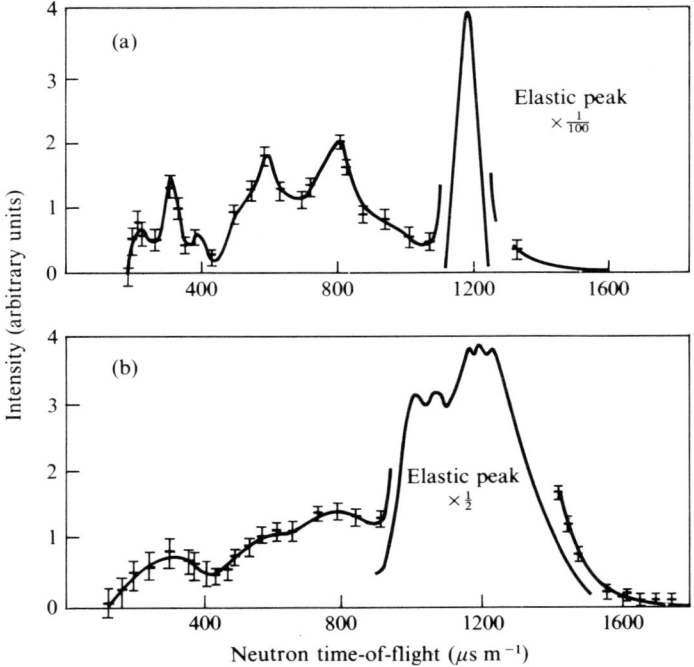

FIG. 3.8. Neutron scattering spectra from p-dichlorobenzene at 290 K using (a) high-resolution, (b) low-resolution instrument. (After Reynolds and White 1969).

at lower temperatures and can be related to the critical points in the density of states for the phonons (see below). The region in energy transfer above approximately 120 cm$^{-1}$ corresponds to molecular vibrations, and the region below is the phonon density of states weighted by the hydrogen amplitudes.

Fig. 3.10 shows the neutron scattering spectrum from the closely related molecule para-di-iodobenzene, again at two angles of scattering. As in Fig. 3.9, a stick diagram underneath the spectrum shows the assignments to molecular modes, classified according to the point-group symmetry of the whole molecule ($D_{2h}$). Well-distinguished peaks appear in the spectrum, even at room temperature.

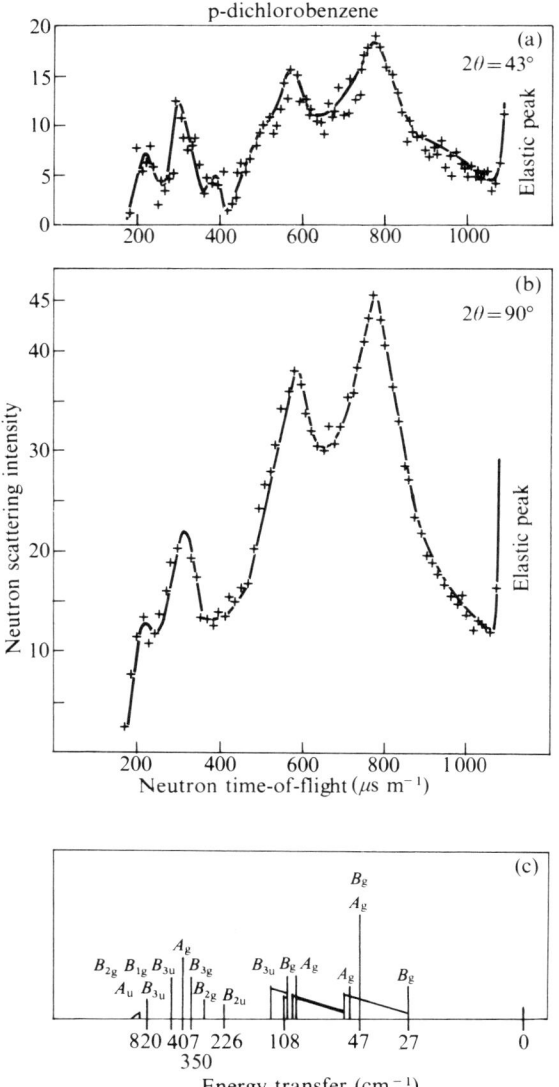

FIG. 3.9. Incoherent neutron scattering spectrum from p-dichlorobenzene at two angles of scattering: (a) 43°, (b) 90°. The stick diagram (c) shows the assignments of the molecular modes. (After Reynolds and White 1969).

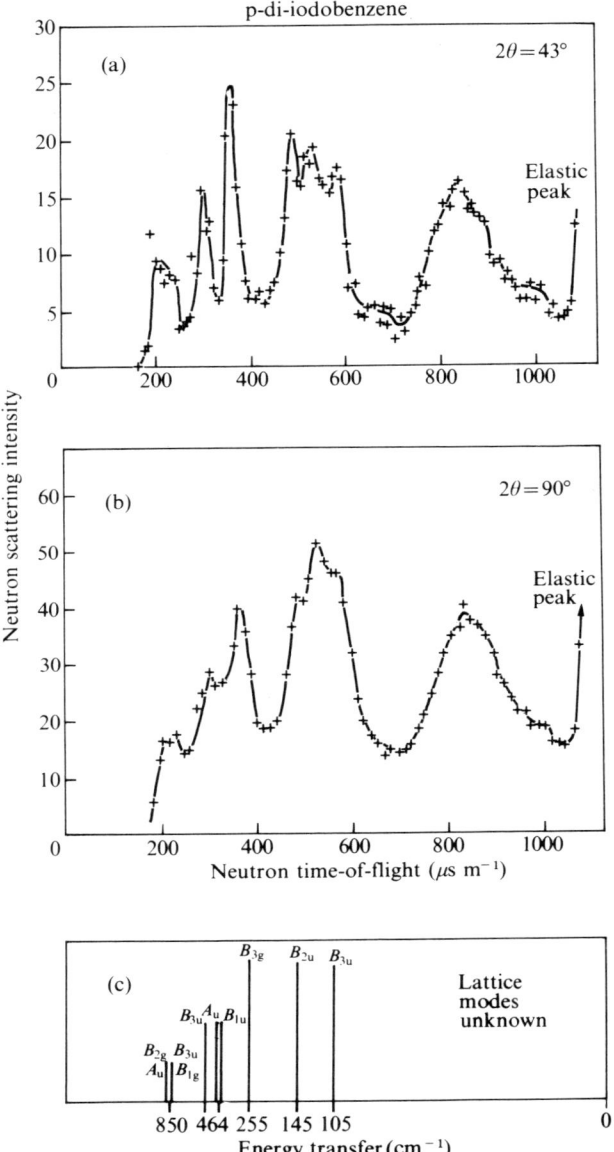

FIG. 3.10. Neutron-scattering spectra (a) and (b) from p-di-iodobenzene at two angles of scattering. The stick diagram (c) shows the assignments of the molecular modes. (After Reynolds and White 1969).

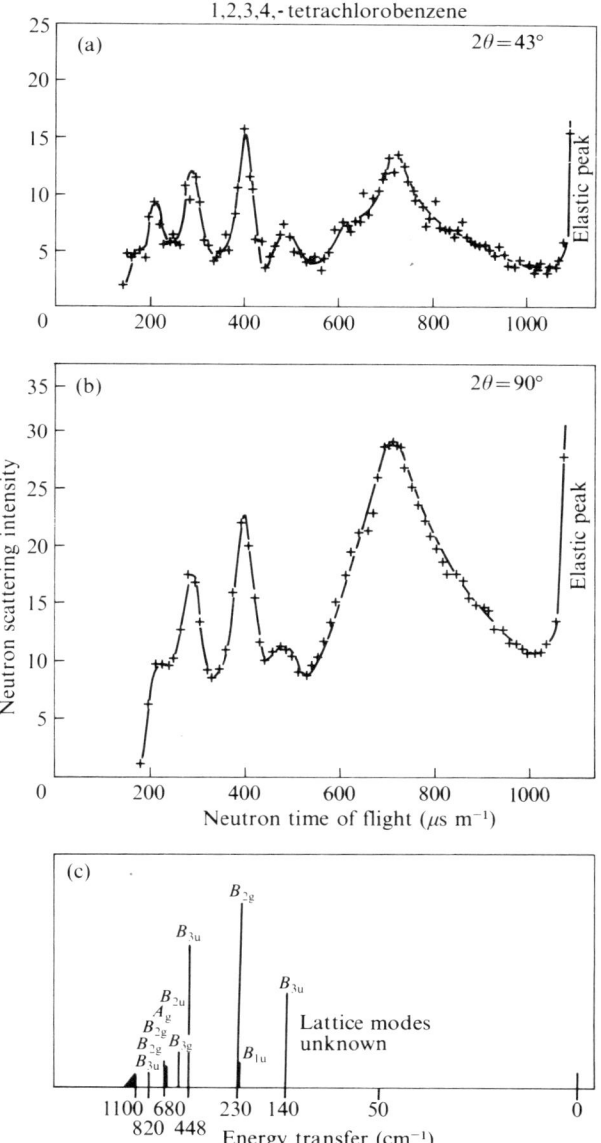

FIG. 3.11. Neutron-scattering spectra (a) and (b) from 1-2-4-5 tetra-chlorobenzene at two angles of scattering. The stick diagram (c) shows the assignments of the molecular modes. (After Reynolds and White 1969).

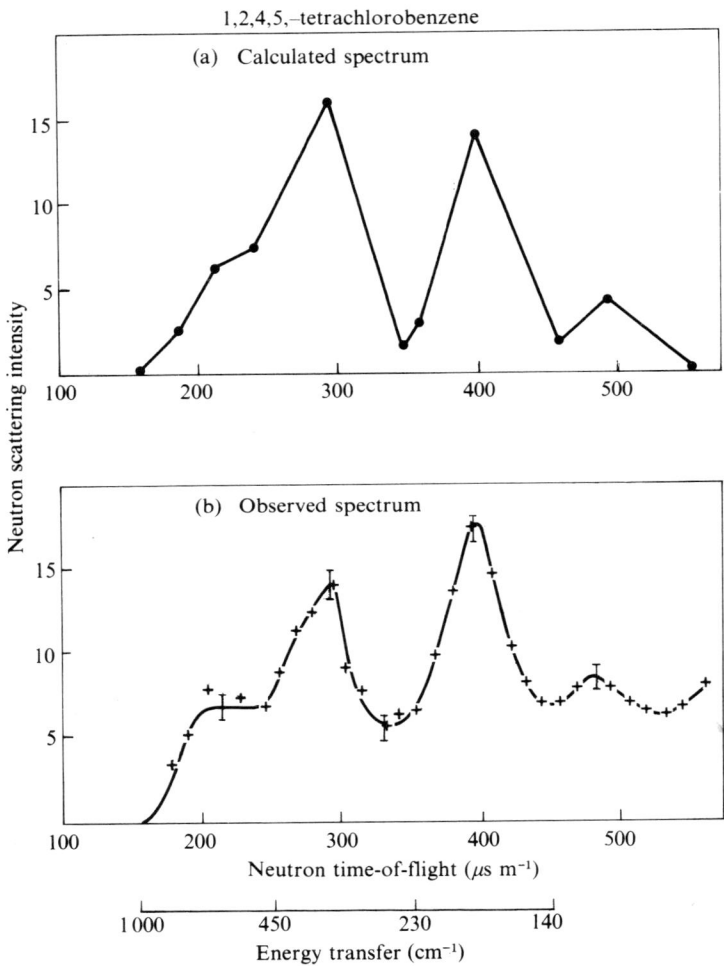

FIG. 3.12. (a) Theoretical intensity for 1-2-4-5 tetra-chlorobenzene, compared with (b) experimental intensity in the molecular scattering region. The scattering angle is 66°. (After Reynolds and White 1969).

The angular dependence in both spectra can be described well by the Debye-Waller factor and $Q^2$ terms in cross-section formula, eqn (3.8). The sharp maxima above 120 cm$^{-1}$ indicate weak dispersion only of the molecular modes. The effect of increasing the mass of the molecule, by substituting iodine for chlorine, can be seen in a strong shift of the maximum in the density of states for the lattice vibrations to lower frequencies in the iodo compound.

Fig. 3.11 shows the spectrum for symmetrical, 1-2-4-5 tetra-chloro benzene. In the molecular region above 150 cm$^{-1}$, four or five sharp singularities can be observed. These persist at all angles of scattering and their intensities are controlled by the Debye-Waller factor and $Q^2$ terms as above. They are well resolved and not obviously obscured by any multi-phonon effects from the lattice vibrational region, nor overlapped particularly by any other weaker modes. The stick diagram beneath Fig. 3.11 shows the assignments given to these modes in the molecular point group.

When one examines the possible vibrations which fall into the frequency region covered by the neutron spectra in Figs. 3.9, 3.10 and 3.11, it is clear that the neutron spectra show far fewer peaks than might have been expected if all molecular modes could be observed with equal ease. The hydrogen-amplitude and $(b^\nu_{\text{incoh}})^2$ weighting of eqns (3.6) and (3.8) again operate as a selection rule to simplify the spectra. To test this assumption and also the gaussian approximation, which is inherent in eqn (3.6) because only the term in $Q^2$ rather than higher terms in $Q^4$, $Q^6$ etc. is taken to determine the neutron scattering intensity, a calculation of the eigen vectors associated with the normal modes of 1-2-4-5 tetra-chlorobenzene was made. These quantities can be related to the Cartesian displacements, $U^s_\rho$, needed in the cross-section formula to predict the scattering intensity. When these were calculated, fitted into the cross-section formula, and the whole convoluted with the resolution function of the instrument, the theoretical curve shown in the top half of Fig. 3.12 was obtained for the molecular scattering region. This is compared in Fig. 3.12 with the experimental spectrum plotted on the same scale and arbitrarily normalized at one point. It can be seen that a good fit to the relative intensities is obtained by this calculation, so that both the theoretical approach and the experimental methods are apparently justified.

## 3.6 Isotopic substitution method for vibrational assignments

The isotopic substitution method, commonly used in infra-red and Raman spectroscopy to make vibrational assignments (Herzberg 1945), can be used for the same purpose in neutron spectroscopy when theoretical methods fail. In optical spectroscopy the method depends upon selective substitution, e.g. of hydrogen by deuterium in the molecule, and in the observation of shifts in the infra-red or Raman spectra consequent upon this change. In neutron scattering not only do these shifts occur, but because of the cross-section difference between hydrogen and deuterium, very large changes in the intensities of the

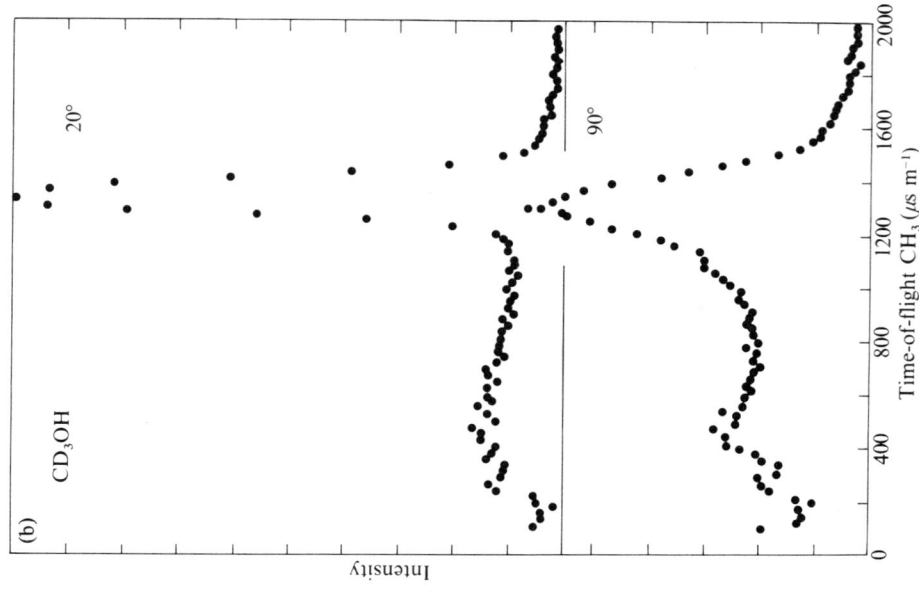
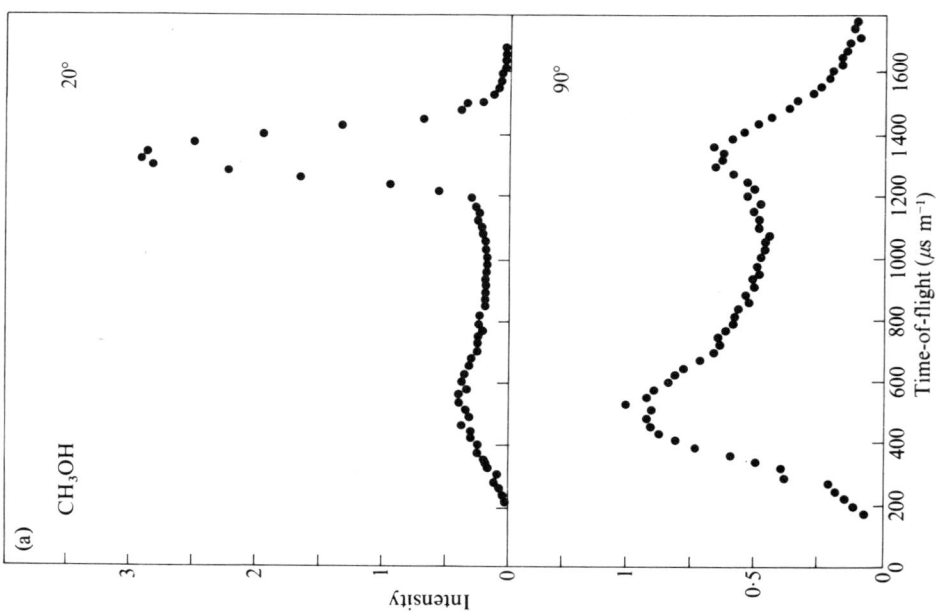

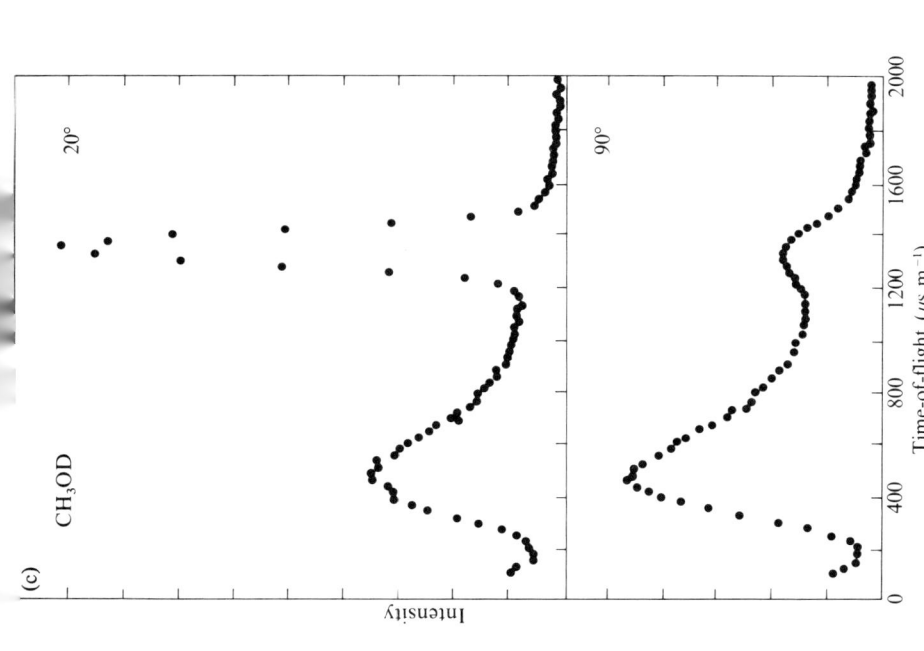

FIG. 3.13. (a) Neutron-scattering spectra from liquid methanol at room temperature for 5·3 Å neutrons scattered through 20° and 90° from the incident beam; (b) same for $CD_3OH$; (c) same for $CH_3OD$. (After White 1968).

spectra occur as well. The modes with large hydrogen amplitudes can be made to disappear almost completely from the neutron scattering spectrum, when deuterium is substituted for the hydrogen atoms involved in the mode.

This technique has been used to assign the hindered rotational motion of the $CH_3$ group in methanol and to find the OH bending motions, which were otherwise obscured by the rotational feature in the neutron spectrum. In Fig. 3.13(a) the spectrum of methanol is shown, and it can be seen that there is a strong sharp elastic peak at low angles of scattering which drops in intensity and becomes very broad at higher angles. At short times-of-flight, there is a broad intense feature rising to a maximum at about an energy transfer of 150 cm$^{-1}$ both at low and at high angles. This spectrum must obviously contain modes of motion due to the $CH_3$ group and to the CH motions. In Fig. 3.13(b) and (c) are shown the spectra of $CD_3OH$ and $CH_3OD$ respectively. The effect of deuterium substitution is to appreciably reduce the relative intensity of the inelastic region compared with the elastic intensity in $CD_3OH$. This intensity appears strongly in the case of $CH_3OD$, and so it can be concluded that the feature is associated with the $CH_3$ group. At the same time, in the spectrum of the $CD_3OH$ it is possible to see a sharp feature at rather short times-of-flight. This corresponds to one of the OH intermolecular modes in the liquid (Aldred, Eden, and White 1967). An elegant use of the isotopic substitution method to assign the H bending mode in $ClHCl^-$ has been made by Stirling, Ludman, and Waddington (1970).

When deuterium substitutions are not available or difficult to perform, studies on molecules closely related in structure, but where the hydrogen atoms have been replaced, for example, by fluorine or other atoms of small scattering cross-sections, can be used to make assignments by an analogous procedure (Aldred *et al.* 1967). This technique has been found useful for understanding the motions of side groups in polymers (Longster and White 1968). The deuterium-substitution technique could obviously play an important role, coupled with the hydrogen-amplitude selection rule, in assigning the vibrational spectra of rather complicated molecules.

## 3.7. Lattice vibrations and the phonon density of states

The acoustic branches of the phonon dispersion curves for molecular crystals are not normally accessible by optical spectroscopy because of the very small region of momentum space explored in optical absorption or scattering experiments. At these small momentum transfers, the acoustic modes have frequencies close to zero. The density of states spectra are also not normally accessible, although partial density-of-states spectra are sometimes built on optical fundamental frequencies in the infra-red (Wilkinson 1969, Dows 1959) and on ultra-violet absorption and fluorescence spectra (Castro and Hochstrasser 1967). This most often happens when the transition is forbidden in the molecular point group for symmetry reasons, so that the mixing with external modes promotes optical activity.

There is no restriction on the momentum transfer range explored in incoherent neutron scattering spectroscopy, except that dictated by the angle of scattering and the energy transfer in the inelastic event; therefore, in principle, phonons throughout the whole of the first and higher Brillouin zones can be excited or de-excited and so contribute to the observed density-of-states spectrum. For hydrogenous compounds, this density of states will naturally be a hydrogen-amplitude-weighted spectrum.

There are two important points of principle to be investigated before density-of-states spectra, measured by incoherent scattering from hydrogenous compounds, can be subjected to an analysis for Van Hove† singularities. It is conceivable that the hydrogen-amplitude-weighted density of states will show different singularities from the unweighted spectrum. Secondly, the true density of states for a system can only be determined by sampling the dispersion curves throughout the whole of energy-momentum space. Because the momentum transfers actually obtained in a neutron spectrum will be limited for the reasons mentioned above, experimentally-determined spectra are those from a volume of momentum space limited by experimental conditions. It is necessary to show that, provided this volume includes certainly the first Brillouin zone and possibly the second and third, the density of states obtained is almost identical with that which could be obtained by a complete averaging of momentum space. These two exercises have been performed for naphthalene by Pawley, Kjems, Reynolds, and White (1971), using Pawley's computer program (Pawley 1967) for calculating the dispersion curves. The phonon density of states was calculated for spherical samples of momentum-transfer space of radii $2\text{Å}^{-1}$, $2.5\text{Å}^{-1}$ and higher values up to a value large enough to give the true density of states for this substance. It was shown that the singularities in the true density of states correspond with those in the spectra from limited volumes of momentum space. When smaller volumes of momentum space were taken, it was found that the singularities became sharper only. The computer program can be used to generate the eigenvectors of the modes as a function of their wave vector, and these data were combined with the dispersion curves to produce a hydrogen-amplitude-weighted density of states for comparison with the 'true' spectrum. This comparison is shown in Fig. 3.14 and once again there is good agreement between the positions of the singularities in both spectra. It can also be seen that the hydrogen amplitude weighting emphasizes certain features with respect to others, as found in the case of the molecular spectra discussed earlier. The weighting can again be used to help make assignments of the singularities to particular symmetry classes of crystal vibrations. For instance, librations can be distinguished qualitatively from translations because of their relatively larger intensities. With these provisos in mind, it is possible to make a qualitative interpretation of the lattice-vibrational region of the spectra for the substituted benzenes shown in Figs. 3.10, 3.11 and

† Singularities in the phonon density of states function $Z(\omega)$ are expected when the dispersion curve, $\omega(\mathbf{q})$ versus $\mathbf{q}$, has zero gradient. These and other singularities have been discussed by Van Hove (1953).

3.12. Taking the gross features of these spectra first, the effect of replacing the chlorines by iodine in para-di-iodo benzene is to shift the centre of gravity in the spectra of the 1-4 compounds to lower frequencies. The lower-frequency peak in the para-dichloro benzene overlaps that at higher frequencies, but in para-di-iodo benzene one component of the high-frequency peak is left at high frequencies and appears as a central line in the lowest-energy triplet. The two outside lines at the energy transfers of 105 and 145Å$^{-1}$ are the $B_{3u}$ and $B_{2u}$ molecular, out-of-plane, bending vibrations. Since the central sharp peak is in approximately the same position that it was in para-dichloro benzene, we may assign it tentatively to the density of states for the libration about the 1-4

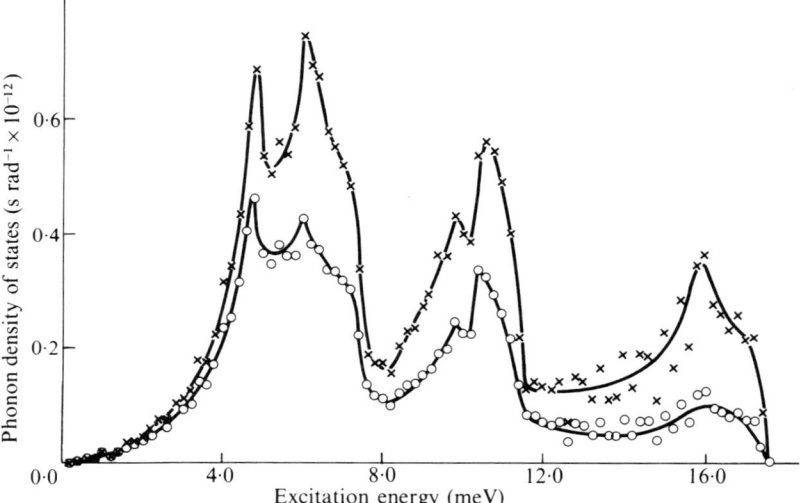

FIG. 3.14. Calculations for naphthalene showing comparison of the hydrogen-amplitude weighted density of states (×) at 296 K with the 'true' density of states (○). (After Pawley et al. 1971).

(halogen-halogen) axis of the molecule. Reference to the spectrum of the 1-2-4-5 tetra-chloro benzene shows that this peak is absent, and the whole density-of-states spectrum peaks around about 50 cm$^{-1}$, indicating that the librations and translations fall more or less in the same region. This seems reasonable since there is now no unique molecular axis of low moment of inertia to give rise to high-frequency librations.

When the spectra of the halogen-substituted benzenes, and of other compounds, are studied at low temperatures and with higher instrument resolution, a number of sharp peaks can be distinguished (Kjems, Reynolds, and White 1972a). These features are undoubtedly Van Hove singularities associated with flat parts of the dispersion curves in these molecular crystals. The fact that they are strongly broadened if the temperature is raised from liquid nitrogen to room temperature is an indication of the anharmonicity of the system, which can be described (Kjems *et al.* 1972b) by a quasi-harmonic approximation.

It must not be inferred that measurements of the kind described above are limited to molecular crystals containing hydrogen. Excellent density of states

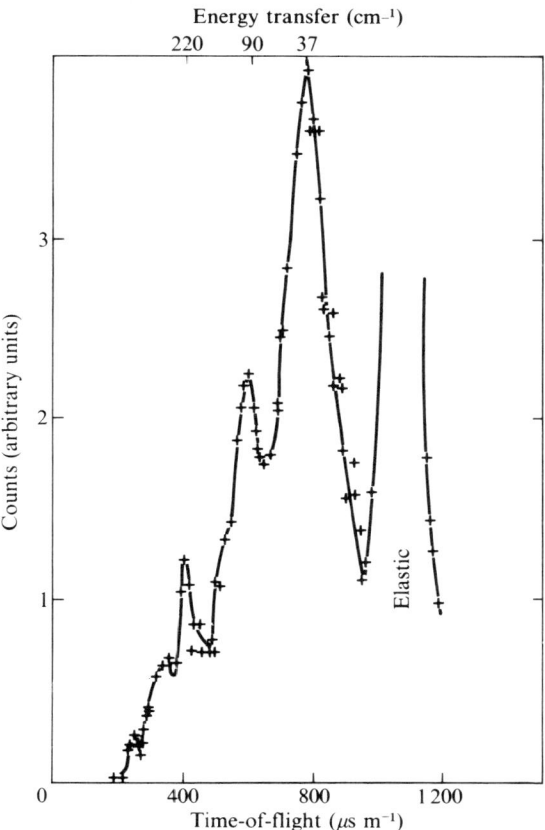

FIG. 3.15. Neutron-scattering spectrum at room temperature of hexachlorobenzene. Incident wavelength is 4·2 Å, and angle of scattering is 81°. (After Kjems *et al.* 1972a).

and molecular scattering spectra have been recorded from fully deuteriated, substituted and unsubstituted, aromatic hydro-carbons (Kjems *et al.* 1972a). The spectrum of hexachlorobenzene shown in Fig. 3.15 illustrates what can be done, even with moderate resolution, for a molecular crystal containing no particularly light atoms. These studies are yet in a very early stage and considerable extensions of both technique and application can be expected.

A very good use for measurements in the lattice dynamical region, as well as in the molecular region, is to obtain a quick appraisal of the extent of dispersion and of the likely complexities to be expected in a full study of the phonon dispersion curves, which must remain as the ultimate goal for a full experimental knowledge of the dynamics of these systems. The density-of-states measurements are a particular attraction, as they can be recorded without great difficulty on polycrystalline materials. At this stage of the subject their value as a survey tool is likely to be particularly high.

## 3.8. Conclusions

This review must not, in any sense, be taken to be complete. Many beautiful molecular systems have been studied by neutron inelastic scattering at low energies. The discussion does not dwell at all heavily on these or on high-energy-transfer measurements, *viz.* those above 500 cm$^{-1}$. An important virtue of neutron experiments, whereby the eigen-vectors as well as the eigen-frequencies are measured for molecular modes, has hardly yet been exploited. What does emerge from the above investigations is the suggestion that the amplitude and cross-section weighting present in neutron scattering spectroscopy can be used as a new type of selection rule for understanding molecular and crystal dynamics, especially in conjunction with optical spectroscopy.

## 3.9. References

ALDRED, B. K., EDEN, R. C. *and* WHITE, J. W. (1967). *Discuss. Faraday soc.* **43**, 169.
BROWN, A. N. (1970). Part II thesis, Oxford University.
BUNCE, L., HARRIS, D. H. C. *and* STIRLING, G. C. (1970). *Rep. U.K. Atom. Energy Auth.* (London) R-6246.
CASTRO, G. *and* HOCHSTRASSER, R. M. (1967). *J. chem. Phys.* **46**, 3617.
DOWS, D. A. (1959). *Spectrochim. Acta* **13**, 308.
EGELSTAFF, P. A. (1961) in *Inelastic scattering of neutrons in solids and liquids.* I.A.E.A., Vienna. p. 25.
EGELSTAFF, P. A., HAYWOOD, B. C. *and* WEBB, F. J. (1967). *Proc. phys. Soc.* **90**, 681.
JONES, S. (1971). Part II thesis, Oxford University.
KJEMS, J., REYNOLDS, P. A. *and* WHITE, J. W. (1972a). *J. chem. Phys.* **56**, 2928.
KJEMS, J., REYNOLDS, P. A., *and* WHITE, J. W. (1972b) in *Inelastic scattering of neutrons (Grenoble meeting)*. I.A.E.A., Vienna.
LONGSTER, G. F. *and* WHITE, J. W. (1968). *J. chem. Phys.* **48**, 5271.
LONGSTER, G. F. *and* WHITE, J. W. (1969). *Molec. Phys.* **17**, 1.
PAWLEY, G. S. (1967). *Phys. Stat. Sol.*, **20**, 347.
PAWLEY, G. S., KJEMS, J., REYNOLDS, P. A. *and* WHITE, J. W. (1971). *Solid St. Comm.* **9**, 1353.

REYNOLDS, P. A. *and* WHITE, J. W. (1969). *Discuss. Faraday soc.* **48**, 131.
SEARS, V. F. (1965). *Proc. phys. Soc.* **86**, 953.
SEARS, V. F. (1967). *Can. J. Phys.* **44**, 1279.
SHERRER, J. R. (1968). *Spectrochim. Acta.* A **24**, 747.
STIRLING, G. C., LUDMAN, C. J., *and* WADDINGTON, T. C. (1970). *J. chem. Phys.* **52**, 2730.
VAN HOVE, L. (1953). *Phys. Rev.* **89**, 1189.
WHITE, J. W. (1968). in *Excitons, magnons and phonons in molecular crystals.* (*editor E. B. Zahlan*). Cambridge University Press.
WHITE, J. W. *and* WRIGHT, C. J. (1970). *Chem. Comm.* 970.
WHITE, J. W. *and* WRIGHT, C. J. (1972). *J. Chem. Soc. Faraday* II, **68**, 1423.
WILKINSON, G. R. (1969) in *Molecular dynamics and the structure of solids.* (*Ed. R. S. Carter and J. J. Rush.*) Nat. Bur. Stand. Special Publication 301. p. 77.

*General*

Standard texts on molecular spectroscopy are:
HERZBERG, G. (1945). *Infra-red and Raman spectra of polyatomic molecules.* Van Nostrand, N.Y.
WILSON, E. B., DECIUS, J. C. *and* CROSS, P. C. (1955). *Molecular vibrations.* McGraw-Hill, N.Y.

# 4 Models for Calculating the Properties of Phonons in Molecular Crystals

By G. S. PAWLEY

Department of Physics, Edinburgh University

## 4.1. Introduction

The purpose of this chapter is to describe the behaviour of molecules undergoing thermal vibration in a crystal. In this introductory section a number of general ideas are presented, followed by a section on the group theory and one outlining the procedure for model calculations. To do these calculations requires some model potential function, and this model can be improved by refining the parameters of the model so as to fit theory to experiment. However, even a simple unfitted model can give insight into the modes of vibration of any crystal under study, and it is advantageous to perform such calculations before doing any experiments.

The worth of model calculations depends on the choice of a realistic potential function from which to obtain all the forces between the components of the crystal structure. Indeed the understanding of the forces between atoms or molecules in both solid and liquid phases is a fundamental theme of the present volume. As the study of molecular crystal dynamics is still in its infancy, we can work with somewhat oversimplified models of molecular interaction and still get useful results.

Let Fig. 4.1(a) represent a molecular crystal, which is typically of fairly low symmetry. The diagram has two-dimensional symmetry $pgg$ (Fig. 4.1(b)), and is analogous to the three-dimensional structure of naphthalene which crystallizes in the monoclinic space group $P2_1/a$. If we have any force model whatever, we can plot a curve (Fig. 4.1(c)) for the potential function $V(\theta)$ in terms of any small displacement co-ordinate $\theta$. The particular choice of $\theta$ in the figure is arbitrary; any other possible rotation or translation would suffice, and for the chosen displacement there must be a minimum in the potential function which can be compared with an experimental result. Two such results that can be readily employed are the sublimation energy and the crystal structure. These determine respectively the $V$ and $\theta$ co-ordinates of the minimum of the calculated $V(\theta)$, and in Fig. 4.1(c) this point is located at the intersection of the axes. It is certainly not difficult to get a model which fits these experimental measurements well, but although necessary this is not a sufficient guarantee that the model is good. Clearly, a realistic model must be stable with respect to all possible vibrational modes, and this condition is far stricter than the conditions which Fig. 4.1

describes. The model must give not just the position of the minimum of $V(\theta)$ but also the correct shape of the function, necessitating experiments where this

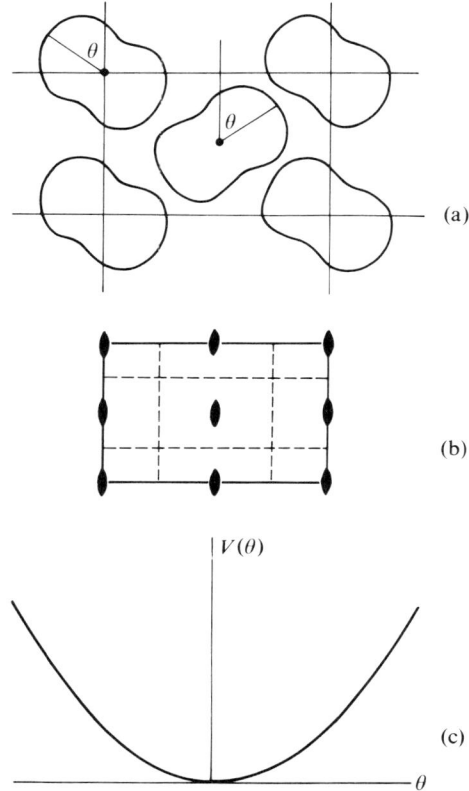

FIG. 4.1. (a) Representation of a molecular crystal in two dimensions having the space-group symmetry *pgg*. (b) The symmetry elements of *pgg*. ♦ is a two-fold point of symmetry; – – – a glide reflexion line. (c) The potential function for the crystal (a) with respect to the single variable $\theta$.

space is explored. Points other than the potential minimum correspond to non-equilibrium states, and these points are explored by molecules in vibration. Phonon frequencies are therefore the requisite measurements. The simplest phonons to consider are those having an infinite wavelength or, in the usual terminology, zero wave vector **q**. For a general wave vector, $|\mathbf{q}| = 2\pi/\lambda$ where $\lambda$ is the wavelength, and **q** is parallel to the propagation direction of the wave. At **q** = 0 all the molecules vibrate in phase and all those on equivalent crystallographic sites have equal amplitudes. In the example of Fig. 4.1, if $\theta$ is the only co-ordinate which can vary in the crystal and is zero at equilibrium, then there will be a mode where all the molecules are displaced by an equal amount $\Theta \cos \omega t$, where $\omega$ is the angular frequency. This is a symmetric mode and is governed by the potential function of Fig. 4.1.

The word 'symmetric' is introduced because a rotation of each molecule by $\theta$ does not destroy the symmetry *pgg* of the crystal. However, it can be seen that the molecules occur in two sets, such that the molecules within any one set are related by the lattice translational symmetry. Within each set the molecules have identical orientation; this orientation differs between the two sets but they are, of course, related. We can extend our concept of the potential function curve as follows; one set of molecules may be displaced by $\theta$ while the other set is displaced by $-\theta$. Immediately, we see that the symmetry *pgg* is broken and only $p\bar{1}$ remains. The potential for the displaced configuration should be above that for the undisplaced configuration as the latter must be an absolute minimum for stability; thus if the potential is plotted for a range of displacements we obtain an approximately parabolic curve with a minimum again at $\theta = 0$. This then governs the antisymmetric librational mode and, because the potential curve is not forced by symmetry to be identical to Fig. 4.1(c), this mode has a different frequency from the symmetric mode and we say they are non-degenerate. The fact that there are two frequencies results from there being two different orientations of molecules in the unit cell, and for the general two-dimensional problem there will be $n$ frequencies if there are $n$ differently oriented molecules in the unit cell.

The above is oversimplified by assuming that $\theta$ is the only structural variable. We know that to position a rigid three-dimensional object in space requires six co-ordinates, three translational co-ordinates for its centre and three angular co-ordinates for its orientation. Consequently, if there are $n$ molecules in the three-dimensional unit cell, there will be $6n$ frequencies for the modes of vibration. Because our consideration at this stage is restricted to $\mathbf{q} = 0$, three of these 'frequencies' must be zero, corresponding to pure translations of the crystal as a whole.

Let us continue with a simple two-dimensional example but extend it so that it involves two parameters. This example contains many of the complications of a realistic model. The two structural variables are now the translations depicted in Fig. 4.2(a). Fig. 4.2(b) is obtained by plotting the potential as a function of the displacements of the single molecule indicated in Fig. 4.2(a), and the resulting curves of constant potential approximate to ellipses. This shows that the direction of easiest motion is along a line near $x = y$, and the most difficult motion is at right angles to this line. These will be the directions of motion in the two normal modes. These directions can be found mathematically by setting up a two-dimensional secular equation from the two equations of motion for the molecule and solving for eigen-vectors (directions of motion) and eigen-values (squared frequencies). For a three-dimensional crystal of rigid molecules we have to do the same but with a $6n$-dimensional secular equation.

We have confined our attention so far to zero wave vector, $q = 0$. It is easy to see that modes with $q = 2\pi/a$ are exactly equivalent to those at $q = 0$, where $a$ is the crystal lattice spacing. Fig. 4.3 shows this equivalence for a monomolecular crystal. Three different functions $\Theta(x)$ are presented describing the amplitude of

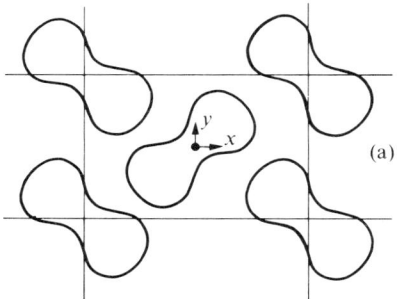

(a)

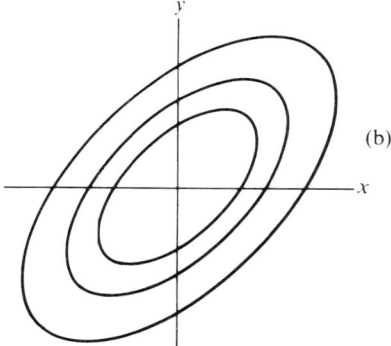

(b)

FIG. 4.2. (a) The same crystal as in Fig. 4.1(a) but assuming two structural variables. (b) Curves of constant potential with respect to these variables.

displacement for the molecule positioned at $x$ and at a given instant of time. The variation with time is then $\Theta(x) \cos \omega t$. The three functions shown are

$$\Theta(x) = \begin{cases} \Theta_0 \\ \Theta_0 \cos(2\pi x/a) \\ \Theta_0 \cos(4\pi x/a). \end{cases}$$

The molecules are represented by large dots in the figure, and it can be seen that, because these dots appear at regular intervals, all the functions $\Theta(x)$ give the same variation. Now the mode frequencies are a function of $q$, and the plot of the frequencies against $q$ is called the dispersion curve diagram. In this diagram we get a periodic repetition of identical results, and all the information can therefore be represented within the first region of this diagram, the first Brillouin zone. However, there are sometimes advantages in presenting this dispersion information over a number of Brillouin zones, as shall be seen later.

Monomolecular crystals are comparatively rare, most molecular crystals having two or four molecules in the unit cell. Let us consider now a crystal with molecules of two different orientations in the unit cell, and represent these by an alternating sequence of dots and circles. The top profile in Fig. 4.4 shows the symmetric mode where $q = 0$, and is therefore constant in space. Moving down through the

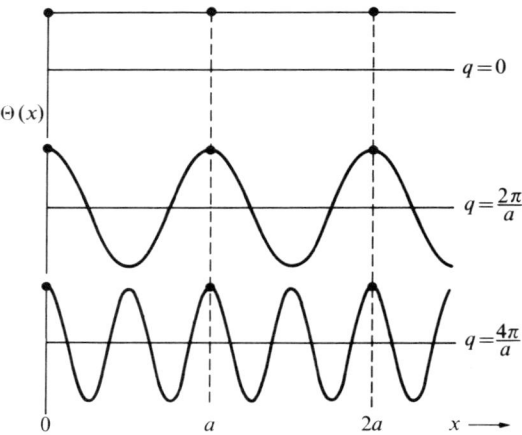

FIG. 4.3. This shows the equivalence of just three of the functions needed to describe a wave of infinite wavelength in a monatomic (or monomolecular) crystal.

figure corresponds to selecting modes with gradually increasing wave vectors, all profiles having the form:

$$\cos(qx), \quad \frac{aq}{2\pi} = 0, \frac{1}{8}, \frac{1}{4}, \frac{3}{8}, \ldots 1.$$

At $q = 0$ the motion is symmetric, and therefore as $q$ increases it traces the so-called symmetric branch of the dispersion curve. As soon as $q$ becomes non-zero the static crystal symmetry is destroyed, and the deviation from this symmetry increases as $q$ increases. When $aq/2\pi = \frac{1}{2}$ the Brillouin zone boundary is reached, as this is halfway to $aq/2\pi = 1$ (see middle profile of Fig. 4.3). We may then expect that the profiles in the bottom half of Fig. 4.3 will correspond to displacement configurations already included in the top half, but the figure shows that this is not so. Continuing along the dispersion branch brings us to $aq/2\pi = 1$, the point which for Fig. 4.2 is equivalent to $q = 0$. Molecules related by the lattice translations are all moving in phase by equal amounts and so the motion corresponds to an infinite wavelength, but the profile differs from that for $q = 0$ by

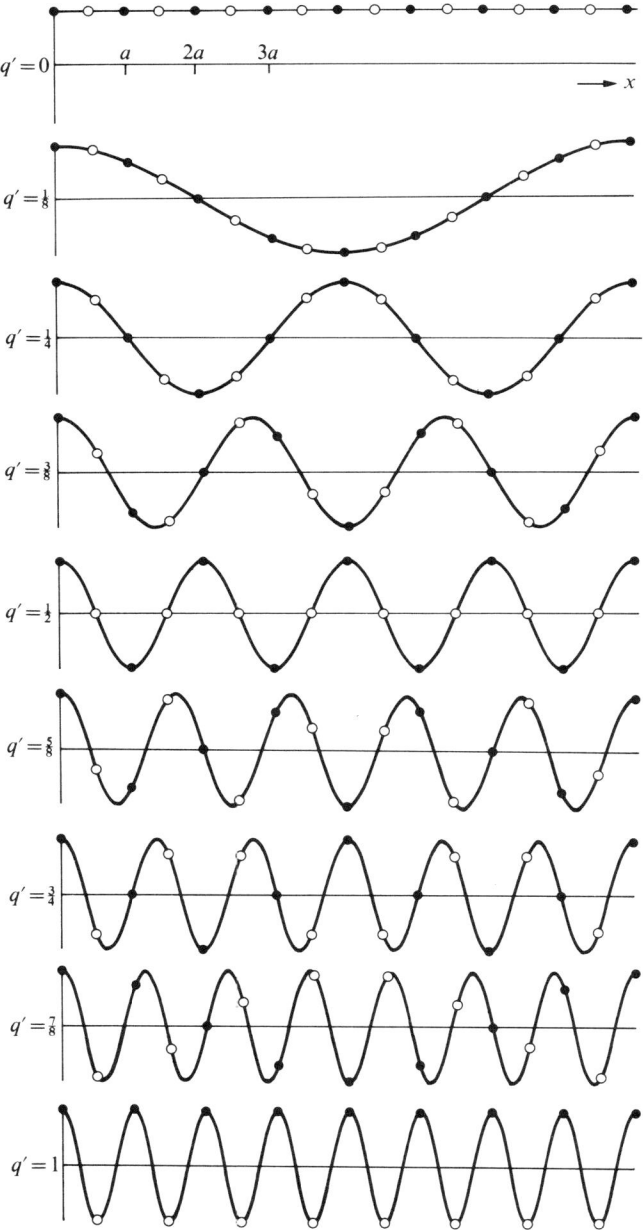

FIG. 4.4. This shows the variation in waveform shape as the wave-vector is varied. The top curve represents a symmetric mode of zero wave-vector and the bottom curve an antisymmetric mode of zero wave-vector. The dots and circles represent molecules in two different orientations. $q'$ is $aq/2\pi$, the reduced wave vector.

## PROPERTIES OF PHONONS IN MOLECULAR CRYSTALS

having a phase difference of $\pi$ between the two sets of molecules. This then is the antisymmetric mode of zero wave vector, and the profiles in the lower half of Fig. 4.3 belong to the antisymmetric branch of the dispersion curves.

A choice now presents itself for the way in which the dispersion curve for this branch can be presented. Either the variation of phonon frequency between $q = 0$ and $aq/2\pi = 1$ can be plotted as a single continuous branch (Fig. 4.5(a))

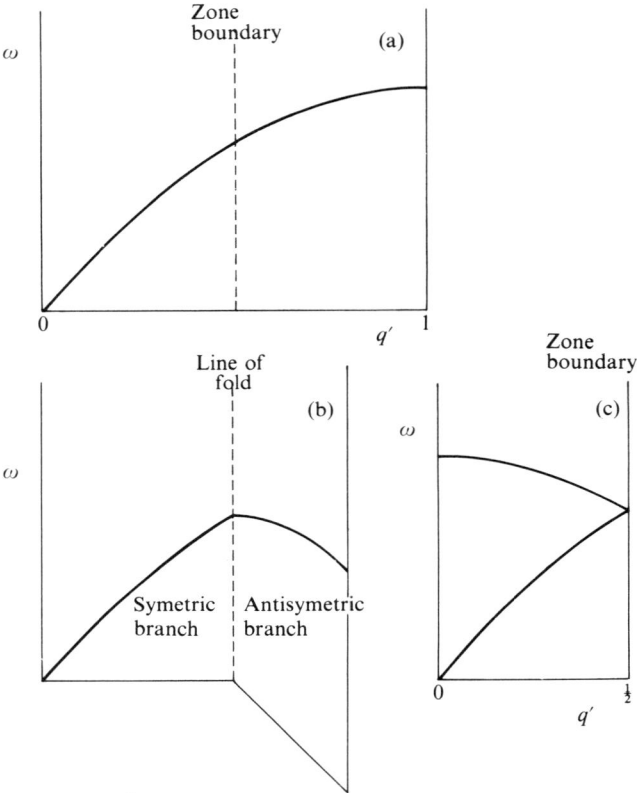

FIG. 4.5. (a) Dispersion curve diagram with the Brillouin zone boundary down the middle. Both ends correspond to zero wave vector, the left and right hand ends being symmetric and antisymmetric modes respectively. (b) Folding of curve (a). (c) Curve (a) completely folded, showing a symmetric and an antisymmetric branch becoming degenerate at the zone boundary.

whereupon the Brillouin zone boundary runs down the middle of the figure, or the antisymmetric branch can be folded over on to the symmetric branch (Fig. 4.5(b)) to give the two branches within the first Brillouin zone (Fig. 4.5(c)). The curves calculated for naphthalene by Pawley and Cyvin (1970), shown in Fig. 4.6, are of the first (unfolded) type. Curves of the second type (Pawley 1967) are presented in Chapter 1, Fig. 1.13. The latter figure is also for naphthalene and both are for wave vectors along the only symmetry direction, the screw diad axis [010]. The symmetric and antisymmetric branches are seen to cross each other in Fig. 1.13, and this is allowed by group theory as they correspond to different representations of the space group. Within the same representation branches may not cross, and this could be seen by unfolding Fig. 1.13, as with

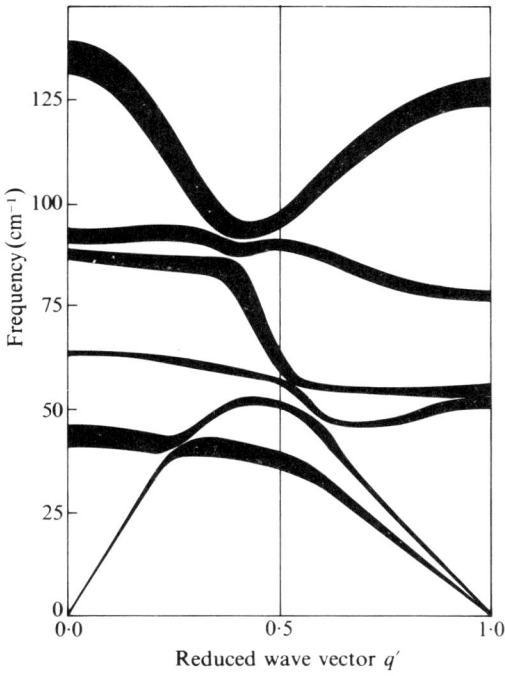

FIG. 4.6. The dispersion of the external modes of naphthalene. $q' = 0$ corresponds to symmetric modes and $q' = 1$ to antisymmetric modes, and the Brillouin zone boundary is down the middle at $q' = 0.5$. The diagram presents two calculations. The tops of the bands give the rigid molecule result, and the bottom the change on allowing the molecule to undergo bending. In three cases the bottom of one band overlaps the top of the band below, but in no cases do the branches for either calculation cross each other. The rigid molecule calculation is also presented (but in a different form: see text) in Fig. 1.13. (After Pawley and Cyvin 1970).

Fig. 4.6. If this were done, none of the branches in the extended zone would cross.

Although the curves of Fig. 4.6 have been calculated, not measured, they give us insight as to the physical nature of the various modes. The naphthalene space group ($P2_1/a$) is satisfied by two molecules in the unit cell, each placed on a centre of symmetry. Rigid-body motion gives rise to the external modes, of which there are twelve for a general $q$. In any mode, the molecular motion can consist of translations in any direction, given by **u**, and rotations about any axis, given by $\theta$. As the centre of symmetry in the molecule itself coincides with the centre in the crystal structure, the phase relationship between **u** and $\theta$ is always $\pi/2$. In other words, the translational velocity is at a maximum when the rotational displacement is a maximum. The modes consist of pure translations or of pure rotations only at $q = 0$, but as a rule the character of the mode does not change rapidly with $q$. Thus a branch which begins as a purely librational† mode at $q = 0$, remains predominantly librational about approximately the same axis as $q$ increases. This can be shown on the dispersion curves by drawing librational branches dashed, the amount of broken character indicating the proportion of libration. However, it is not feasible to indicate on the dispersion diagram the direction of either **u** or $\theta$.

Although the above statement referring to the slow variation in mode character is generally true, this is not always the case. It is easy to find an example in the dispersion curves where two branches of the same representation approach each other in a way that indicates that they would cross, but then they curve away from each other. These branches 'seem to want to cross' but cannot do so, as otherwise they would be breaking the rules of group theory. A hypothetical case is given in Fig. 4.7(a), where an acoustic mode 'A' approaches the librational optic mode 'C' as $q$ increases. The observer readily imagines Fig. 4.7(b), in which AB illegally crosses CD. If we examine the mode eigen-vectors we find a close similarity between that at A and that at B, and also a similarity between C and D. We can argue that B is an acoustic mode because of the former similarity. An acoustic mode is defined as one having molecular motion similar to that produced by a sound wave passing through the crystal. The point A on the dispersion curve corresponds exactly to a sound wave, and the molecular motion has no librational character. The eigen-vector at B is clearly much more consistent with that at A than is the librational mode at D. It is illogical to call D an acoustic mode, as some authors do, simply because it has the lower frequency. In the region where the branches avoid crossing the variation in the mode character with $q$ can be very rapid. The molecular motion becomes a complicated mixture of eigenvectors of modes similar to A and C. This 'mixing of mode character' is physically unavoidable, and occurs in calculations and in reality. At this point we must be careful not to use the expression 'mixing (or interaction) of modes' for this suggests that the phenomenon is a result of anharmonic effects, which it is not. We will return to the problem of 'branches wanting to cross' at the end of this section.

Some of the other branches in Figs. 4.6 or 1.13 can be readily understood. At $q = 0$ the modes are either purely translational or purely librational, symmetric $\mathcal{S}$ or antisymmetric $\mathcal{A}$. The librational modes can be studied by Raman scattering. In early papers, the libration was described to be about the principal axes of inertia of the molecules, with the $\mathcal{S}$ and $\mathcal{A}$ modes of close, but not degenerate,

† The word 'libration' combines the meaning of 'rotation' with 'vibration'.

frequency. Although there is no theoretical reason for the motion to be exactly about the principal axes, calculations do suggest that this description is fairly good and worthwhile in some cases. Modern experimental techniques can now usually resolve the $\mathscr{S}$, $\mathscr{A}$ doublets and the relative assignments can be made with more certainty. There is no reason to assume that the $\mathscr{A}$ mode has a higher frequency than the $\mathscr{S}$ mode. The other modes at $q = 0$ are optic translational modes, whose frequencies can be measured rather indirectly in the far infra-red (see, e.g., Harada and Shimanouchi 1966).

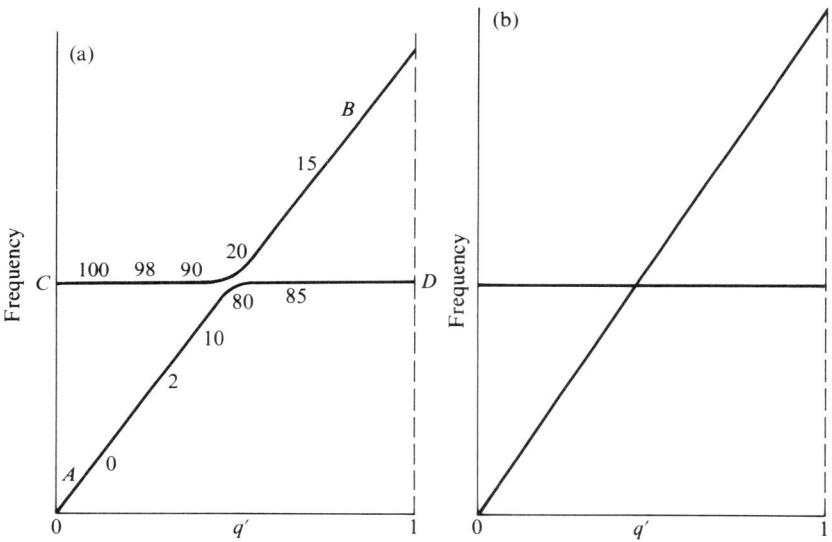

FIG. 4.7. (a) Two branches of the same representation. The numbers give the percentage of librational character in the eigen-vector in a hypothetical example. (b) Interpolation of the modes as if they cross. This can be used for scattering cross-section calculations.

A model calculation is of great benefit in the interpretation of an experiment. Continual interpretation, throughout the course of an experiment using neutron coherent inelastic scattering, can greatly increase efficiency. When we have found evidence for a phonon at **Q**, **q**, $\omega$ (**Q** is the scattering vector), the next step is to find out its mode eigen-vector. Assigning this mode to a certain model eigen-vector, we can use this eigen-vector to calculate intensities of scattering from the mode at different **Q**'s and then check these by further experiments. Modes assigned in this way allow us to improve the model, using the experimental frequencies in a refinement of the potential function parameters. Having identified a certain phonon branch, we then want to find how the branch frequencies vary with **q**. Measurements are made for a different value of **q** and the intensity compared with calculated values. We expect the new mode eigen-vector to be very similar to that of the first measurement, and this will be true unless the model incorrectly predicts the close approach of a non-crossing branch or fails to predict this when it actually happens. In these circumstances the calculated intensity will be considerably in error. It would be better in this case to use a model calculation where the eigen-vectors in this region are got by interpolating as between A and B (in the example of Fig. 4.7(a)), that is, as if the branches actually do cross. The model should then agree with experiment in all regions where there are no close approaches of non-crossing branches in the actual crystal dispersion curves.

## 4.2. Group theoretical assignments

Formal group theory is used in good measure throughout lattice dynamics, and it is unfortunate that there are many different notations used for describing the same things. The purpose of this section is to show what it is that is being described, without using much more than the standard space-group symbolism. The only extension of this symbolism that is necessary in this discussion is that widely used to describe the Shubnikov space groups. These groups, otherwise known as black-and-white space groups or magnetic space groups, will be seen to have some application in understanding the properties of phonon dispersion curves.

Let us consider a molecular crystal in the space group No. 26, $Pmc2_1$. This group is chosen because there are only four general positions in the unit cell and because all the symmetry operations are of different types. If there were only two general positions in the unit cell the problem would be trivial, whereas a number greater than four would increase the complication but introduce no further concepts in the analysis. Fig. 4.8(a) shows the symmetry elements of this space group in the orientation given in the *International Tables for X-ray Crystallography,* Vol. 1. Instead of circles for the atoms, arrows have been added to the picture, and these are meant to show the direction of movement of the atoms in the symmetric mode at $q = 0$. All the arrows are related by the full symmetry of the crystal, a necessary condition for the symmetric mode, and the thick arrows are related to the thin arrows by operations which add half a $c$-axis translation; the thicker arrows are nearer the reader by $c/2$.

The actual pattern of movement of the atoms in the symmetric mode depends, of course, on the potential function, and in Fig. 4.8(a) the arrows are free to take any orientation in the three-dimensional space. Consequently, there are three symmetric modes as the arrows have three degrees of freedom. Within a unit cell there are four arrows representing four molecules, giving a total of twelve degrees of freedom for translation. Therefore, there must be nine other translational modes still to be classified. These could all be classed as antisymmetric, but we shall see that they are antisymmetric in different ways.

The antisymmetric mode discussed earlier in this chapter consisted of half

G. S. PAWLEY

of the molecules in motion exactly out of phase with the other half. This condition will now be applied, and it is easily seen that there are three ways of dividing the molecules into pairs. The simplest way to describe these is to explain Figs. 4.8(b), (c), and (d). In all these figures there is one arrow which is the same, namely the upper left arrow in the central grouping. Fig. 4.8(b) retains only the plane of symmetry perpendicular to the $a$-axis, and this is the only symmetry symbol appearing in bold black as in Fig. 4.8(a). This symmetry operation relates

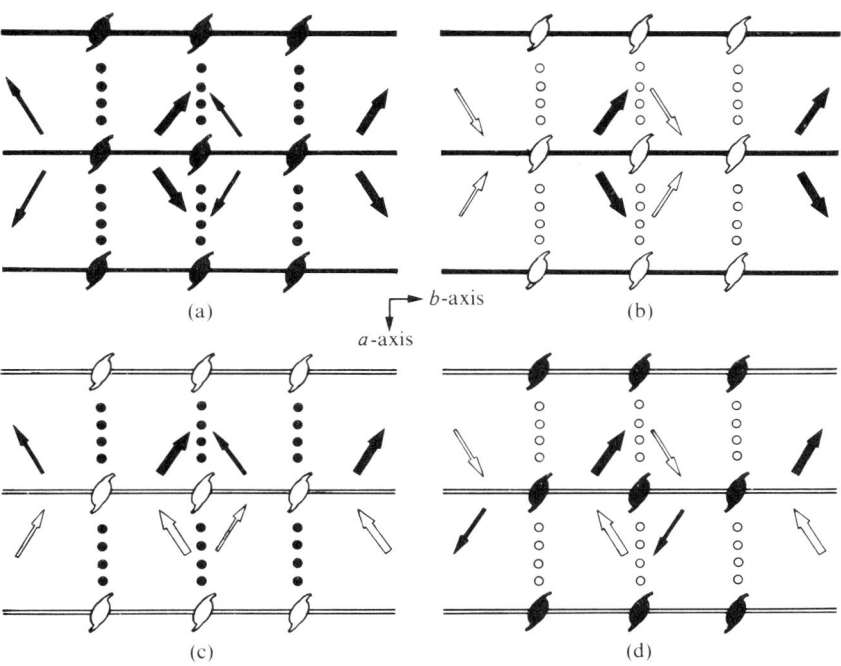

FIG. 4.8. (a) The symmetry elements of the space group $Pmc2_1$. The arrows represent general displacements of the atoms or molecules in the general positions of the space group. The thicker arrows denote displacement towards the reader of half a $c$-axis lattice translation. (b) The symmetry elements of the group $Pmc'2_1'$. The symmetry elements drawn empty are the antisymmetry operations, and they relate full arrows to empty arrows in such a way that the direction of the latter is reversed. (c) The symmetry elements of the group $Pm'c2_1'$. (d) The symmetry elements of the group $Pm'c'2_1$.

the two arrows drawn thickly. The other two arrows have had their directions reversed, and have been drawn empty. These two arrows are related to each other by the plane of symmetry, and as there is no symmetry relating them to the first two arrows the symmetry of the picture is $Pm^{**}$. Here an asterisk appears where a symmetry element has been lost. Similar considerations result in the symmetries $P^*c^*$ and $P^{**}2_1$ for Figs. 4.8(c) and (d), respectively.

The statement in the last paragraph, that there is no symmetry relating the arrows drawn full and those drawn empty, immediately provokes the thought that the latter arrows are determined uniquely from the former in a way consistent with the definition of antisymmetry. The symmetry elements lost in going from $Pmc2_1$ to $Pm**$, $P*c*$, and $P**2_1$ can now be replaced by antisymmetry operations, and the resulting groups are Shubnikov groups. In these groups an antisymmetry operation has the effect of interchanging black and white colourings, or in another context it causes the reversal of a spin. In our usage the operation reverses the sign of a vector displacement. In the figures, any operation which has this property has been drawn empty, and if we follow the convention of priming the antisymmetry elements of the space group we get the symbols $Pmc'2_1'$, $Pm'c2_1'$, and $Pm'c'2_1$ for Fig. 4.8(b), (c), and (d), respectively. The

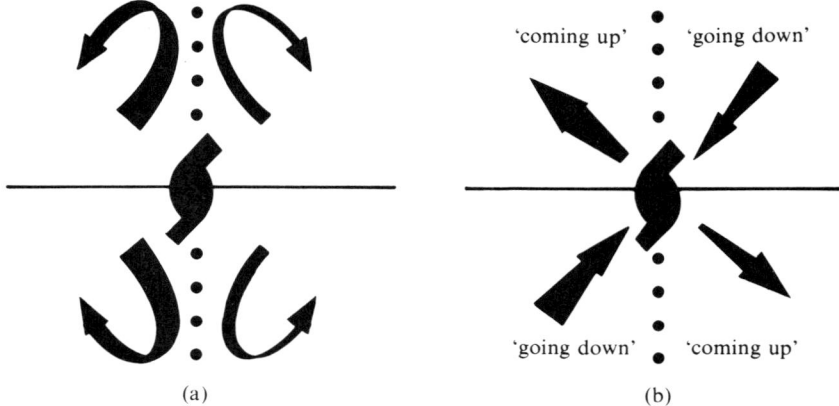

FIG. 4.9. (a) Molecular rotations corresponding to the symmetry $Pmc2_1$. (b) The rotations of (a) are represented by axial vectors, which are drawn positively along the rotation axis in a right-handed sense. The arrows considered as ordinary (polar) vectors clearly do not obey the symmetry.

argument leading to the conclusion that there are three symmetric modes corresponding to Fig. 4.8(a) now holds good for the three other diagrams. In each diagram one vector has three degrees of freedom and all the other vectors are related to that one; therefore, there are three modes of vibration corresponding to the symmetry of each of the diagrams of Fig. 4.8. Thus all the twelve translational modes have been found.

So far, in this example, the librational modes have been ignored. There are again twelve such modes, and these occur in four symmetry types each with three modes. As there are no centres of symmetry in the space group chosen, here it turns out that the four symmetry types are the same as those just derived. If a molecule appears on a centre of symmetry in the crystal, then all librational modes are such that the local centrosymmetry is preserved, whereas this is never the case with translational modes. Hence when the molecules are on centres of symmetry in the crystal, the translational and librational modes necessarily have different symmetries, but this extra complexity is avoided in the present example. However, if we do try to represent rotations vectorially, we get a problem of a different nature. It is clear that the schematic representations of rotations given in Fig. 4.9(a) could be used without difficulty throughout Fig. 4.8, but difficulty arises when we replace a right-handed rotation about an axis by a positive vector along that axis (known as an axial vector). Fig. 4.9(b) is an attempt to portray this set of vectors related to the rotations of Fig. 4.9(a). The symmetry is not obeyed by these vectors, but it is the vectors which are at fault. They are not vectors in the true sense, but are infinitesimal rotations and have transformation properties different from true vectors. Under proper symmetry operations (axes and screw axes of symmetry) the behaviour of both vectors is the same, but under improper operations (planes, glide planes, and inversions) the infinitesimal rotation vector transforms as the true vector plus a sign reversal. For this reason the infinitesimal rotations should be used with caution.

Most measurements by optical techniques are confined to the special case of $q = 0$ just considered. Neutron measurements, however, range over the whole of the Brillouin zone, so let us continue the analysis of the $Pmc2_1$ structure to modes with non-zero $q$. If we choose a general $q$, there is only one symmetry representation; all the modes belong to this symmetry and there is nothing more of interest. In special directions there are different representations, so let us analyse the problem for $q_a$, $q_b$, and $q_c$, that is, for wave vectors along the $a$, $b$, and $c$-axes respectively.

For a wave with wave vector $q_a$ all the molecules related by lattice translations and lying in a plane perpendicular to the $a$-axis move in phase. Molecules in other planes move with a different phase and are therefore unrelated by symmetry. Consequently, the symmetry pertaining to these modes is that within the plane of constant phase. In choosing $q_a$ we have to analyse the symmetry in the $b$-$c$ plane, resulting in the group $p*c*$. The asterisks here now represent symmetry elements which are definitely missing, and these must be all those planes of symmetry not containing the $a$-axis, and all axes of symmetry not parallel to the $a$-axis. The allowed operations then comprise a form of plane group as they transform the planes of constant phase into themselves. A small $p$ is used in these symbols to denote the primitive two-dimensional Bravais lattice, following the usage of the International Tables. By the process just outlined we have finished with the group $p*c*$ starting from the group $Pmc2_1$, the symmetry of Fig. 4.8(a), and by applying this procedure to the symmetries of Fig. 4.8(b), (c), and (d) we obtain $p*c'*$, $p*c*$ and $p*c'*$ respectively. The generated groups are seen to fall into pairs, and following the same procedure for $q_b$, pairing is found again but in a different combination. This is indicated in Table 4.1. No pairings occur for $q_c$ as none of the symmetry elements are disallowed. What then is the significance of this pairing? At $q = 0$ the modes belong to the four symmetry representations of Fig. 4.8. As we move away from this point we are investigating the phonon branches which, for the particular directions of $q$ given in Table 4.1, belong to the symmetry representations of the plane groups tabulated. Thus the branches starting with the symmetry (a) or (c) at $q = 0$ are of the same representation for $q_a$, and those branches starting with symmetry (b) or (d) are likewise of the other representation. (By the 'symmetry (a)' we mean the symmetry of Fig. 4.8(a)). The members of the pairs (a) and (c), and (b) and (d) are said to be compatible with one another for $q = q_a$. The compatibility relations for $q = q_b$ are different, as indicated in the table, and when $q$ is in the direction of the $c$-axis the branches fall into four representations and the compatibility relations are trivial. This last fact is easy to understand as the largest number of representations is always associated with the highest symmetry direction in the crystal point group, which is obviously the unique axis $c$.

TABLE 4.1

*The symmetry groups for zero wave vector (1st col.) and for wave vectors in the directions of $q_a$, $q_b$, and $q_c$, and their compatibility relations.*

| Symmetry of Fig. 4.8 | Symmetry of $q_a$ mode | Symmetry of $q_b$ mode | Symmetry of $q_c$ mode |
|---|---|---|---|
| (a) $Pmc2_1$ | $p*c*$ | $pm**$ | $Pmc2_1$ |
| (b) $Pmc'2_1'$ | $p*c'*$ | $pm**$ | $Pmc'2_1'$ |
| (c) $Pm'c2_1'$ | $p*c*$ | $pm'**$ | $Pm'c2_1'$ |
| (d) $Pm'c'2_1$ | $p*c'*$ | $pm'**$ | $Pm'c'2_1$ |
| Compatibilities | (a) and (c)<br>(b) and (d) | (a) and (b)<br>(c) and (d) | no<br>pairings |

The way the dispersion curves join up at $q = 0$ has now been determined, along with the number of representations governing the branches. Finally, we need to know how these branches behave at the Brillouin zone boundary. Let us use labels A, B and C to denote the boundary points reached in the directions of the $a$, $b$ and $c$-axes respectively. At A and B there is no degeneracy, but at C there is degeneracy between the $pmc2_1$ and $pmc'2_1'$ branches, and a separate degeneracy between the $pm'c2_1'$ and $pm'c'2_1$ branches. To understand this let us look closely at how the $\mathscr{S}$ and $\mathscr{A}$ branches became degenerate in the example of the previous section. In this case the two different molecule sets were situated on lattices related to each other by a half lattice translation in the direction of $q$. The phase difference between the motion of the molecules in these two sets started at zero and increased as $q$ increased to reach $\frac{1}{2}\pi$ at the zone boundary, and then continued to increase until it reached $\pi$ at the next $q = 0$ point. This gradual transition from $\mathscr{S}$ to $\mathscr{A}$ is only possible because of the half lattice translation in the direction of $q$. In the $Pmc2_1$ example, the only half lattice translation is in the direction of $q_c$, giving degeneracy at the point C only. The molecules on the same level up the $c$-axis are related by the mirror symmetry or mirror antisymmetry, and therefore no phase difference can build up between these two molecules. Thus the branch p$mc2_1$ becomes p$mc'2_1'$, the $m$ symmetry remaining throughout; in the other case the $m'$ symmetry is always present.

The symmetry representations, degeneracies and compatibility relations have now been determined for dispersion curves in P$mc2_1$. There is a formal jargon of group theory used for expressing these facts, and in many cases more than one nomenclature. On phonon dispersion curves will be found various letters of the Greek and Roman alphabets, usually with numerical suffices, and these convey to the diligent reader the representation for any branch. This is very similar in its usage to the spectroscopists jargon of $A_{2u}, B_{1g}$ etc., except that the latter symbolism conveys its meaning to the reader without recourse to any tabulation.

## 4.3. Model calculations

It would be out of place in the present chapter to give the fine details of model calculations when these are available in the general literature (Venkataraman and Sahni, 1970; Pawley 1972). Instead it is much more pertinent to give an outline of the procedure. A model for which calculations are possible must be in one of two forms. Either there is an explicit analytic crystal potential function, or all the force constants necessary for the calculation must themselves be parameters of the model. The force constants are in essence the double differentials of the crystal potential function with respect to possible displacements of the constituent molecules or atoms (cf. eqns (3.1) and (3.2)). They are compounded in the standard way to build up the dynamical matrix, from which the eigen-vectors for the phonon modes and their eigen-values are determined.

The procedure used by Cochran and Pawley (1964) for hexamethylene-tetramine used the force constants as parameters. In this case it was feasible to adopt this approach. The very high symmetry of the crystal structure caused many of the force constants to have identical values, and the number of independent parameters was quite small. Such simplification was not possible for the naphthalene structure (Pawley 1967), because of the very low symmetry. The molecules are held together in the crystalline state by the relatively weak van der Waals forces, and as these forces are of a fairly short range it is only necessary to consider interactions over distances of up to 5 Å.

How then do we introduce the interaction? It is clear that steric effects are of great importance in governing the crystal structure, and it is plausible that the interactions between the hydrogen atoms of neighbouring molecules are of dominant importance. Assuming a secondary role for interactions involving carbon atoms, we follow Kitaigorodskii (1966) and write the potential $V(r)$ for one pair of atoms a distance $r$ apart as

$$V(r) = -\frac{A_i}{r^6} + B_i \exp(-\alpha_i r)$$

where the atoms in the pair belong to different molecules. In this equation the index $i$ denotes the specific chemical pairing, and Kitaigorodskii's values are given in Table 4.2. Because of its functional form this potential function is often called the '6-exp' potential, and is otherwise known as the Buckingham pair-potential function. The interactions between any pair of molecules can be found by summing over all the atom-atom interactions within a radius of 5 Å from each constituent atom, as shown schematically in Fig. 4.10. In this way all the molecule-molecule interactions are determined by the nine parameters of Table 4.2, of which the three H . . . H parameters will be the most important.

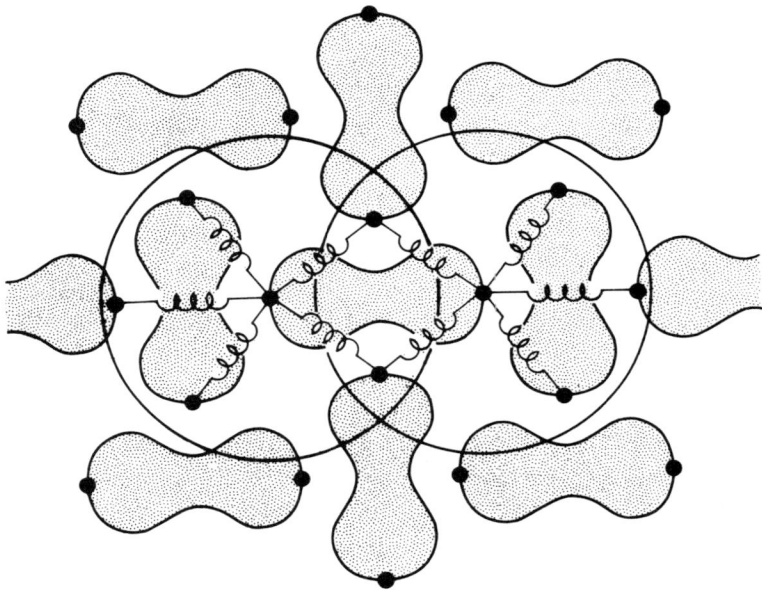

FIG. 4.10. Molecules are represented as shapes with just two atoms each. Molecular interaction is assumed to be between atoms in different molecules, and the interactions for the molecule shown are found by summing over all the atom-atom interactions within the two circles drawn. These interactions are represented as springs.

The potential function of the crystal is an analytic function of the nine parameters, and it is possible to perform all the lattice dynamical calculations following analytic procedures. This has recently been done in some detail by Pawley (1972), who shows how to set up the dynamical matrix for a general wave vector. If it is desired to have the result for a wave vector in a special direction, then substitution of the special value of $q$ must yield the correct result as regards the group theory of the problem. The analysis of the previous section is then very useful in interpreting the results, and such an interpretation is essential for assigning a calculated phonon frequency with a measurement. All measurements must be assigned to a particular symmetry representation, and this is best done by comparing with a rough calculation of neutron cross-sections.

When the measurements are complete, fitting must be done in earnest. It is usual to have recourse to a least-squares procedure, but whatever method is used the fitting is necessarily iterative. The dynamical matrix has to be set up many times and finding the solutions can be very time-consuming. Most measurements

TABLE 4.2

*Kitaigorodskii's values for the pair potential function.*

| Atom pair, $i$ | $A_i$, (kcal/mole)Å$^6$ | $B_i$, kcal/mole | $\alpha_i$, Å$^{-1}$ |
|---|---|---|---|
| H...H | 57 | 42 000 | 4·86 |
| H...C | 154 | 42 000 | 4·12 |
| C...C | 358 | 42 000 | 3·58 |

are made in symmetry directions, mainly because they are easier to assign and interpret, but this means that, in solving the general dynamical matrix for each measurement and each iteration, we are doing much more calculation than is necessary. By using the known symmetry of the problem the dynamical matrix can be block diagonalized, whereupon each block, when solved separately, gives solutions corresponding to just one representation. However, in view of the complicated nature of the compatibility relations discussed in the last section, it is clear that no single scheme of block diagonalization exists for extensive measurements, making the writing of a general computer program almost impossible. Each problem requires its own program, tailored to the possible measurements. Although this problem is by its very nature complicated, recourse to numerical methods should not be necessary (Pawley 1972).

The models discussed so far have been restricted to molecular motion of a rigid-body type. Many molecules are however far from being rigid, and the assumption of rigidity may be rather poor. To test this for naphthalene, Pawley and Cyvin (1970) did the following calculation. The interaction between the molecules was again assumed to be of the atom–atom sum type, but the molecules were treated as an assembly of atoms. The integrity of each molecule was obtained by including the interactions between the atoms within the molecule, this interaction being simply the molecular force field obtained from fitting the frequencies and mean amplitudes for the molecule in the free state. As all the interactions are now between atoms, the calculation is of the original Born–von Kármán type, and the dynamical matrix has a dimension of three times the number of atoms in the unit cell. The results of the calculation are given in Fig. 4.6, where each branch is presented as a band. The lower edge of the band gives the result from the calculation where the molecule is allowed to deform as governed by the molecular force field, whereas the top edge corresponds to a rigid molecule result. The latter is obtained from the same program as the former but the molecular force field is increased by a large factor making the molecule essentially rigid.

The thickness of the band therefore gives an indication of the error which is being introduced by accepting a rigid-molecule model.

None of these calculations has required consideration of electrical interactions. Molecules with a strong dipole moment have interactions in excess of the van der Waals forces, and thus need special attention. Higher electrical multipoles are of lesser importance and may therefore be erroneously neglected. In hexamethylenetetramine the lowest multipole is octopolar, and it may well be that the interaction between these multipoles can explain a puzzling fact. The molecules are roughly spherical in shape, that shape being similar to adamantane. Adamantane has a transition in the solid state to a high temperature plastic crystalline phase which has no parallel in hexamethylenetetramine. In such a phase there is very little resistance to molecular reorientation and the molecules interchange frequently between two orientations. It is surmized that the lower octopole moment on the adamantane molecules causes this difference in physical behaviour, and calculations are in progress to test this.

It is fitting to end this chapter by pointing out the importance of the plastic-crystal phase transition, owing to its widespread occurrence. Molecular reorientation cannot be treated within the framework of the models described in this chapter, as the latter are all based on the validity of the harmonic approximation. With large librations and reorientations a completely different theoretical approach is called for.

## 4.4 References

COCHRAN, W. *and* PAWLEY, G. S. (1964). *Proc. R. Soc.* A**280**, 1.
HARADA, I. *and* SHIMANOUCHI, T. (1966). *J. chem. Phys.* **44**, 2016.
KITAIGORODSKII, A. I. (1966). *J. Chim. phys.* **63**, 6.
PAWLEY, G. S. (1967). *Phys. Stat. Sol.,* **20**, 347.
PAWLEY, G. S. (1972). *Phys. Stat. Sol.,* (b) **49**, 475.
PAWLEY, G. S. *and* CYVIN, S. J. (1970). *J. chem. Phys.* **52**, 4073.

*General*
COCHRAN, W. *and* PAWLEY, G. S. (1964). The theory of diffuse scattering of X-rays by a molecular crystal. *Proc. R. Soc.* A**280**, 1-22.

   The phonon dispersion relations are calculated in this paper for the external modes of hexamethylenetetramine: this is the first molecular crystal for which detailed phonon calculations were carried out.

VENKATARAMAN, G. *and* SAHNI, V. C. (1970). External vibrations in complex crystals. *Rev. mod. Phys.* **42**, *no.* 4, 409-470.

   This is a review of the theoretical and experimental studies of the dispersion relations for the external modes of molecular crystals.

# 5 Neutron Scattering Studies of the Dynamics of Polymer Chains

*By* G. ALLEN

*Chemistry Department, Manchester University*

## 5.1. Background

*5.1.1. Polymer molecules*

A polymer molecule consists of a large number of repeating sub-units 'A' strung together in a long chain, e.g. A—A—A— or $(A)_n$. In practice, $n$ generally ranges from 100 to 100 000 units. To a theoretical physicist, it may be sufficient just to define the length ($c.$ $10^3$ to $10^5$ Å) and breadth ($c.$ 5 Å) of these molecules, and subsequently to treat them as long thin pieces of string arranged either in ordered or random internal and external configurations. For the chemist and chemical physicist, it is necessary to consider the detailed chemical structure of the molecule, this being a three-dimensional structure, the simplest of which is polyethylene,

$(CH_2)_n$ i.e.

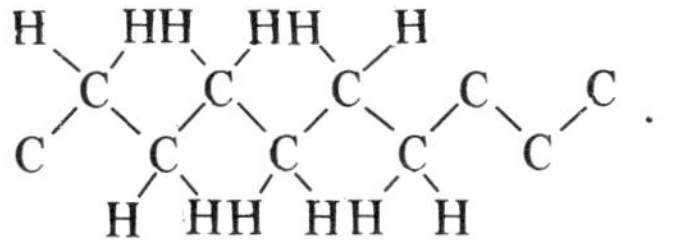

```
    H   HH  HH   H
     \ / \ / \ / \ / \
      C   C   C   C   C  .
     / \ / \ / \ / \ / 
    C   C   C   C   C
    H  HH  HH   H
```

Each carbon atom subtends four bonds in a tetrahedral configuration, and the conformation of the molecule as drawn above represents its pulled-out form. By virtue of internal rotation about the C—C bonds, the molecule can take on millions of different conformations, there being three stable rotational isomers, separated by approximately 120°C for the rotation about each bond. Another simple polymer molecule is poly(methylene oxide) $(CH_2O)_n$,

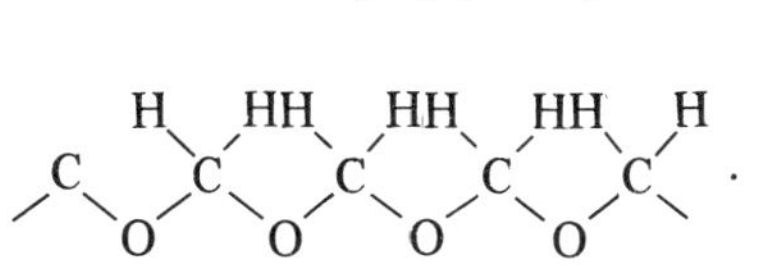

```
   H   HH  HH  HH   H
    \ / \ / \ / \ / \
     C   C   C   C   C   .
      \ / \ / \ / \ /
       O   O   O   O
```

Many polymer molecules have side groups attached to the repeat unit, e.g., poly(propylene oxide) which has a repeating unit

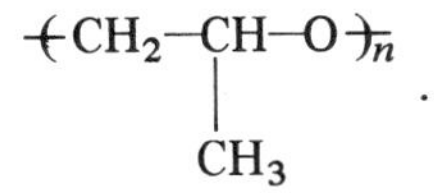

$(CH_2-CH-O)_n$
    |
    $CH_3$

# THE DYNAMICS OF POLYMER CHAINS

A more complicated polymer, poly(methyl methacrylate), Perspex, has the structure

$$\mathrm{-(CH_2-\underset{\underset{COOCH_3}{|}}{\overset{\overset{CH_3}{|}}{C}})_n}.$$

Polyethylene and poly(methylene oxide) both crystallize because each polymer chain has an inherently regular structure, but the molecules having side

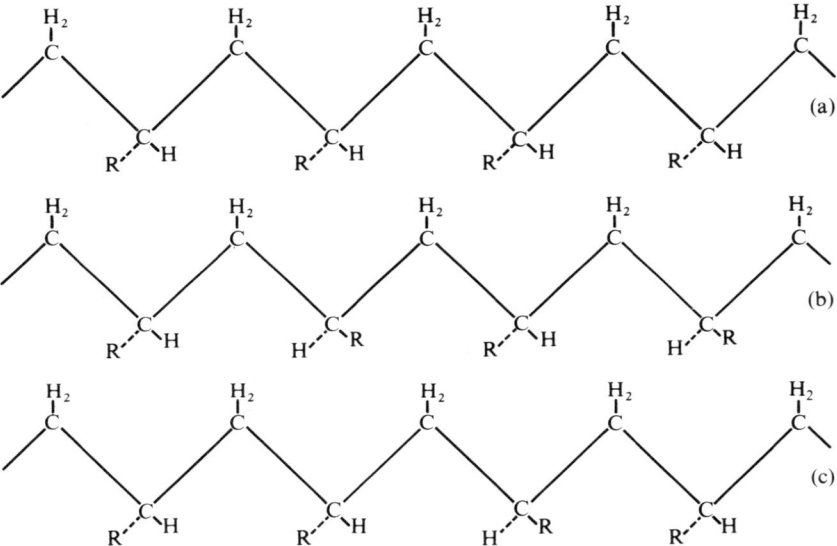

FIG. 5.1. Isotactic (a), syndiotactic (b), and atactic (c) forms of a poly (α-olefin), e.g.
$\mathrm{-(CH.CH_2)_n}$.
|
$CH_3$

groups only crystallize when the side groups are attached to the main chain in a regular fashion. There are two alternatives shown in Fig. 5.1. The isotactic polymer chain is one in which all the side groups appear on the same side of the chain; the syndiotactic molecule has the side groups alternating regularly from one side of the backbone to the other. Careful selection of the polymerization catalyst allows the chemist to make polymer chains containing, predominantly, one or the other of these isomers. However, many catalysts operating on the same monomer produce polymer chains with randomly alternating side groups—atactic structures (Fig. 5.1(c)). Such materials are completely amorphous. They do not crystallize and they are either glassy or rubbery substances. For example, atactic polypropylene is a rubbery material at room temperature, and freezes to a glass at $-20°C$. Isotactic polypropylene, on the other hand, crystallizes very efficiently, and is a hard material, melting at $170°C$. It is used a great deal in the motor industry in the interior trim of vehicles.

If the chain has no side groups, then the material is only encountered as a rubbery or crystalline substance: it does not exist as a glass. If the polymer chain possesses side groups, then the stereo regularity of the molecules determines whether the material is crystalline or amorphous. It is possible then to have the same chemical substance exist in two completely separate and non-interchangeable forms. Polymer science is concerned firstly with properties of the material which are conferred predominantly by the nature of the main chain, and secondly with properties which arise from the nature and disposition of the side groups.

So far, we have assumed that the polymer chains are essentially linear, but, of course, side reactions do occur during the synthesis of polymer chains, and these may lead to branched molecules. Polyethylene, for example, is very susceptible to chain branching. Although it is possible to tell from the physical properties of polymer melts when the molecules have branched structures, it is impossible to determine precisely the topology of the chains. In the case of rubbers, chemical cross-links are deliberately introduced between individual chains to tie the molecules into three-dimensional networks, thus preventing the rubber from flowing and adopting a permanent set when it is subjected to large strains. Natural rubber, for example, is not a good engineering material unless its polymer chains are crosslinked into a three-dimensional network. The cross-links are introduced approximately once every 100 units per chain, so that there is still sufficient length of chain between cross-links to undergo the coiling and uncoiling function which is the origin of rubber elasticity in a molecular network.

*5.1.2. The states of polymerized matter*

Basically, polymerized substances exist in one of three states of condensed matter.

(a) The rubbery state is really the liquid-like state of polymeric material. The molecules undergo long-range coiling and uncoiling, with frequencies in the region of $10^8$-$10^9$ Hz, as they wriggle through the various molecular conformations available. The rate of diffusion of the long chains is very low indeed, owing to the low probability of obtaining a co-operative translation simultaneously of all the segments in a chain. This is the high-temperature state of all polymers which do not decompose before melting.

(b) The glassy state is the state of frozen disorder which is obtained by cooling geometrically-regular polymer chains, or by the very rapid chilling of systems composed of regular structures. The molecules exist in a variety of conformations, but they no longer have sufficient thermal energy to change their individual conformations. The glass-to-rubber transition is one in which the molecules acquire sufficient thermal energy to begin their long range coiling and uncoiling motions.

(c) The crystalline state of a polymer is usually the state in which the material is only partially crystalline. The actual extent of crystallization depends on the stereo regularity of the molecules, the linearity of the chains and the thermal history of the sample. The non-crystalline regions may be either glass-like or rubbery, depending on the temperature of the crystalline polymer. The material is really a composite of tiny crystallites embedded in an amorphous matrix, and is isotropic provided that the melt is undeformed during crystallization. If the melt is strained either by stretching or by extrusion through a die, then orientated crystallization occurs, and an anisotropic material (e.g., a fibre) will be produced, in which the crystallites are aligned relative to the direction of the strain. Nylon fibre is a typical anisotropic material.

## 5.2. The dynamics of the main polymer chains in crystals

§§ 5.2, 5.3, and 5.4 give the basic background for discussing the dynamics of polymer chains, and the information that can be obtained from neutron scattering. A variety of studies is possible. First of all, we can study crystalline polymers in the same way as the organic crystals discussed in the review article by Venkataraman and Sahni (1970) (see General References, Chapter 4 (p. 96)). We can investigate the intra-molecular vibrations within the polymer chains and the inter-molecular vibrations between chains in the crystalline regions. Infra-red and Raman spectroscopy studies will be required to supplement the neutron scattering results. Secondly, the dynamics of side chains in glassy or partially crystalline polymers is of interest. When the main backbone of the polymer chain is frozen, the side groups can still continue to rotate and do not freeze until lower temperatures are reached. These motions can give rise to mechanical and dielectrical energy loss phenomena. Thirdly, in the rubbery state, we can study the overall diffusion of the polymer chains by measuring the Doppler broadening of the quasi-elastic neutron scattering (see p. 124); but although we can get information about the translational motion, it is very difficult to obtain detailed information about the internal dynamics of the polymer chains in the rubbery state. We know a considerable amount about the time-scales of these motions, from pulsed n.m.r. measurements for example, and we know a fair amount about the energetics involved from thermodynamic measurements on rubbers, but we still have a long way to go in understanding the dynamic properties of the main chain in terms of the internal rotation of coupled asymmetric rotors.

In this section we shall consider the dynamics of the main chain in crystalline polymers. The side-group motion in polymeric glasses is discussed in § 5.3, and the self diffusion of polymer chains in the rubbery state in § 5.4.

## 5.2.1. The normal modes of vibration of polymer chains in crystals

The molecular dynamics of a *small* non-linear molecule containing $N$ atoms is described in terms of the $3N - 6$ normal modes of vibration.† The vibrational frequencies can be calculated from a normal-co-ordinate analysis, provided the potential function given by eqn (3.1)

$$2V = 2V_0 + 2 \sum_{i=1}^{3N} (\partial V/\partial q_i) q_i + \sum_{i=1, j=1}^{3N} (\partial^2 V/\partial q_i \partial q_j) q_i q_j + \dots, \quad (5.1)$$

and the geometry of the molecule are known. It is usual to consider the molecule as a set of point masses coupled by harmonic forces, thus truncating the Taylor expansion in eqn (5.1) at the second-order term. The first-order term vanishes as the energy must be a minimum for all the co-ordinates $q_i = 0$; if the energy of the equilibrium configuration is taken to be zero, $V_0$ also is zero. A linear transformation from the geometrical co-ordinates $q_i$ to a set of normal co-ordinates $Q_i$, so that the cross terms $i, j$ are eliminated in the potential energy, gives the potential energy ($V$) and the kinetic energy ($T$) of the molecule as the sums of series of squared terms,

$$2V = \sum_{i=1}^{3N-6} \lambda_i Q_i^2, \qquad 2T = \sum_{i=1}^{3N-6} \dot{Q}_i^2. \quad (5.2)$$

The coefficients $\lambda_i$ are related to the frequencies $\nu_i$ of the normal vibrations by

$$\lambda_i = 4\pi^2 \nu_i^2. \quad (5.3)$$

In a normal vibration, all the atoms move in phase with the same frequency $\nu$ such that the cartesian co-ordinates of displacement change sinusoidally. The overall vibrational motion of the molecules is obtained by superimposing all $3N - 6$ normal vibrations.

Now let us consider, as the simplest model of a polymer chain, an extended set of $N$ (where $N$ is now large, $> 10^3$) identical point masses $m$, separated by distances $a$ (as shown below) and joined by weightless springs which obey Hooke's law and have a force constant $f$. When the $n$th mass is displaced from its

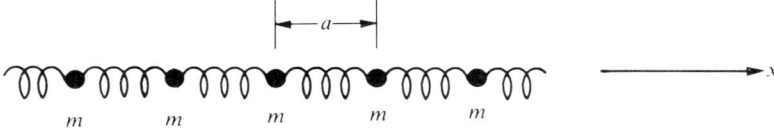

equilibrium position, along the internuclear axis, the equation of motion for this mass is

$$m\ddot{x}_n = f(x_{n+1} + x_{n-1} - 2x_n), \quad (5.4)$$

† For a *linear* molecule there are $3N - 5$ normal modes of vibration.

# THE DYNAMICS OF POLYMER CHAINS

where it is assumed that only nearest neighbours are effective. The solution of eqn (5.4), describing a wave travelling along the internuclear axis with angular frequency $\omega_i (= 2\pi \nu_i)$, is

$$x_n = A \exp[i(\omega_i t - qna)], \tag{5.5}$$

where $q$ is the wave number. Differentiation of eqn (5.5) and substitution of the result in eqn (5.4) gives a dispersion relationship between $\omega_i$ and $q$

$$\omega_i = \pm (4f/m)^{\frac{1}{2}} \sin(qa/2), \tag{5.6}$$

which is periodic in $2\pi/a$. It describes a wave propagated through the chain of wavelength $\lambda = 2\pi/q$. For $q = 0$, all particles move exactly in-phase with the same amplitude and this corresponds to a translation of the whole chain along the internuclear axis. At $q = \pi/a$, alternate particles move exactly out of phase with each other and constitute a standing wave of wavelength $\lambda = 2a$. In general, the phase difference $\phi$ between successive particles is

$$\phi = 2\pi a/\lambda, \tag{5.7}$$

and the region between $q = \pi/a$ and $q = -\pi/a$ defines the first Brillouin Zone, which contains $N$ independent and discrete values of $q$ for this degree of vibrational freedom. Values of $q$ outside this range simply describe motions already included within these limits. (There are, in addition, $2N$ modes of vibration corresponding to the two degrees of transverse vibrational freedom: their frequencies can be described by the same expression (5.6) as for the longitudinal motion but with different force constants $f$.)

Next consider the longitudinal motion (shown below) of a one-dimensional chain with two types of atoms of masses $m_1$ and $m_2$, again coupled by bonds of

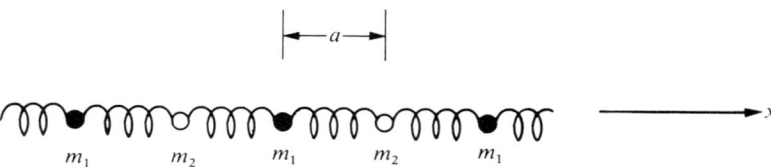

force constant $f$ at an internuclear distance $a$. For each value of $q$, two values of $\omega_i$ are obtained (see Kittel 1966), given by

$$\omega_i^2 = f(m_1^{-1} + m_2^{-1}) \pm f[(m_1^{-1} + m_2^{-1})^2 - 4\sin^2 qa/(m_1 m_2)]^{\frac{1}{2}}. \tag{5.8}$$

Thus there are now two branches of the dispersion relation containing a total of $N$ modes:

(a) the *acoustical* modes, so-called because, when $q = 0$, adjacent particles move together with equal amplitude, as when a *sound* wave is transmitted along the chain;

(b) the *optical* modes, in which adjacent particles move against one another for small $q$, but the centre of mass in the pair does not move.

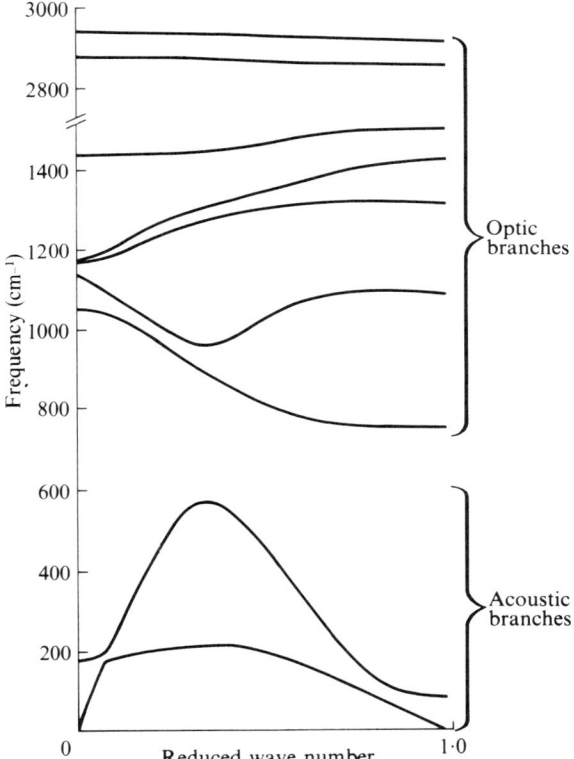

FIG. 5.2. Calculated dispersion curves for the polyethylene crystal. The abscissa is $q/q_{max}$ where $q_{max}$ is the wave number at the Brillouin zone boundary. (After Tasumi, Schimanouchi, and Miyazawa 1962).

The shapes of the branches of the optical and acoustical components can only be predicted if the molecular force field is known. The shape reflects the intramolecular coupling. If the branches are flat and show little phase dependence the coupling along the chains is weak. Strongly curved, phase-dependent branches are indicative of strong coupling. A typical set of dispersion curves is drawn in Fig. 5.2 for $+CH_2 \}_n$, polyethylene. These curves were calculated assuming an infinitely extended chain. They show two strongly curved acoustic branches and relatively flat high-frequency optical branches.

## THE DYNAMICS OF POLYMER CHAINS

*5.2.2. Incoherent inelastic neutron scattering*

The selection rules governing the spectral activity for electromagnetic radiation are

$\left|\dfrac{\partial \mu}{\partial Q_i}\right| > 0$ and $\phi = 0, \pi/2$ for infra-red activity (where $\mu$ is the dipole moment of the molecule and $Q_i$ is a normal co-ordinate),

$\left|\dfrac{\partial \alpha}{\partial Q_i}\right| > 0$ and $\phi = 0, \pi/2, \pi$ for Raman activity (where $\alpha$ is the polarizability).

In incoherent inelastic neutron scattering all modes are active and all phase relations are allowed. The intensity of the mode is related to the mean-square displacement of the nuclei in the normal vibration, and the inelastic spectrum is dominated by the movement of the protons in the structure because of their large incoherent scattering cross-sections (see p. 8).

An important consequence of the relaxed selection rule concerning phase relations in the vibrational spectra obtained from inelastic neutron scattering is that modes which show no dispersion appear as sharp lines and modes which are dispersed produce broad lines. This contrasts with the vibrational bands observed in the infra-red and Raman spectra, which should be sharp because of the well-defined selection rules.

From the potential function of the molecule, the number of vibrational lines in the frequency interval $\omega$ to $(\omega + d\omega)$ can be computed, i.e. $Z(\omega)\, d\omega$. Since the transition probability depends only on the wave function of the initial and final states and the amplitudes of vibration of the nuclei, and is independent of $\mu$ and $\alpha$, the curve $Z(\omega)$ should represent the observed intensities in the incoherent inelastic spectrum. Thus intensity relations in the inelastic neutron scattering spectrum should lend themselves to quantitative analysis.

*5.2.3. Experimental results from incoherent inelastic scattering*

Consideration of the experimental results for polyethylene will serve to illustrate the most comprehensive application to date of neutron scattering to the molecular dynamics of a polymer chain.

The infra-red and Raman spectra of high density (virtually linear) polyethylene are remarkable for the fact that there are very few vibrational bands below 500 cm$^{-1}$. They have been interpreted using a normal co-ordinate analysis of an infinitely extended isolated molecule. The dispersion relations account for the optical branch of lowest frequency having a limiting value of 720 cm$^{-1}$, and for two acoustical branches with limits below this frequency. One is designated the carbon-skeletal torsional mode,

$$\diagup^{C}\diagdown_{C}\diagup^{C}\diagdown_{C}\diagup^{C}\diagdown_{C}\diagup^{C}\diagdown_{C\downarrow}$$

with a calculated frequency limit of 190–200 cm$^{-1}$; the other is described as the skeletal stretching mode (sometimes called the 'accordion' mode) with a limiting frequency of about 500 cm$^{-1}$. The one distinct feature at 76 cm$^{-1}$ in the far-infra-red spectrum is assigned as a lattice mode of vibration, i.e. a vibration in which the chain moves as a rigid unit.

Incoherent inelastic neutron scattering spectroscopy is peculiarly suitable for the observation of fundamental vibrational modes of low frequency, and particularly so in this case because the modes are not optically active. A clearly defined peak at 195 cm$^{-1}$ was readily detected in the earliest neutron scattering

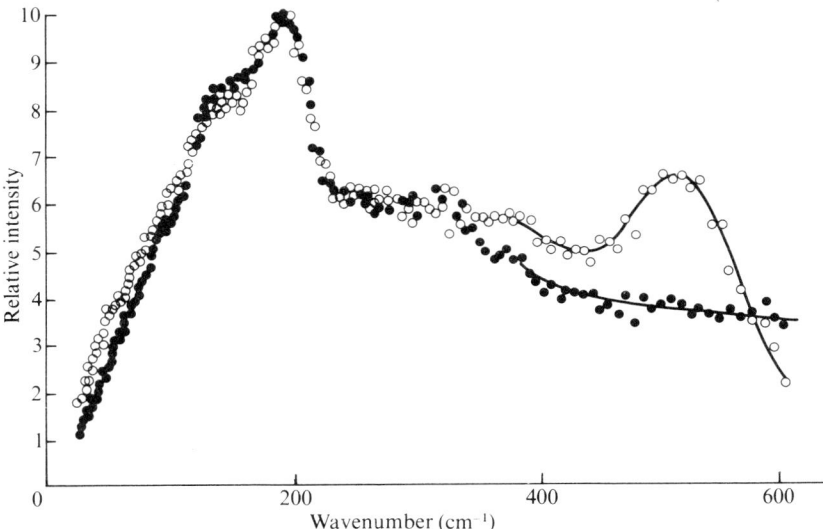

FIG. 5.3. Inelastic neutron scattering curves for oriented polyethylene, normalized at the 190 cm$^{-1}$ peak. ○ Momentum transfer parallel to chain axis; ● momentum transfer perpendicular to chain axis. (After Myers, Summerfield, and King 1966).

measurements (Danner *et al.* 1964; Myers *et al.* 1965) which used an unoriented polyethylene specimen. Later, more precise measurements by Myers, Summerfield and King (1966) identified the second peak at 525 ± 15 cm$^{-1}$. More important is the fact that these measurements were carried out, using a triple-axis spectrometer, on a specimen which had been uniaxially oriented by stretching. The results are shown in Fig. 5.3. When the sample is oriented such that the momentum transfer occurs in a longitudinal direction, i.e. along the chain axis, the 525 cm$^{-1}$ peak is intense, but when the momentum transfer occurs in a transverse direction, i.e. perpendicular to the chain axis, the intensity of this mode is greatly diminished in accord with its assignment to the longitudinal stretching mode which should have no transverse component. Fig. 5.4 shows the predicted $Z(\omega)$ curve for the two orientations and gives further evidence for the assignment. The calculations were carried out for a model of the crystal which included intermolecular force constants, rather than for the isolated polymer chain. Inspection of Fig. 5.4 shows an additional peak at about 80 cm$^{-1}$ which is related to the infra-red band at 76 cm$^{-1}$ previously assigned as a lattice vibration. There are, however, two unsatisfactory and unexplained aspects of the experimental results. First, it will be noted that although the torsional transverse peak at 195 cm$^{-1}$ has been somewhat reduced in intensity in the longitudinal spectrum it is still pronounced, whereas it would be expected to have no longitudinal component. This is not obvious from Fig. 5.3 because in the figure the raw data have been normalized at 195 cm$^{-1}$. Second, there is a suggestion of subsidiary weak peaks at c. 150 and 250-300 cm$^{-1}$ which are not

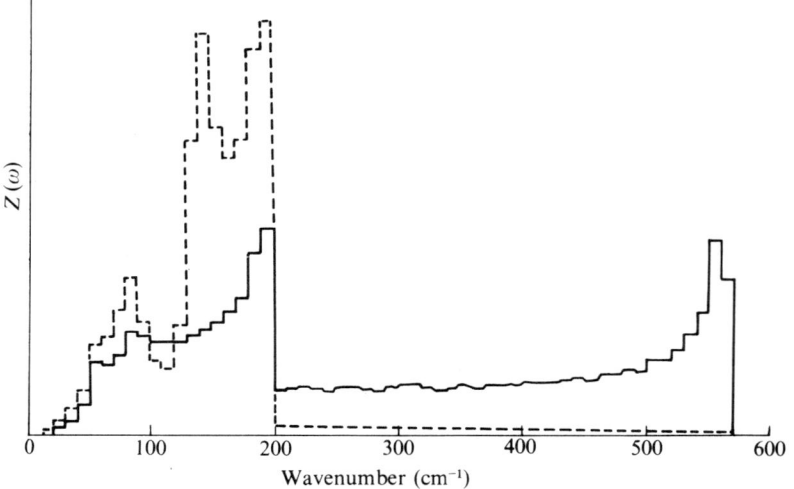

FIG. 5.4. Calculated density of states function $Z(\omega)$ for oriented polyethylene. —— Momentum transfer parallel to chain axis; – – – momentum transfer perpendicular to chain axis. (After Kitagawa and Miyazawa 1967).

accounted for on the basis of our present knowledge. Nevertheless, a more complete picture of the molecular dynamics of the chain has been obtained and, furthermore, one can deduce that the interchain forces must be very weak since the torsional and longitudinal modes do not interact appreciably.

Thus the vibrational spectrum of the polyethylene chain can only be comprehensively explored by including incoherent inelastic neutron scattering as an additional experimental technique. In principle, similar results can be obtained on other polymers. There are, however, practical difficulties in obtaining specimens which are sufficiently well oriented. Also, where the polymer can exist in several tactic forms, a specimen of relatively pure stereo-regularity must be used since each form will have a different molecular vibrational spectrum. Finally, the calculation of $Z(\omega)$ is a major task requiring considerable computer time and a fairly good knowledge of a suitable potential function.

### 5.2.4. Coherent inelastic neutron scattering

In polymers containing hydrogen atoms, the strong incoherent scattering from the protons usually masks completely the relatively small coherent scattering. However, the study of coherent scattering is complementary to the incoherent spectra discussed in the foregoing section, because it provides a direct method for the determination of the dispersion curves. The application to organic crystals such as naphthalene is discussed in Chapter 4. In order to reduce the incoherent scattering to a manageable level, it is customary to use perdeutero samples at the present time. Hopefully, in the future, the development of neutron polarization analysis† will remove the necessity for complete deuteriation.

Coherent scattering of a monochromatic incident beam of radiation will produce a Bragg diffraction pattern from an oriented polymer sample. The elongated, regularly-arranged polymer chains behave, essentially, as an assembly of one-dimensional diffraction gratings. Extra periodicities are present in the system, because the atoms in the polymer chain are vibrating and because there are also lattice modes of vibration. The phonon waves corresponding to the optical and acoustical normal modes produce inelastic peaks displaced from the incident frequencies by $\pm \nu_p$, where $\nu_p$ are the phonon frequencies occurring as in Raman scattering. $\nu_p$ can be measured from the observed energy transfer. Furthermore, the wavelengths $\lambda_p$ of the phonon waves can be determined at selected inelastic scattering angles. Thus by careful alignment of an oriented polymer sample with respect to the incident beam of radiation, the coherent scattering allows the dispersion curve to be explored; that is to say, $\nu_p$ can be obtained as a function of $q(= 2\pi/\lambda_p)$ for each vibrational mode.

The particular power of coherent inelastic neutron scattering spectroscopy lies in the fact that all normal modes are active. Optical spectroscopy is limited by the stringent selection rules mentioned in § 5.2.2. From the experimental data, information concerning intramolecular and intermolecular forces can be obtained. Again, polyethylene serves as a good example. Using a triple-axis spectrometer and an oriented $+CD_2+_n$ specimen, the dispersion of the acoustical modes corresponding to torsion of the C–C backbone $(\nu_9)$ and the accordion mode $(\nu_5)$ have been studied by Trevino and Boutin (1967) and by Feldkamp, Venkataraman, and King (1968). The results are plotted in Fig. 5.5. As for the incoherent scattering spectrum, these results can be compared with the theoretical dispersions calculated for an isolated chain using a single force constant per $CH_2$ to describe the chain vibrations.‡ Only for $\nu_5$ are the experimental results

---

† In this technique, the polarizations of both the incident and scattered beams are determined, leading to a separation of the coherent and spin-incoherent cross-sections.

‡ Allowance is made in the comparison for the lowering of frequencies following replacement of H by D.

## THE DYNAMICS OF POLYMER CHAINS

sufficiently extensive to allow detailed comparison. A single-force-constant, individual chain model gives a very good representation of the experimental data. Two important conclusions are that the interchain forces are very small relative to intramolecular forces, as one would expect from a knowledge of covalent and van der Waals forces in alkane systems, and that the Young's Modulus of the polyethylene crystal measured along the direction of the chain axis is calculated from these data to be $3 \cdot 5 \times 10^{11}$ $Nm^{-2}$. The latter figure is in general agreement with direct experimental estimates of the Young's Modulus. But the direct measurement is bedevilled by the fact that a macroscopic piece of polyethylene is not a single crystal and consequently assumptions have to be made regarding the stress–strain relations between the amorphous and heterogeneous regions.

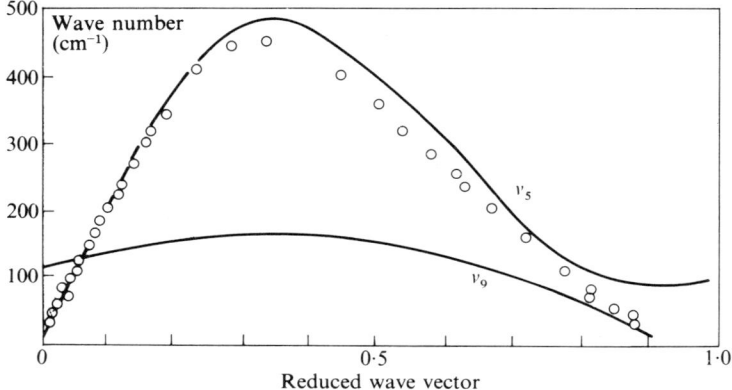

FIG. 5.5. Dispersion curves for the torsional $\nu_9$ and the accordion $\nu_5$ acoustical modes of polyethylene. — calculated, OOO experimental (After Feldkamp, Venkataraman, and King 1968).

Precise information on the intermolecular forces can be obtained from coherent neutron scattering studies on the appropriate lattice modes of vibration. Altogether, neutron scattering spectroscopy can provide a detailed picture of the intra and intermolecular dynamics of polymerized materials. The considerable advantage over optical methods is seen by recalling the Raman spectroscopic study of the dispersion of $\nu_5$ in the long-chain molecule $CH_3(CH_2)_{92}CH_3$. Because of the selection rule requiring a change in molecular polarization only, the fundamental and every second overtone are active. Shimanouchi and Schaufele (1967) have observed these Raman bands but even so, only a small part of the Brillouin zone can be explored ($q < 0 \cdot 1 q_{max}$) because of the long wavelength of the incident radiation. The effective wavelengths of the neutron beams used are two orders of magnitude smaller and so allow the whole of the zone to be explored.

## 5.3. Side-group motion in polymers

In the introduction it was pointed out that, in the glassy and crystalline states, the polymer-chain backbone is frozen into rigid conformations and the long-range motion responsible for rubber-like elasticity disappears. The constituent atoms then vibrate about their lattice sites and the molecular dynamics can be explored, in principle and, often, in practice, by the methods outlined above. If, however, side groups are subtended by the polymer chain, these groups may still undergo rotation about the chemical bond which attaches them to the main chain at temperatures below which the conformation of the main chain has ceased to change. These side-group motions are important because they are responsible for some of the energy-loss mechanisms which are manifest in mechanical, dielectrical and nuclear magnetic relaxation studies of polymer properties in the glassy state.

Consider, for example, the $-CH_3$ group in poly(propylene)

$$\begin{array}{c} CH_3 \\ | \\ +CH_2-CH\!\!+_n \end{array}$$

or in poly(propylene oxide)

$$\begin{array}{c} CH_3 \\ | \\ +CH_2-CH-O\!\!+_n. \end{array}$$

The torsional oscillations of the $CH_3$ groups are not likely to be very strongly coupled to the main chain vibrations and even less to each other; the torsional frequencies are expected, by analogy with similar small molecules, to lie near 200 cm$^{-1}$. Unfortunately, such vibrational modes, though active, are very weak in both infra-red and Raman spectra, and their vibrational bands have not been detected. Incoherent inelastic neutron scattering spectroscopy, on the other hand, is a very effective technique for the detection of $-CH_3$ group torsional frequencies. This is because the dispersion of the torsional vibration is small, and consequently a sharp inelastic peak is observed, and, also, because the amplitude of motion of the protons in the torsional mode is relatively large and so the inelastic peak is intense. The incoherent inelastic spectrum for amorphous poly(propylene oxide) is shown in Fig. 5.6. The strong peak at 228 cm$^{-1}$ is assigned as the torsional band and this is confirmed by the fact that the peak is absent from the corresponding spectrum of poly(propylene oxide) $-d_3$

$$\begin{array}{c} CD_3 \\ | \\ +CH_2-CH-O\!\!+_n \end{array}$$

(Deuterium has a much smaller incoherent scattering cross-section than hydrogen.)

## THE DYNAMICS OF POLYMER CHAINS

The rotation of the methyl group by an angle $\phi$ is hindered by a potential barrier $V(\phi)$ periodic in $\tfrac{2}{3}\pi$, which originates in the repulsion between non-bonded atoms and which can be approximated as the first term in a Fourier series

$$V(\phi) = \frac{V_N}{2}(1 - \cos N\phi), \qquad (5.9)$$

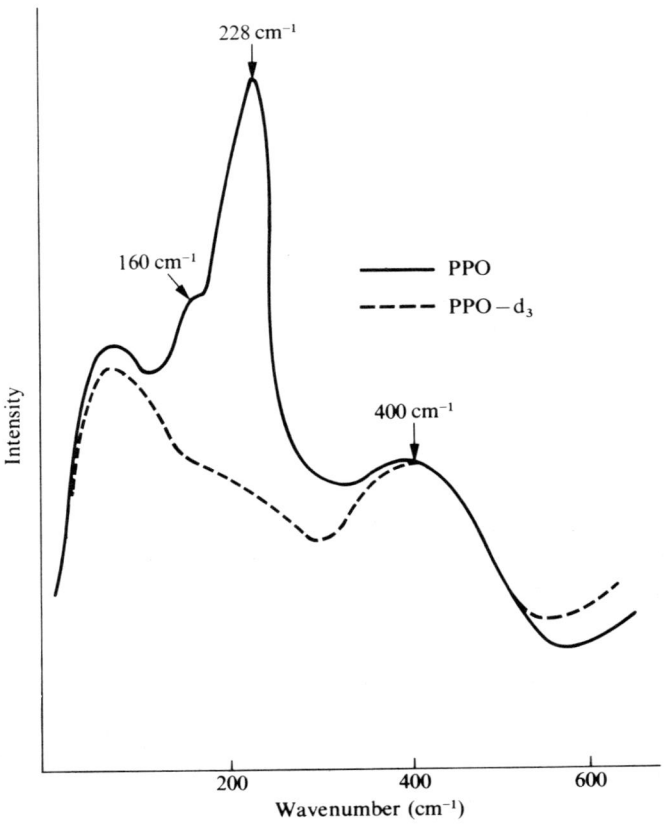

FIG. 5.6. Incoherent inelastic neutron scattering spectra of poly(propylene oxide) $+\!\!\operatorname{CH_2-CH-O}\!\!+_n$ and of poly(propylene oxide)–$d_3$
       $|$
      $\mathrm{CH_3}$
$+\!\!\operatorname{CH_2-CH-O}\!\!+_n$. (After Higgins, Allen, and Brier 1972).
            $|$
           $\mathrm{CD_3}$

where $N = 3$. For the C–$\mathrm{CH_3}$ group undergoing rotation about the C–C bond, the Schrödinger wave equation is

$$\left[-F\frac{\partial^2}{\partial\phi^2} + V(\phi)\right]\psi_{v\sigma}(\phi) = E_{v\sigma}\psi_{v\sigma}(\phi) \qquad (5.10)$$

where $v$ is the torsional quantum number, $\sigma$ specifies the periodicity of the wave functions $\psi$, and $F$ is a geometrical factor representing the reduced rotational constant of the methyl group. $E$ is the torsional eigen-value. This equation can be transformed into Mathieu's equation

$$\frac{\partial^2 y}{\partial x^2} + (b - s\cos^2 x)y = 0, \tag{5.11}$$

by substituting

$$\left. \begin{array}{ll} 2x = 3\alpha + \pi, & y_{v\sigma}(x) = \psi_{v\sigma}(\phi) \\ V_3 = 9\,Fs/4, & E_{v\sigma} = \dfrac{9Fb_{v\sigma}}{4}. \end{array} \right\} \tag{5.12}$$

In eqn (5.11), $s$ and $b$ are dimensionless parameters corresponding to the barrier height and the torsional energy levels respectively. For each value of $s$, there is a corresponding set of eigen-values $b_v$, and these can be tabulated for a range of values of $s$. The experimental datum obtained from the incoherent inelastic neutron spectrum is the fundamental torsional frequency $\nu_0$ which in turn gives the energy of the $v = 0 \rightarrow v = 1$ transition. (There are in fact two sets of solutions to the equation, one for which $\sigma = 0$ and is periodic in $2\pi/3$, and another for which $\sigma = \pm 1$ and is periodic in $\pi$. In the present application the separation between the two sets is too small to be detectable.) Thus we have

$$E_1 - E_0 = h\nu_0$$

and from eqn (5.12)

$$b_0 - b_1 = 4\omega_0/9F.$$

Since the local geometry of the molecule is known, $F$ and hence $b_0 - b_1$ can be calculated. The value of $s$ corresponding to the change in $b$ is obtained from the tables of $b_v$ as a function of $s$ and finally the barrier height $V_3$ is calculated from the relation given in eqn (5.12).

In this way the observed value $\nu_0 = 228$ cm$^{-1}$ leads to a barrier to internal rotation of the methyl group $V_3 = 15$ kJ mole$^{-1}$. The result is particularly interesting because measurements of the nuclear spin lattice relaxation time $T_1$ at 30 MHz and the corresponding relaxation time in the rotating frame $T_{1\rho}$ at 18KHz show that the methyl group relaxation process observed at low temperatures has an activation energy of $17 \pm 3$ kJ mole$^{-1}$. It is most probable that hindered methyl rotation is the origin of this relaxation process, and if so, $V_3$ provides the major contribution to the energy barrier.

Perspex, i.e.

$$\begin{array}{c} \mathrm{CH_3} \\ | \\ \mathrm{\!-\!(CH_2C\!)\!_{\mathit{n}}} \\ | \\ \mathrm{COOCH_3} \end{array}$$

has two methyl groups which undergo internal rotation. The inelastic spectra of

$$\begin{array}{c} CH_3 \\ | \\ +CH_2-C\!\!\!+_n \\ | \\ COOCD_3 \end{array} \quad \text{and} \quad \begin{array}{c} Cl \\ | \\ +CH_2-C\!\!\!+_n, \\ | \\ COOCH_3 \end{array}$$

all having similar degrees of tacticity, have enabled the torsional frequencies for the $C\!\!\!-\!\!\!CH_3$ and $O\!\!\!-\!\!\!CH_3$ groups to be assigned at 340 cm$^{-1}$ and 120 cm$^{-1}$ respectively and consequently the two methyl groups are hindered by barriers of different heights, respectively 34 and 4 kJ mole$^{-1}$ as shown by Higgins, Allen and Brier (1972). These results are paralleled by relaxation studies which indicate that the activation energy for $C\!\!\!-\!\!\!CH_3$ rotation is about 24 kJ mole$^{-1}$.

The torsional motions only of side groups which have a degree of symmetry, e.g. $-CH_3$, $-CCl_3$, $-C_6H_5$, can be studied in this way. Asymmetrical groups such as

$$-CH_2Cl \quad \text{or} \quad -C\!\!\!\stackrel{\displaystyle\nearrow O}{\searrow\!\!OCH_3}$$

require the determination of more than one potential constant in the expansion of $V(\phi)$, e.g. $V_1$, $V_2$ and $V_3$. At the present time there is no really satisfactory way of dealing with coupled asymmetrical rotors, and, in fact, a survey of the problem by Cunliffe (1970) suggests that it may prove very difficult to collect, with a suitable precision, the required experimental results.

## 5.4 Molecular motion in rubbers

Investigation of the long-range internal rotational motion of the main polymer chain in a rubber, by spectroscopic methods, presents a difficult problem. However, neutron scattering studies of the Doppler broadening of the quasi-elastic peak can provide information on segmental diffusion, provided that the time scale of segmental motion lies in the appropriate range to give detectable broadening. There is no detailed application to a high molecular-weight polymer system at present in the literature, and indeed it is possible that the Doppler broadening in such systems, at accessible temperatures, will be too small to be detected by present-day equipment (Allen et al. 1972). Nevertheless, a brief description of some preliminary results for low molecular-weight precursors of poly(dimethyl siloxane) will serve to illustrate the power of the method.

The spectra of neutrons scattered at 18° and 90° from a dimethyl siloxane chain

$$CH_3\text{-}(Si\text{-}O)_3\text{-}Si\text{-}CH_3$$
with $CH_3$ groups on each Si,

and the corresponding 4-membered ring

$$(Si\text{-}O)_4,$$
with $CH_3$ groups on Si,

are shown in Fig. 5.7. The spectra were recorded on a time-of-flight spectrometer and the neutron counts have been corrected to give the incoherent cross-section

$$\frac{d^2\sigma_{incoh}}{d\Omega\, d\omega}.$$

The broadening of the quasi-elastic scattering region is noticeable at 90° in scattering angle. The Doppler broadening is even clearer in Fig. 5.8. This shows the shapes of the quasi-elastic peaks at each angle, compared with the scattering from a standard vanadium sample which gives the actual instrumental resolution function. The comparison of the widths of the peaks is made through the incoherent scattering law $S_s(\mathbf{Q}, \omega)$ where $\hbar Q$ is the magnitude of the momentum transfer and $\hbar\omega$ is the energy gain experienced by the neutrons (see eqn (1.25d) of Chapter 1):

$$\frac{d^2\sigma_{incoh}}{d\Omega\, d\omega} = Nb_{incoh}^2 \frac{k'}{k} S_s(\mathbf{Q}, \omega).$$

The function $\tilde{S}$ plotted in Fig. 5.8 is the symmetrical scattering law†

$$\tilde{S}(\mathbf{Q}, \omega) = S(\mathbf{Q}, \omega) \exp(-\hbar\omega/2k_B T).$$

The theory of diffusional broadening predicts a Lorentzian shape for the quasi-elastic peak, i.e. in general

$$S(\mathbf{Q}, \omega) = \frac{1}{\pi} \cdot \frac{p(Q)}{\omega^2 + [p(Q)]^2}$$

where $p(Q)$‡ is some function of $Q$. Thus the broadening at half peak height is given by

$$\Delta\omega = 2p(Q).$$

† So called because, in accordance with the expression (1.26a) for $S(\mathbf{Q}, \omega)$, $\tilde{S}(\mathbf{Q}, \omega)$ is an *even* function in both $\mathbf{Q}$ and $\omega$.
‡ The nature of this function is discussed in Chapter 6 for several diffusion models.

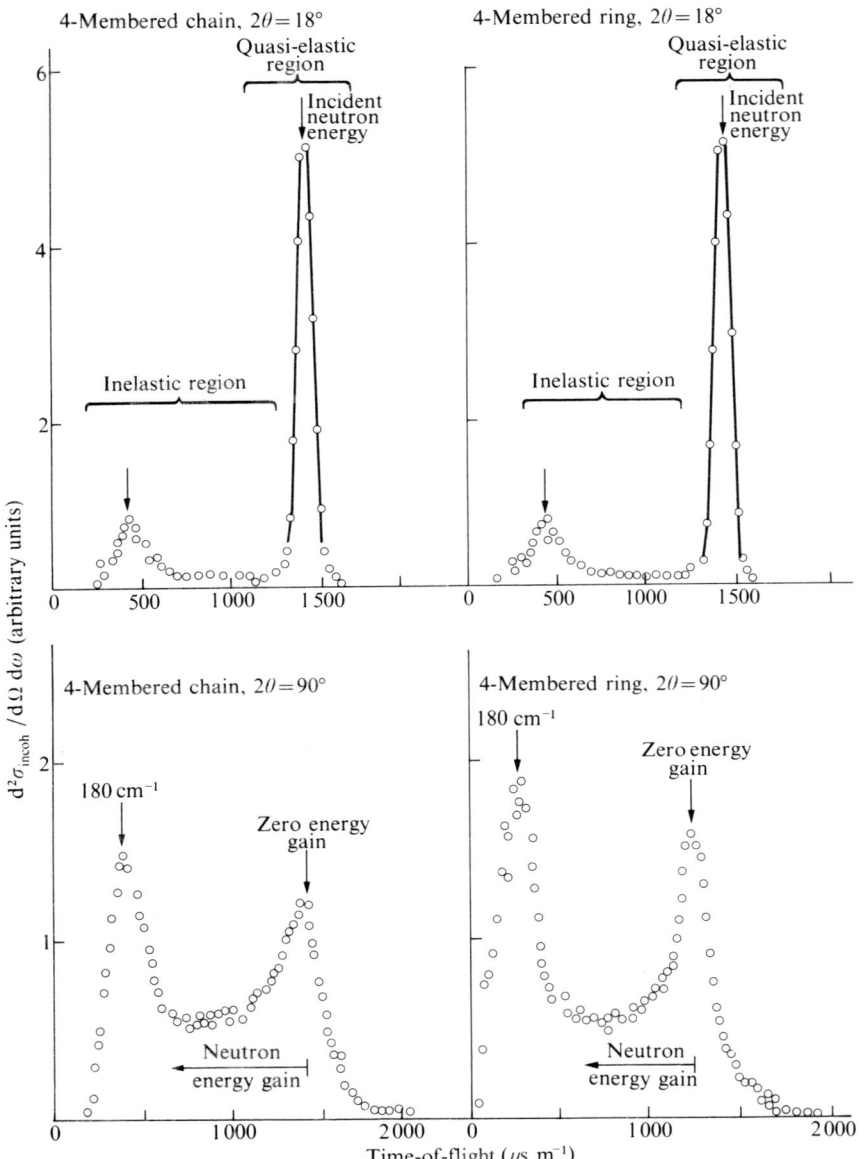

FIG. 5.7. Time-of-flight spectra at scattering angles of 18° and 90° for linear and cyclic dimethyl siloxanes.

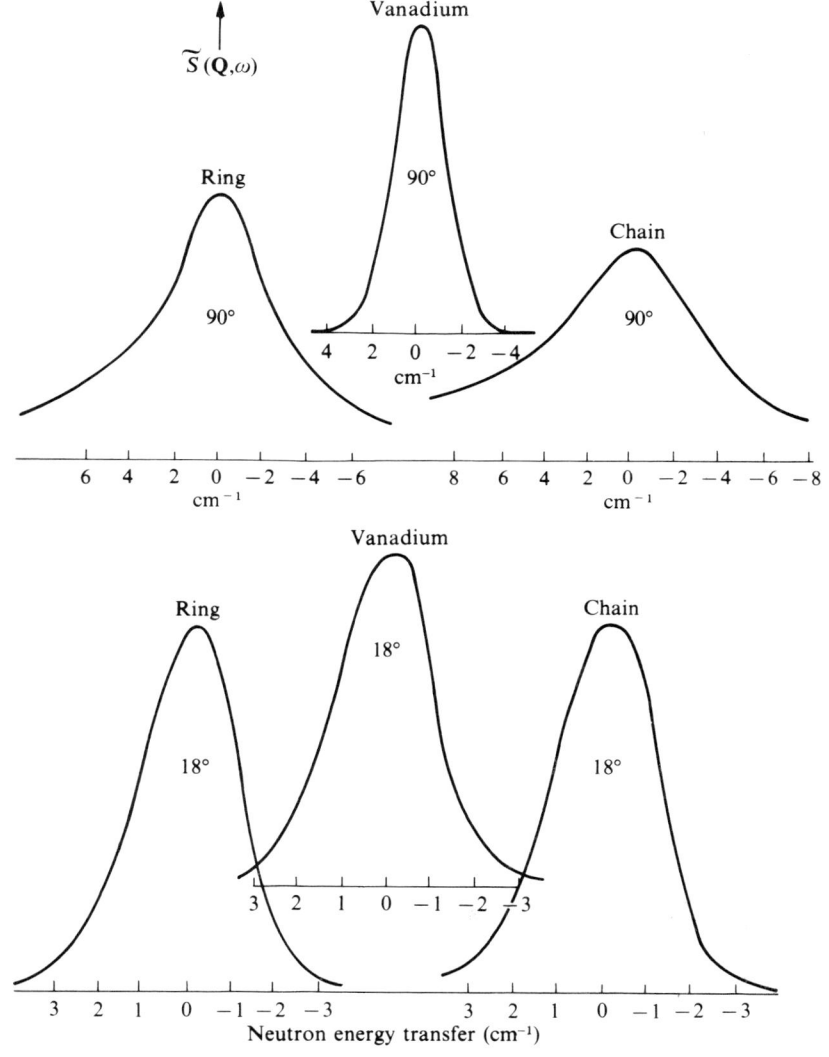

FIG. 5.8. Quasi-elastic peaks of linear and cyclic dimethyl siloxanes at 18° and 90° compared with unbroadened peak from a standard vanadium sample.

In practice, the Lorentzian curve is convoluted with the actual resolution function which approximates to a Gaussian. Thus the curves shown in Fig. 5.8 are Voigt functions. The difficulty in analysing the Doppler broadening lies in the deconvolution not only of the Voigt function, but also of the low-energy components of the inelastic spectrum which lie under the quasi-elastic peak. In the present example we shall assume that the deconvolution has been satisfactorily performed and that $\Delta\omega$ can be obtained from the results shown in Fig. 5.8.

## THE DYNAMICS OF POLYMER CHAINS

For a model in which simple diffusion (governed by Fick's law for macroscopic diffusion) is the main cause of Doppler broadening

$$\Delta\omega = 2DQ^2$$

where $D$ is the coefficient of self diffusion (see §6.2). $D$ can be obtained directly from Fig. 5.9 in which $\hbar\Delta\omega$ is plotted against $Q^2$. The results are linear in $Q^2$

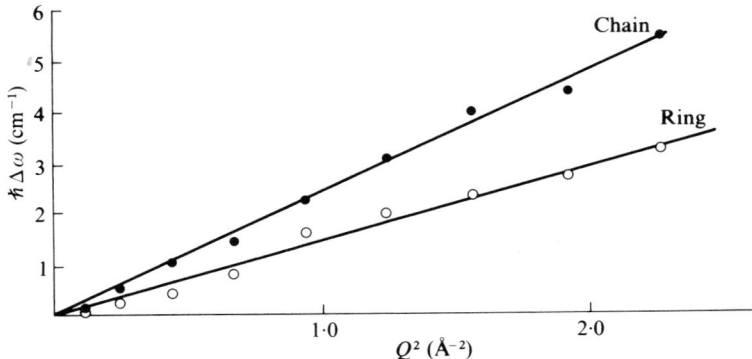

FIG. 5.9. Full-width at half-maximum of deconvoluted quasi-elastic peak for linear and cyclic dimethyl siloxanes, plotted as a function of $Q^2$.

for both samples, but it will be seen that $D$ is very different for the linear and cyclic molecules:

$D$ (m$^2$s$^{-1}$ at 25°C)
chain    2·3 . 10$^{-9}$
ring     1·4 . 10$^{-9}$.

More work must be done to establish the absolute values of $D$ but the relative values are considered reliable. The model treats diffusion as a rate process, so that

$$D = A \exp(-E/k_B T),$$

and measurements at different temperatures will allow the frequency factor $A$ and the activation energy $E$ to be determined. Until these parameters are separated it is not satisfactory to speculate on the different values of $D$.

Measurements on longer chains will allow the trend in the character of diffusional motions of the chain segments to be studied under conditions approaching the behaviour in long molecular chains. There are, of course, other methods for the measurement of $D$, e.g. tracer methods and spin-echo n.m.r. studies. Comparison of the results from different techniques will be of great interest because of the very different time scales involved, and may well shed light on the relative contributions of rotational and translational motions to the self diffusion. The scope of neutron scattering studies in this field has yet to be firmly established.

## 5.5. References

ALLEN, G., BRIER, P. N., GOODYEAR, G., *and* HIGGINS, J. S. (1972). *Faraday Symp. Chem. Soc.* **6**, 169.
CUNLIFFE, A. V. (1970). *J. molec. Struct.* **6**, 9.
DANNER, H. R., SAFFORD, G. H., BOUTIN, H. *and* BERGER, M. (1964). *J. chem. Phys.* **40**, 1417.
FELDKAMP, L. A., VENKATARAMAN, G. *and* KING, J. S. (1968). *Neutron inelastic scattering*, vol. II. I.A.E.A., Vienna p. 159.
HIGGINS, J. S., ALLEN, G., BRIER, P. N. (1972). *Polymer*, **13**, 157.
KITIGAWA, T. *and* MIYAZAWA, T. (1967). *J. chem. Phys.* **47**, 337.
KITTEL, C. (1966). *Introduction to Solid State Physics (3rd Ed.)*. Wiley, New York, p. 142.
MYERS, W. R. DONOVAN, J. L. *and* KING, J. S. (1965). *J. chem. Phys.* **42**, 4299.
MYERS, W. R. SUMMERFIELD, G. C. *and* KING, J. S. (1966). *J. chem. Phys.* **44**, 184.
SHIMANOUCHI, T. *and* SCHAUFELE, R. F. (1967). *J. chem. Phys.* **47**, 3605.
TASUMI, M., SCHIMANOUCHI, T. *and* MIYAZAWA, T. (1962). *J. molec. Spectrosc.* **9**, 261.
TREVINO, S. *and* BOUTIN, H. (1967). *J. macromol. Sci.* A1(4), 723.

*General*

The following are general texts on polymers: they are not specifically concerned with neutron studies.

McCRUM, N. G., READ, B. E. *and* WILLIAMS, G. (1967). *Anelastic and dielectric effects in polymer solids*. J. Wiley and Sons, London.
TRELOAR, L. R. G. (1970). *Introduction to polymer science*. The Wykeham science series, London and Winchester.
ZBINDEN, R. (1964). *Infra-red studies on high polymers*. Interscience, New York.

# 6 Atomic and Molecular Motion in Liquids by Thermal Neutron Scattering

By J. G. POWLES

Physics Laboratories, University of Kent

## 6.1. Introduction to the liquid state

The problem of the molecular motion in a liquid is a fundamental one in the sense that if we knew the position and velocity† of every atom in the liquid as a function of time for any initial conditions and for any external constraints then we would know almost all there is to know about a liquid. In fact we do not have this information, although an approach to this situation has been achieved in recent years, for somewhat idealized systems, by using computers to study various aspects of the molecular dynamics of liquids. Even if we did have such detailed knowledge it would still be a major operation to reduce it to a form in which we could predict the results of experiments which could actually be performed. In practice, what we need are various averages over the motions (or over ensembles) which are related to various measurable macroscopic properties of the liquid. If the number and variety of the averaged quantities which we can measure and calculate is sufficiently great, we may hope to say something significant about the molecular motion.

The study of liquids is much frustrated by the lack of a suitable 'zero-order approximation' from which to approach the behaviour of an actual liquid. This situation contrasts with that for crystalline solids. Here we have the harmonic approximation for a perfect lattice, leading to the idea of expressing the translational motion of the atoms or molecules in terms of phonons. At the other extreme we have a perfect gas where the positions and motions of the atoms or molecules are completely disordered and the motion is relatively simple. Liquids, it turns out, are very poorly described by either approximation. They have no long-range order and any sort of crystalline approximation is dubious. Nevertheless, a cell model of liquids has been used with a little success. Perhaps its gravest defect is the contortions, which are necessarily resorted to, to allow for the fact that the atoms are not actually confined to cells—in particular, the communal entropy problem. The approach from the perfect-gas end is equally hazardous. The corrections to the perfect-gas limit are all analogous to the virial expansion which expresses the pressure, $P$, as a power series in the density, $\rho$:

$$P/\rho k_B T = 1 + B_2(T)\rho + B_3(T)\rho^2 + \ldots \tag{6.1}$$

† We shall be concerned only with 'classical' liquids.

Although the virial coefficients $B_n(T)$ can be found—with increasing difficulty for increasing $n$—it is thought that for liquid densities the series almost certainly diverges. The same is probably true of virtually all other liquid properties, both equilibrium and non-equilibrium, which are calculated in this way.

More recently it has become apparent that the so-called hard-sphere approximation may be a good zero-order approximation, at least for atomic liquids. In this model (Fig. 6.1(a)) the interaction between the particles corresponds to representing them as spheres which show no attraction for each other but show infinite repulsion when the separation between the centres of two spheres is equal to their diameter. It seems likely that perturbation of this approximation, for instance towards a more realistic pairwise interaction potential, $u(r)$, such as the Lennard-Jones 6-12 potential (Fig. 6.1(b)) and to non-pairwise interactions, will give a reasonable interpretation of liquid properties. However, this approach

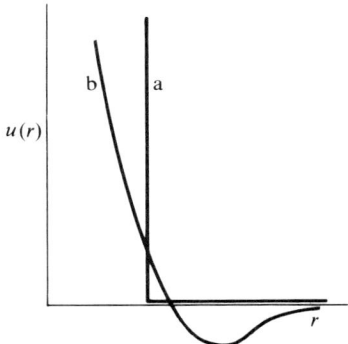

FIG. 6.1. Interatomic potentials: (a) hard-sphere; (b) Lennard-Jones.

is considerably frustrated by the lack of exact analytical theories for the hard-sphere approximation. It is as if, in ideal crystal theory, we were unable to solve the equation for the harmonic oscillator. A zero-order approximation which is not itself analytically simple leads to obvious difficulties in generalization.

In view of the present impossibility of any exact treatment, models are widely used. Although in a sense they are the only way one can get a physical grasp of liquids, models are subject to their usual drawbacks of being limited in range of application, difficult to disprove, and attracting personal loyalty. Molecular motion can be studied in a variety of ways. We shall concentrate our attention on the slow-neutron scattering method but try to bear in mind other methods which also provide important, complementary and, in some cases, more readily digestible information.

## ATOMIC AND MOLECULAR MOTION IN LIQUIDS

### 6.2. 'Incoherent' translational motion

In an atomic liquid we are concerned with the translational motion of the atoms. We can never measure the actual motion of an individual atom and so we may ask what averages are significant. One average which proves to be useful arises from asking: if an atom is at a point $r_0$ at time $t_0$ what is the probability of the *same* atom being at $r_0 + r$ at time $t_0 + t$? This probability is called $G_s(r, t)$ since for a liquid in equilibrium (in contrast to a crystal) it is evidently independent of both $r_0$ and $t_0$. On the other hand, we might consider the relative motion of different atoms as a function of time. Thus if one atom is at $r_0$ at $t_0$, what is the probability of finding a *different* atom at $r_0 + r$ at $t_0 + t$? This probability is called $G_d(r, t)$. Evidently the probability of finding *any* atom at $r_0 + r$ at $t_0 + t$ is $G(r, t)$, where

$$G(r, t) = G_s(r, t) + G_d(r, t). \tag{6.2}$$

The separation on the right-hand side can only be made, of course, if the atoms are distinguishable, i.e. for classical systems.

Evidently $G_s(r, t)$ is simpler than $G_d(r, t)$ and we look at it first in this section. It is well established that for long enough times (and distances) the motion of an atom is represented by the diffusion equation (or Fick's law). $G_s(r, t)$ is then the appropriate solution of the diffusion equation for $p(r, t)$, the probability of finding an atom at $r$ at time $t$:

$$\frac{\partial p}{\partial t}(r, t) = D \nabla^2 p(r, t) \tag{6.3}$$

Here $D$ is the self-diffusion coefficient and $\nabla^2 = \frac{1}{r^2} \frac{\partial}{\partial r} \left( r^2 \frac{\partial}{\partial r} \right)$ for isotropic diffusion. For an atom at the origin at time $t = 0$, we find, using eqn (6.3), that

$$G_s(r, t) = \frac{1}{(4\pi D |t|)^{3/2}} \exp(-r^2/4D|t|). \tag{6.4}$$

In particular, this means that the mean-square distance travelled in time $t$, $\langle r^2(t) \rangle$, is given by

$$\langle r^2(t) \rangle = 6Dt. \tag{6.5}$$

(This relation follows immediately from eqn (6.10) below.) On the other hand, for very short times an atom must move with effectively constant velocity, since the forces have no time to accelerate it.† Thus $r = vt$ and

$$\langle r^2(t) \rangle = \langle v^2 \rangle t^2 = (3 k_B T/M) t^2, \tag{6.6}$$

† Contrary to what is sometimes implied, the forces are certainly there, but they do not have time to affect the motion.

where $M$ is the atomic mass. We have used the equipartition relation for the velocity, $\frac{1}{2}mv_x^2 = \frac{1}{2}k_BT$, which applies to a classical liquid. There must, therefore, be a region of time when the behaviour changes over from eqn (6.6) to eqn (6.5), and that is the important region where the initimate details of the atomic motion play their part. The atomic motion is, of course, complicated at all times but is simple if averaged over either short times (eqn (6.6)) or over long times (eqn (6.5)). Since the diffusion equation is not obeyed for short times, the result (6.5) needs modification and so we might expect for long times to have, rather crudely, a short time contribution to $\langle r^2 \rangle$, say $\langle r_0^2 \rangle$, and a long time part, say $6D(t - t_0)$, so that

$$\langle r^2(t) \rangle = 6Dt + C. \tag{6.7}$$

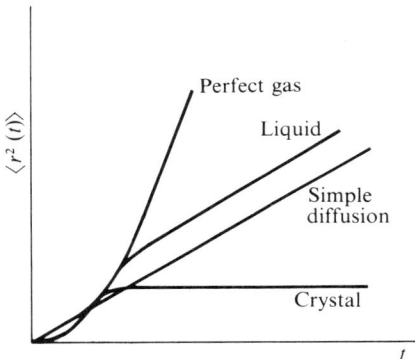

FIG. 6.2. Mean-square distance moved in time $t$ for various types of translational motion.

This form is obtained for instance for one model, the Langevin equation, for which it is supposed that the forces acting on an atom of mass $M$ consist of a viscous part $M\gamma d\mathbf{r}/dt$ and a fluctuating force $\mathbf{f}(t)$, i.e.

$$M\frac{d^2\mathbf{r}}{dt^2} + M\gamma \frac{d\mathbf{r}}{dt} = \mathbf{f}(t), \tag{6.8}$$

where there is reason to believe that $\gamma = k_BT/MD$. It can be shown that eqn (6.8) leads to eqn. (6.7) but with $C$ negative (but see §6.3, eqn (6.28) *et seq.*). The Langevin equation will also be of interest later. A satisfactory empirical calculation of $\gamma$ is not known. The simplest type of behaviour for $\langle r^2(t) \rangle$ is given by a smooth join of the asymptotic results (6.6) and (6.7). According to molecular dynamics calculations, this is true for computer argon at 85·5K with a Lennard-Jones interatomic potential, as shown in Fig. 6.3,† but this apparent simplicity is in large measure due to the severe averaging implicit in $\langle r^2(t) \rangle$. The result for

† All the figures are diagrammatic and should not be used as sources of data without reference to the originals.

## ATOMIC AND MOLECULAR MOTION IN LIQUIDS

a liquid may be contrasted with that for a perfect gas and for a harmonic crystal where $\langle r^2(t) \rangle$ rises to a constant value (see also eqn 6.15), as illustrated in Fig. 6.2.

We can estimate the time of the changeover in $\langle r^2(t) \rangle$ by introducing the plausible notion that an atom must surely move by at least an atomic radius before its motion is sufficiently averaged to be described by a macroscopic parameter like $D$. If an atomic radius is, say, $a = 2$ Å and a typical value of $D$ is $2 \cdot 10^{-9}$ m$^2$ s$^{-1}$ we have for this limit,

$$t > a^2/6D \sim 3 \times 10^{-12} \text{ s}. \tag{6.9}$$

Fig. 6.3 suggests that an average over such a time is quite sufficient. By considering the forces acting on an atom due to a neighbour at the mean interaction distance, we can show that the velocity of an atom changes appreciably after about $10^{-13}$ s.

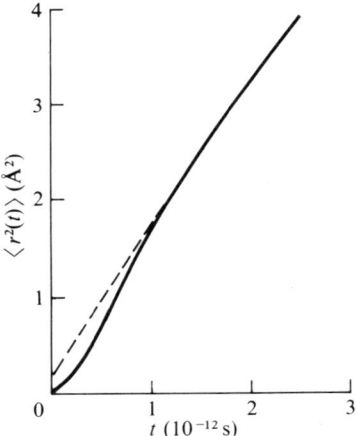

FIG. 6.3. Computed mean-square distance moved for Lennard–Jones potential: cf. 'liquid' curve in Fig. 6.2. (After Rahman 1964).

In a sense, the time interval between about $10^{-13}$ s and about $10^{-12}$ s after the beginning of observation of an atom is the important range for describing molecular motion, and this time interval arises repeatedly in the consideration of the dynamics of liquids. We can see immediately, for instance, why slow neutrons may be a useful probe for investigating liquids. The neutrons will 'see' the interesting changes in the liquid most readily if they are in the region of an atom (4 Å) for a time of order $4 \times 10^{-13}$ s. Hence a suitable velocity would be of order 4 Å/($4 \times 10^{-13}$ s) = $10^3$ m s$^{-1}$. Such a neutron has a de Broglie wavelength $\lambda$ given by $\lambda = h/m_n v \simeq 4$ Å, which is a typical wavelength for a slow neutron. Thus slow neutrons have not only the right 'size' for atoms, which is important for sensitivity to the liquid structure, but also have the right velocity for sensitivity to the molecular motion.

The quantity $\langle r^2(t) \rangle$ cannot be measured directly by any known method but the quantity $G_s(r, t)$ can be obtained, in principle, by appropriate scattering experiments (see below), and $\langle r^2(t) \rangle$ can then be derived from

$$\langle r^2(t) \rangle = \int_0^\infty r^2 G_s(r, t) \cdot 4\pi r^2 \, dr. \tag{6.10}$$

We shall therefore leave $\langle r^2(t) \rangle$ for consideration later and now look at the less-averaged quantity $G_s(r, t)$.

In order to see $G_s$ only and not $G$ (or $G_d$) we need an incoherent scatterer, and in this respect neutrons are unique in the possibility of the scattering being at least partially incoherent. Protons and a few other nuclei scatter almost entirely incoherently. We can obtain $G_s$ in principle using the inverse Fourier transform relation (see Appendix III),

$$G_s(r, t) = \frac{1}{2\pi} \iint S_s(Q, \omega) \exp[-i(\mathbf{Q} \cdot \mathbf{r} - \omega t)] \, d\mathbf{Q} \, d\omega. \tag{6.11}$$

$S_s(Q, \omega)$ is the self scattering function (the incoherent scattering law) for momentum change $\hbar \mathbf{Q}$ and energy change $\hbar \omega$ in the scattered particle. It is essentially the double differential scattering cross-section $d^2\sigma/d\Omega \, d\omega$ i.e. it is determined from the scattered intensity per unit solid angle, $\Omega$, and per unit frequency (or energy) range $\omega$ (or $\hbar\omega$). $S_s(Q, \omega)$ is seldom measured over sufficient range of $\mathbf{Q}$ and $\omega$ to make the four-dimensional $\mathbf{Q}$-$\omega$ Fourier transform in eqn (6.11) give a reasonably faithful representation of $G_s(r, t)$.† Consequently, it is more usual to put in a theoretical or model expression for $G_s$ and to deduce what $S_s(Q, \omega)$ should be. One then compares this theoretical $S_s(Q, \omega)$ with that part of $S_s(Q, \omega)$ which has been measured, and one hopes to be able to decide to what extent the model works. For instance, if the atomic motion were entirely controlled by macroscopic diffusion, then $G_s$ is given by eqn (6.4) and, using the Fourier transform relation,

$$S_s(Q, \omega) = \frac{1}{2\pi} \iint G_s(r, t) \exp[i(\mathbf{Q} \cdot \mathbf{r} - \omega t)] \, dr \, dt, \tag{6.12}$$

the corresponding $S_s(Q, \omega)$ is

$$S_s(Q, \omega) = \frac{1}{\pi} \cdot \left( \frac{DQ^2}{\omega^2 + (DQ^2)^2} \right) \exp(-\langle u^2 \rangle_\mathbf{Q} Q^2). \ddagger \tag{6.13}$$

† Those who have never actually done numerical Fourier transforms, especially in (r, t), will be unlikely to appreciate the difficulty in getting a reliable transform from data which are necessarily restricted in range even for the spatially isotropic situation encountered here.

‡ The derivation of eqn (6.13) from eqn (6.4) and eqn (6.12) is given in Appendix III (p. 301).

## ATOMIC AND MOLECULAR MOTION IN LIQUIDS

We have here included a simplified form of the Debye-Waller factor $\exp(-\langle u^2 \rangle_\mathbf{Q} \times Q^2)$, suitable for liquids, which results from vibrational motion as in a crystal (see Chapter 1); without it this formula is rather unrealistic, and is still unrealistic for large energy transfers $\hbar\omega$. Values for $\langle u^2 \rangle_\mathbf{Q}$, the component in the direction of $\mathbf{Q}$ of the vibrational part of $\langle r^2 \rangle$, can be found from the total quasi-elastic intensity at fixed $Q$, $S_s(Q)$, where

$$S_s(Q) = \int S_s(Q, \omega) \, d\omega. \tag{6.14}$$

For $S_s(Q, \omega)$ given by eqn (6.13) we have,

$$S_s(Q) = \exp(-\langle u^2 \rangle_\mathbf{Q} Q^2), \tag{6.15}$$

and a log-plot of the quasi-elastic part of $S_s(Q)$ *versus* $Q^2$ should be linear. Such a plot is shown for liquid sodium (at 102°C) in Fig. 6.4 (Randolph 1968). The

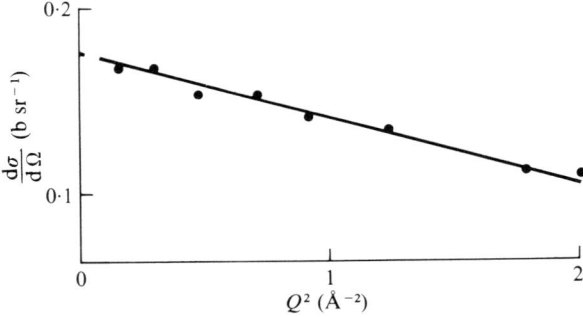

FIG. 6.4. Incoherent scattering by liquid sodium at 102°C, illustrating eqn (6.15). (After Randolph 1968).

deduced value of $\langle u^2 \rangle$ is 0·22 Å$^2$, much the same as for the solid, and this result is typical, suggesting that an atom in a liquid has a vibrational motion similar to that in a crystal. This motion does, of course, contribute to $\langle r^2(t) \rangle$.

For given $Q$, the Debye–Waller factor is constant and we look next at the $\omega$-dependence of $S_s(Q, \omega)$. According to eqn (6.13), this should be a Lorentzian curve of full-width at half-height, $\Delta\omega$, for the Simple Diffusion model given by

$$\Delta\omega_{SD} = 2DQ^2. \tag{6.16}$$

Experimental results indeed show such a peak but usually the width is not given other than roughly by $2DQ^2$. The width is less than $2DQ^2$ for atomic liquids, and the peak is usually standing on another contribution to $S_s(Q, \omega)$. It is customary to say that the broader background peak is due to 'inelastic' scattering, which is in part where the Debye–Waller intensity reduction goes, whereas the relatively sharp peak at the middle is due to 'quasi-elastic' scattering (see Fig. 6.5). The separation into the two contributions is plausible for small $Q$ but less so for large

$Q$, which is really the more interesting region. Small $Q$ corresponds to large wavelengths and so implies averaging over large distances, i.e. we tend to see macroscopic effects and so the behaviour ought to be described by macroscopic equations such as eqn (6.3). This idea is further reinforced by the observed widths of the peaks, which for small $Q$ are sharp in $\omega$. Small $\omega$ in $S_s(Q, \omega)$ tends to correspond to long times in $G_s(r, t)$, so again a macroscopic picture is acceptable. On the other hand, for very large $Q$ we only see over small distances and, since the quasi-elastic peak is broad in $\omega$, for short times. For this situation the perfect-gas equation for $G_s$ should apply,

$$G_s(r, t) = \frac{1}{\pi^{3/2} v_0^3 |t|^3} \exp(-r^2/v_0^2 t^2), \tag{6.17}$$

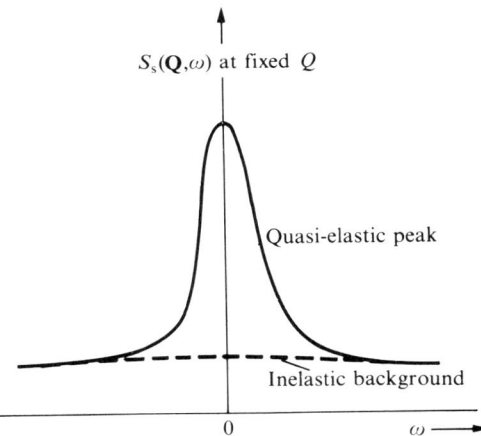

Fig. 6.5. Sketch illustrating the quasi-elastic peak standing on an inelastic background.

leading to the incoherent scattering law

$$S_s(Q, \omega) = \frac{1}{\pi^{\frac{1}{2}} Q v_0} \exp[-\omega^2/Q^2 v_0^2], \tag{6.18}$$

where $v_0^2 = 2k_B T/M$. Thus for large $Q$, the full-width at half-height for the Perfect Gas model should approach

$$\Delta\omega_{PG} = (8 \ln 2)^{\frac{1}{2}} (k_B T/M)^{\frac{1}{2}} Q. \tag{6.19}$$

The observed behaviour of $S_s(Q, \omega)$ for protons in water is shown in Fig. 6.6 (Sakamoto *et al.* 1962), where, however, it must be remembered that we have the complexities of a molecular liquid to account for. Larsson (1965) has also measured water and expresses his results as a width $\Delta\omega$ *versus* $Q^2$ as shown in Fig. 6.7. This sort of departure from a linear relation between $\Delta\omega$ and $Q^2$

ATOMIC AND MOLECULAR MOTION IN LIQUIDS

is found quite generally and is due to the detailed nature of the molecular translational motion. Almost any model for the actual nature of the translational motion gives a breakdown of the eqn (6.16) for large $Q$. For instance,

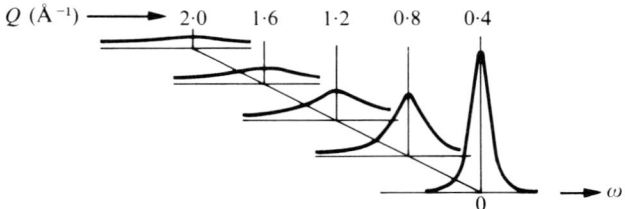

FIG. 6.6. $S_s(Q, \omega)$ for liquid water at 75°C, plotted as a function of $\omega$ at the fixed values of $Q$ indicated. (After Sakamoto *et al.* 1962).

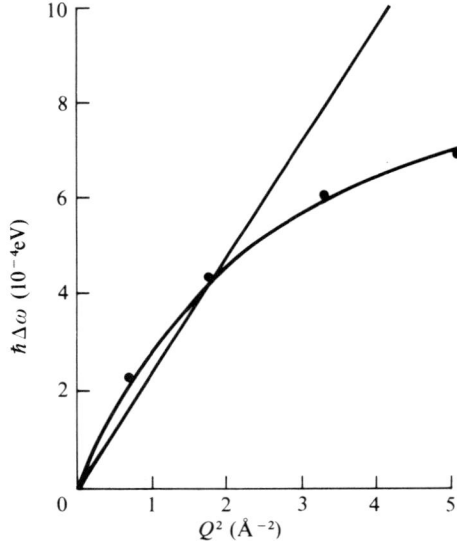

FIG. 6.7. Quasi-elastic scattering width for water at 20°C, plotted as a function of the square of the momentum transfer. The straight line is for simple diffusion, eqn (6.16). (After Larsson 1965).

if the translation is supposed to be by finite 'jumps' of mean length $l_0$ at time intervals $\tau_0$, one can show (Chudley and Elliot 1961) that

$$S_s(Q, \omega) = \frac{1}{\pi} \left( \frac{\Delta\omega/2}{\omega^2 + (\Delta\omega/2)^2} \right), \tag{6.20}$$

where

$$\Delta\omega = \tau_0^{-1} [1 - (1 + Q^2 l_0^2)^{-1}]. \tag{6.21}$$

Thus for small $Q$, $\Delta\omega \to (l_0^2/\tau_0)Q^2$ so that, by comparison with eqn (6.16),

$$D = \tfrac{1}{2}(l_0^2/\tau_0). \tag{6.22}$$

For large $Q$, $\Delta\omega \to \tau_0^{-1}$ and so is independent of $Q$. If one *assumes* that this model is valid and fits the limiting value of $\Delta\omega$, one finds $\tau_0 \sim 10^{-12}$ s (and, using the known $D$ in eqn (6.22), $l_0 \sim 2$ Å) for a surprisingly wide variety of liquids (see Table 6.1). It is perhaps more realistic to say that the time scale

TABLE 6.1

*Characteristic times and distances for atomic motion in a liquid (Egelstaff 1967)*

| Liquid | Temperature | $D/(10^{-9}$ m$^2$ s$^{-1})$ | $\tau_0/(10^{-12}$ s) | $l_0/($Å$)$ |
|---|---|---|---|---|
| Sodium | 108 to 198°C | 4 to 8 | 1·0 to 1·6 | 2·5 |
| Pentane | −35 to +25°C | 3 to 5 | 1·0 to 1·2 | 1·5 |
| Argon | 85 K | 1·6 | 1 | 1 |
| Hydrogen | 15 to 18 K | 4·7 to 7·5 | 2·6 to 1·1 | 2·5 |

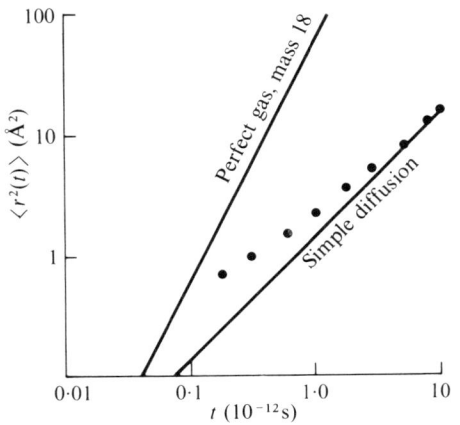

FIG. 6.8. Mean-square distance moved by water molecules at 25°C, showing the simple diffusion limit at long times and the perfect gas limit at short times. (After Sakamoto *et al.* 1962).

for the microscopic process of translation in a liquid is $10^{-12}$ s, as we have already surmised, rather than that the jumping model is necessarily valid.

Since we have discussed $\langle r^2(t) \rangle$, it is of interest to obtain this quantity experimentally from the scattering measurements; this is possible, at least in principle, since $S_s(Q, \omega)$ determines $G_s(r, t)$ and hence (from eqn (6.10)) $\langle r^2(t) \rangle$. Results for water presented in this form are shown in Fig. 6.8, which shows the expected

ATOMIC AND MOLECULAR MOTION IN LIQUIDS

behaviour. For $t > 10^{-12}$ s the diffusion equation holds; deviations from this equation start at about $3 \times 10^{-13}$ s; and the behaviour is gas-like (for motion of the whole molecule rather than of protons) for about $10^{-13}$ s.

The fact that $G_s(r, t)$ has the same $r$-dependence in the two extremes [eqns (6.4) and (6.17)] led Vineyard (1958) to suggest that this form be postulated for all times, i.e.

$$G_s(r, t) = (4\pi w(t))^{-3/2} \exp(-r^2/4w(t)) \qquad (6.23)$$

where $w(t)$ determines the width of the Gaussian distribution. Experimental results fit this quite well so that they can be expressed in terms of $w(t)$ which is then (from eqn (6.10)) the same as $\frac{1}{6} \langle r^2(t) \rangle$. This is the so-called Gaussian approximation. It receives appreciable theoretical support and there are many models which are used to predict $w(t)$. Comparison with experiment is often made via the intermediate scattering function $I_s(Q, t)$, which is defined by

$$I_s(Q, t) = \int S_s(Q, \omega) e^{i\omega t} \, d\omega = \int G_s(r, t) e^{i\mathbf{Q} \cdot \mathbf{r}} \, d\mathbf{r} \qquad (6.24)$$

and is therefore 'intermediate' between experiment, $S_s(Q, \omega)$, and theory, $G_s(r, t)$. For $G_s(r, t)$ given by eqn (6.23), we have the particularly simple form,

$$I_s(Q, t) = \exp(-Q^2 w(t)). \qquad (6.25)$$

We shall not review here the extensive literature on $w(t)$ (see General References).

## 6.3. The velocity auto-correlation function

A somewhat less averaged quantity than $\langle r^2(t) \rangle$ but more averaged than $G_s(r, t)$ is the velocity auto-correlation function. If the velocity of an atom is $\mathbf{v}(t_0)$ at time $t_0$ and the velocity of the *same* atom is $\mathbf{v}(t_0 + t)$ at time $t_0 + t$, we can consider the quantity $\langle \mathbf{v}(t_0) \cdot \mathbf{v}(t_0 + t) \rangle$; in equilibrium, this is independent of $t_0$ so that we can write

$$\langle \mathbf{v}(t_0) \cdot \mathbf{v}(t_0 + t) \rangle = \langle v^2 \rangle \, \phi(t). \qquad (6.26)$$

The actual form of $\phi(t)$, the normalized auto-correlation function, for 'computer argon' is shown in Fig. 6.9 (Rahman 1964).

The macroscopic self diffusion constant, $D$, is related to $\phi(t)$ as follows,

$$D = \tfrac{1}{3} \langle v^2 \rangle \int_0^\infty \phi(t) \, dt = \tfrac{1}{3} \langle v^2 \rangle \tau_v. \qquad (6.27)$$

The second step represents the usual definition of the correlation time $\tau_v$ of a time correlation function. Since $\langle v^2 \rangle = 3k_B T/M$, we have $\tau_v = DM/k_B T \sim 5 \times 10^{-13}$ for a typical liquid. It is not unexpected that the characteristic time for a liquid appears again. The velocity correlation function determines $w(t)$ of eqn (6.23) via

$$w(t) = \tfrac{2}{3} \int_0^t dt'(t - t') \langle \mathbf{v}(t_0) \cdot \mathbf{v}(t_0 + t') \rangle, \qquad (6.28)$$

so that we make contact immediately with the discussion of motion in §6.2. For instance, using the computer result of Fig. 6.9 in eqn (6.28), we find that eqn (6.7) is obeyed with $C$ positive, as illustrated by the broken line in Fig. 6.3.

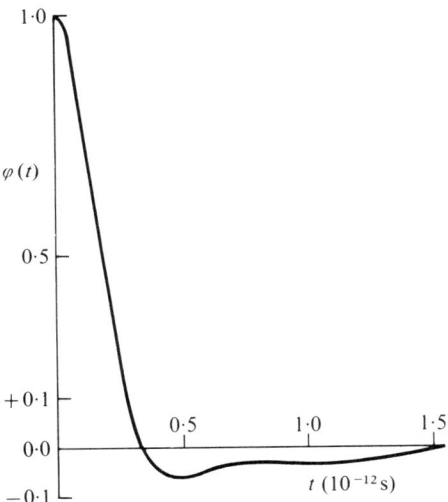

FIG. 6.9. Computed velocity auto-correlation function for Lennard–Jones potential at 85 K. (After Rahman 1964).

Of course, every type of motion—vibration, diffusion, jumps—contributes to $\langle \mathbf{v}(t_0) \cdot \mathbf{v}(t_0 + t) \rangle$. In practice, it is more convenient, and rather closer to experiment, to consider the spectrum of this quantity which is defined by

$$Z(\omega) = \frac{1}{3\pi} \int_0^\infty \langle \mathbf{v}(t_0) \cdot \mathbf{v}(t_0 + t) \rangle \cos \omega t \, dt. \qquad (6.29)$$

Clearly, using eqn (6.27), we have

$$Z(0) = D/\pi. \qquad (6.30)$$

From the definition of $Z(\omega)$ we can show (Egelstaff 1961) that it is related to the 'measured' quantity $S_s(Q, \omega)$ as follows,

$$Z(\omega) = \omega^2 \operatorname*{Lt}_{Q \to 0} (S_s(Q, \omega)/Q^2). \qquad (6.31)$$

## ATOMIC AND MOLECULAR MOTION IN LIQUIDS

This is one of several such limits which arise in this field. We should note that experimentally $\mathcal{L}t_{Q \to 0}$ is not easy to achieve, if only because small $Q$ means low scattering angles and/or very slow neutrons. This formula applies, in particular, to the one-phonon cross-section† for a crystal for which (for high enough temperature)

$$S_s(Q, \omega)\big|_{\text{crystal}} = \frac{k_B T}{M} \frac{Q^2}{\omega^2} Z(\omega)(e^{-2W}), \tag{6.32}$$

where $Z(\omega)$ is the phonon density of states function and $W$ is the Debye-Waller exponent (see eqn (6.13) *et seq.*). We expect a remnant of the crystal phonon spectrum to be retained in the liquid since the atoms take a time of order $3 \times 10^{-12}$ s to diffuse an atomic radius (see eqn (6.9)), whereas the period of the most abundant crystal phonons is of order $5 \times 10^{-13}$ s. There is therefore plenty of time for an atom to vibrate if the structure of the liquid permits it to do so.

From the definition (6.29), or from eqn (6.31), we can find $Z(\omega)$ for various motions.

For a perfect gas, the velocity is perfectly correlated, i.e. $\phi(t) = 1$, so that

$$Z(\omega) = (k_B T/M)\, \delta(\omega). \tag{6.33}$$

For a perfect Einstein solid, the atoms all vibrate at the Einstein frequency $\omega_E$ so that $\phi(t) = \cos \omega_E t$ and

$$Z(\omega) \propto \delta(\omega - \omega_E). \tag{6.34}$$

For a harmonic solid, $Z(\omega)$ is just the vibration spectrum as is seen by inserting eqn (6.32) in eqn (6.31) (For the harmonic solid, $D = 0$ so that $Z(0) = 0$). For a classically diffusing liquid, using eqn (6.12) in eqn (6.31), we find

$$Z(\omega) = D/\pi, \text{ for all } \omega, \tag{6.35}$$

which, of course, is impossible because the total intensity of any spectrum must be finite (eqn (6.36)). The effect of the 'short time' behaviour of the translational motion affects mainly the high-frequency parts of $Z(\omega)$, and in fact for all models $Z(\omega)$ falls with increasing $\omega$ in a way consistent with the normalization,

$$\int_0^\infty Z(\omega)\, d\omega = k_B T/2M. \tag{6.36}$$

---

† The multi-phonon cross-section is usually regarded as a correction; it is assumed that this correction to experimental data has been made, although this would appear not to be an entirely trivial matter.

For the Brownian motion (Langevin) model (see eqn (6.8)), one finds

$$Z(\omega) = \frac{k_B T}{\pi M} \frac{\gamma}{\omega^2 + \gamma^2}, \qquad (6.37)$$

which is illustrated in Fig. 6.10.

What do we expect for a liquid? We must have diffusion, and so the low-$\omega$ Brownian part must be present. We must also have a phonon spectrum since a liquid certainly supports longitudinal ultrasonic ($\sim 10^8$ Hz) and even hypersonic ($\sim 10^{10}$ Hz) waves: hence some combination of the two may be expected in the spectrum with appropriate redistribution of the degrees of freedom. For a

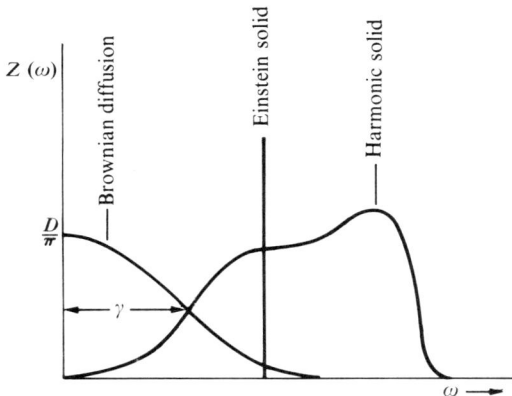

FIG. 6.10. Diagram illustrating the spectrum of the velocity auto-correlation function for various types of motion.

molecular liquid one also usually gets contributions from the reorientational molecular motion.

In practice, it is rather difficult to measure $Z(\omega)$. Most nuclei do not scatter entirely incoherently so that the coherent scattering has to be allowed for, for instance by assuming that it is relatively small at small $Q$ (see §6.4). Also there is the difficulty of extrapolation, which is not trivial. The analysis has been performed for liquid sodium (Cocking and Randolph 1965), and the observed curve (Fig. 6.11) shows roughly the expected form, with the dip at intermediate values of $\omega$ indicating in some measure the separation between the diffusive and vibrational modes of motion. A somewhat different curve, Fig. 6.12, is given by Larsson (1968), who shows nicely the difference between the spectra for solid and liquid sodium. As expected, for the liquid the transverse vibrational modes are not present, and for the solid the diffusional modes are absent. In

Figs. 6.11 and 6.12, $Z(\omega)$ is normalized so that $Z(\omega)\,d\omega = 1$ rather than as in eqn (6.36). The broken line in Fig. 6.11 shows the Brownian motion $Z(\omega)$ from

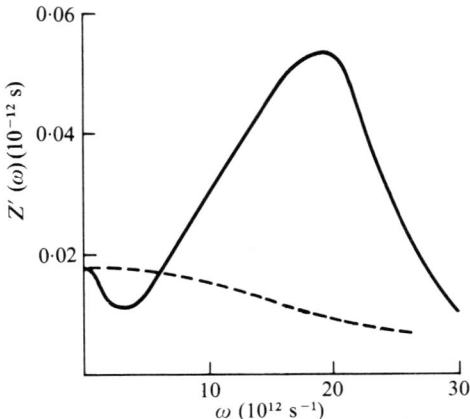

FIG. 6.11. The spectral density function for liquid sodium at 108°C. (After Egelstaff 1966).

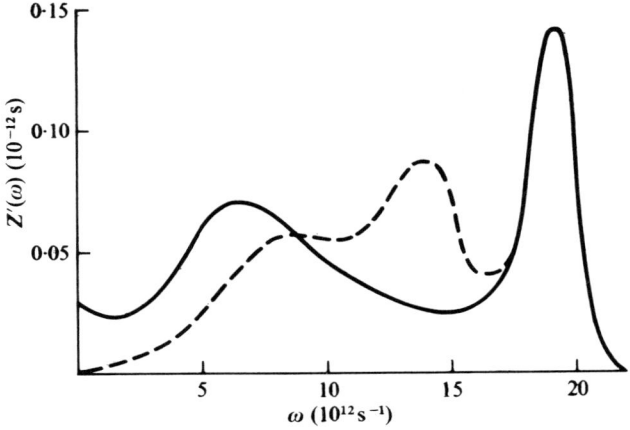

FIG. 6.12. Frequency spectrum for liquid sodium (full line) compared with that for solid sodium (broken line). (After Larsson 1968).

eqn (6.37), which is clearly inadequate. It is possible that a larger $\gamma$ is appropriate, and this has been taken to mean a larger effective mass $M$ and is closely related to the number of degrees of freedom of diffusional motion.

The most interesting case would be liquid argon but it is not easy to separate the coherent from the incoherent scattering.† Dasannacharya and Rao (1965) have claimed that this can in effect be done by separating $G$ into $G_s$ and $G_d$ 'by eye' for time regions short enough ($< 10^{-12}$ s) for them not to overlap appreciably (see Fig. 6.15). However, the molecular dynamics calculation (Rahman 1964) gives the result for liquid computer argon in Fig. 6.13. This result shows how this simplest of liquids departs from naive expectation and is a salutary warning to those who apply simple ideas to liquids and particularly to molecular liquids.

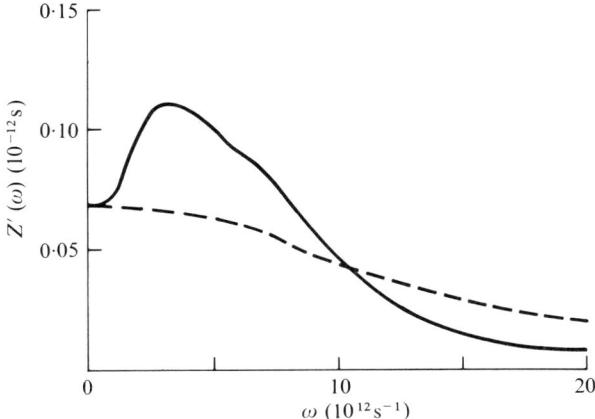

FIG. 6.13. Computed frequency spectrum for Lennard-Jones potential at 85 K (full line) compared with the Langevin result (broken line) based on eqn (6.37). (After Rahman 1964).

## 6.4. 'Coherent' translational motion

Most nuclei are at least partially coherent scatterers (e.g. argon). It so happens that the nuclei of most of the liquid metals which have been studied are coherent scatterers. We must now therefore consider coherent scattering.

For a coherent neutron scatterer the scattering pattern $S(\mathbf{Q}, \omega)$ depends on $G(\mathbf{r}, t)$ i.e. on both the self term $G_s(\mathbf{r}, t)$ which we have discussed at some length in §§6.2 and 6.3 and on the 'distinct' term $G_d(\mathbf{r}, t)$. The results of coherent scattering are, therefore, inherently more complicated to interpret, because of this and also because the motion of two different atoms relative to one another is surely more complex than that of a single atom. Whereas the 'self structure factor' $S_s(\mathbf{Q})$ in eqn (6.14) is rather simple, being virtually independent of $\mathbf{Q}$, the coherent structure factor‡

$$S(\mathbf{Q}) = \int S(\mathbf{Q}, \omega) \, d\omega \qquad (6.38)$$

† For argon $\sigma_{\text{coh}} = 0\cdot5$ barns and $\sigma_{\text{incoh}} = 0\cdot4$ barns.

‡ The 'structure factor' in neutron work is called the 'intensity' or 'interference function' in X-ray work.

## ATOMIC AND MOLECULAR MOTION IN LIQUIDS

is a complicated function of $Q$ (Fig. 6.14) but, of course, not as complicated as for a crystal since we have only short-range order in a liquid. The Fourier transform of $S(\mathbf{Q})$ is just the pair-distribution function $g(r)$,† which is defined as the probability of finding an atom at $r$ if there is an atom at the origin, and is normalized to unit probability at large $r$. The behaviour of $g(r)$ shows the strong repulsion of atoms that are close together $[g(r) \to 0$ for $r \to 0]$ and the independence of atoms which are far apart $[g(r) \to 1$ for $r \to \infty]$. $S(\mathbf{Q})$ and $g(r)$ are fairly well understood for atomic liquids but very poorly understood for anything else (see Chapter 8). The nature of $G_d(\mathbf{r}, t)$ is not well understood even for atomic liquids.

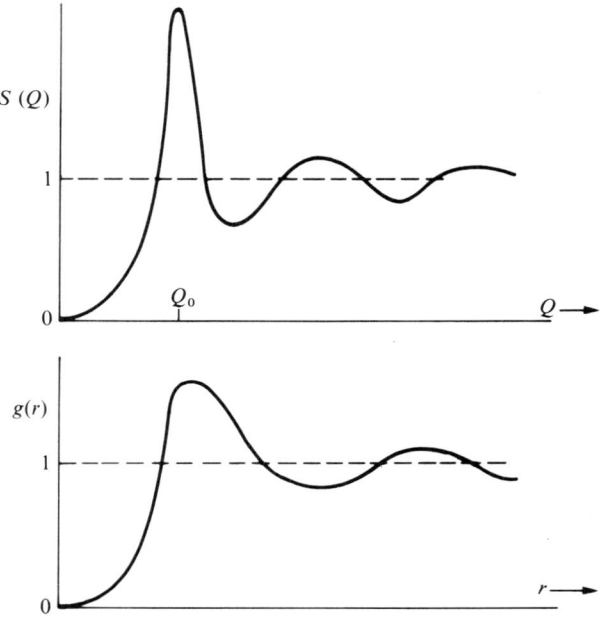

FIG. 6.14. $S(Q)$ and $g(r)$ for a liquid.

It is clear that

$$\mathop{\text{Lt}}_{t \to 0} [G_d(\mathbf{r}, t)] \to \rho g(r) \tag{6.39}$$

and that

$$\mathop{\text{Lt}}_{t \to \infty} [G_d(\mathbf{r}, t)] \to \rho, \tag{6.40}$$

where $\rho$ is the macroscopic number density of atoms. It is also to be expected that, because the important time scale for the self motion is $10^{-12}$ s, this is also

† $g(r)$ determines all time-independent (equilibrium) quantities which are not our direct concern (see Chapter 8).

likely to be important for the distinct motion. The general behaviour of $G_s$ and $G_d$ (van Hove 1954) is shown in Fig. 6.15.

It is worth noting here that the van Hove space-time correlation function $G(\mathbf{r}, t)$, while being sufficient to describe scattering experiments, is not in general adequate for other types of experiment. For instance, in nuclear magnetic resonance a different function is required (Powles 1968) which cannot be expressed in terms of $G(\mathbf{r}, t)$ and so essentially different information may be obtained by other than scattering methods. One should be aware, therefore, that even if we understood $G(\mathbf{r}, t)$ perfectly, our knowledge of molecular motion would still be imperfect in the sense that the results of many experiments would still not be explained.

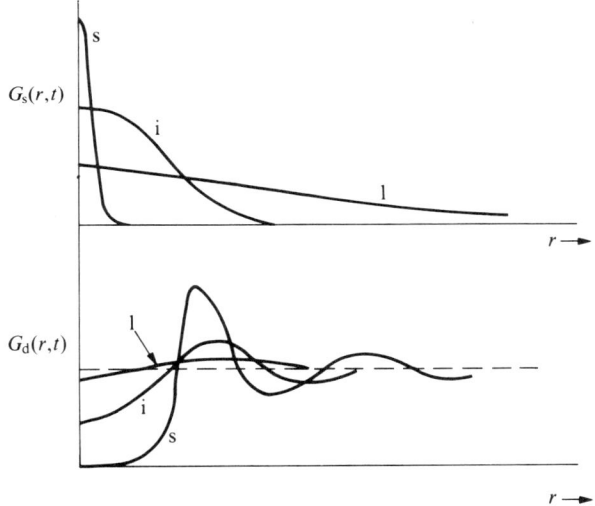

FIG. 6.15. The dependence of the self and distinct van Hove correlation functions on the distance $r$ for short (s), intermediate (i) and long (l) times.

$G_d(\mathbf{r}, t)$ is also sensitive to any co-operative motions, which in some degree and in some approximation may often be associated with phonons. A phonon, represented by a decaying wave motion, relates the motion of an atom at $\mathbf{r}_0 + \mathbf{r}$ to one at $\mathbf{r}_0$ if the mean free path of the phonon is larger than $r$. Thus $G_d(\mathbf{r}, t)$ will have a contribution of the form $\cos[\mathbf{q} \cdot \mathbf{r} + \omega_p t] \exp(-\Gamma t)$ (where $\Gamma$ is a characteristic relaxation frequency) for each phonon (sound wave) which is well defined. This leads to peaks in $S(\mathbf{Q}, \omega)$ at $\omega = \pm v_s Q$ and of width $2\Gamma$, where $v_s = \omega_p/q$ is the velocity of sound. These are the Brillouin lines of light scattering (Fig. 6.16) but they are not readily observed by neutron scattering, because of the experimental difficulty in getting sufficiently small values of $Q$ together with sufficient resolution in $\omega$. The effect may be regarded as the scattering of the neutron by a phonon, which is created or annihilated, whose momentum must be $\pm\hbar Q$ and whose energy is $\hbar v_s Q$. Alternatively, the effect may be considered as Bragg scattering of the neutron wave by the density variations corresponding to a phonon of the appropriate wavelength.† This wavelength must be $2\pi/Q$ and since the 'phonon grating' is moving at velocity $v_s$ the scattered neutron is Doppler shifted by $\Delta\omega = v_s Q$. This only applies to coherent scattering since the phonon is a collective mode necessarily involving more than one nucleus. When the phonons have a short lifetime these particle-collision-type selection rules are relaxed, if not meaningless, and a whole range of $\omega$ for given $Q$ leads to scattering much as for incoherent scattering. In other terms, the grating only has a few 'planes' and the diffraction peaks are not sharp. In fact, for large enough $Q$, say comparable to the quantity $\mathbf{Q_0}$ discussed below, and large

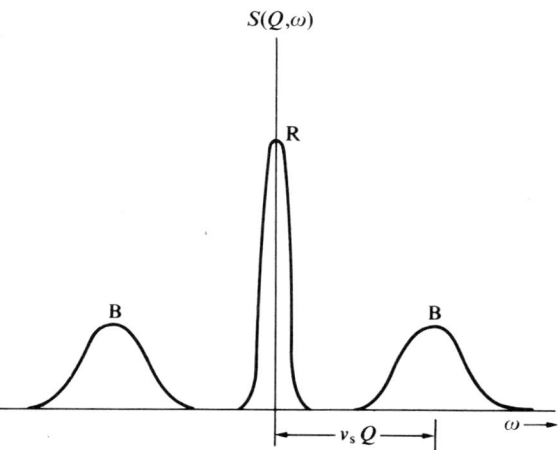

FIG. 6.16. Diagram illustrating the form of $S(Q, \omega)$ as a function of $\omega$ for a very small, fixed value of $Q$. R is the Rayleigh line and BB are the Brillouin lines.

enough $\omega$, so that short time behaviour is implicitly seen, the $S(\mathbf{Q}, \omega)$ for a liquid is rather similar to that for the corresponding poly-crystal.

Larsson, Dahlborg and Jovic (1965) have compared liquid and polycrystalline aluminium, which is a coherent scatterer. For a polycrystal the spectrum is broad because various phonons can be scattered for various orientations of the crystal. This should not be taken to mean that the liquid is just a mobile poly-crystal, since the similarity between the $S(Q, \omega)$ in the two states only exists for limited ranges of $Q$ and $\omega$. It is probably reasonable to say, however, that the short-range order is little changed on melting. Again the low-$\omega$ parts of the spectra for solid and liquid tend to be rather different. For very small $Q$,

† Indeed all coherent scattering may be regarded as due to density fluctuations since $G(\mathbf{r}, t) = \rho^{-1} \langle \rho(\mathbf{r_0}, t_0) \rho(\mathbf{r_0} + \mathbf{r}, t_0 + t) \rangle$.

$S(Q, \omega)$ also has a peak at $\omega = 0$ (see Fig. 6.16) due to non-propagating temperature (strictly speaking, entropy) fluctuations.

At the other extreme of large $Q$, the neutrons can only 'see' individual nuclei and so the scattering is, in effect, incoherent and the result is that $S(Q, \omega)$ is like $S_s(Q, \omega)$, i.e. is a Gaussian of width $(8\ln 2)^{\frac{1}{2}}(k_B T/M)^{\frac{1}{2}} Q$, as for a perfect gas. As $Q$ increases from small values, the behaviour becomes less 'macroscopic' and the Brillouin peaks become broader and less distinct and the velocity $v_s$, in so far as it is well defined, is no longer constant. In other words, we no longer have a linear relation between $\omega$ and $Q$ and we get 'phonon dispersion' as for a crystal. This is certainly expected to occur when the wavelength associated with the scattering becomes short enough (i.e. $Q$ is large enough) to see the 'structure' of the liquid. This effect will be appreciable when $Q \sim Q_0$ where $Q_0$ is the value

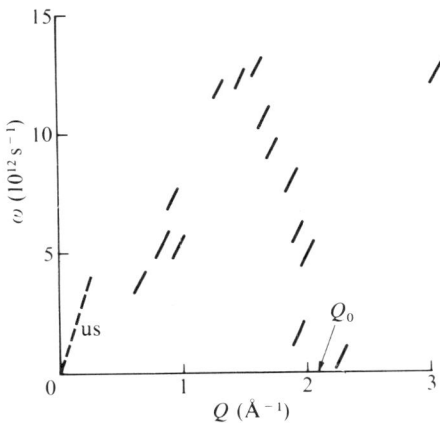

FIG. 6.17. Frequency–wave number plot for the peaks in $S(Q, \omega)$ for liquid lead (thick strokes), showing a possible dispersion relation and a Brillouin zone effect. The broken line us is calculated from $\omega = v_s Q$, where $v_s$ is the ultrasonic velocity. (After Egelstaff 1966).

of $Q$ for the peak in $S(Q)$ (see Fig. 6.14). Since it corresponds to a diffraction peak, $Q_0$ in the liquid plays a role rather similar to that of a reciprocal lattice vector in a crystal. In an approximate sense the momentum conservation law $\mathbf{Q} = \mathbf{Q}_0 + \mathbf{q}$, where $\mathbf{q}$ is a phonon wave vector, may be applied. When $Q < Q_0$, $q$ ranges from $Q_0 - Q$ to $Q_0 + Q$ and the frequency ranges correspondingly.

Fig. 6.17 for liquid lead (Egelstaff 1966) shows the $\omega$-peaks in $S(Q, \omega)$, each measured at fixed $Q$. This is very reminiscent of the Brillouin-zone type dispersion of waves in a crystal (see Chapter 4). For a crystal one would say that the first Brillouin zone ends at $Q \sim 1 \cdot 1$ Å$^{-1}$. The negative of the slope at the end of the second Brillouin zone ($Q \sim 2 \cdot 2$ Å$^{-1}$), gives a reasonable value for the velocity of sound consistent with that observed directly. We shall not explore here this rather complicated aspect of motion although the range of validity of the phonon concept in a liquid is an interesting question. The general conclusion from both incoherent and coherent scattering experiments is that the phonon spectrum of a liquid near its melting point is much the same as that for the corresponding poly-crystalline solid, except that the lifetime of the longitudinal phonons is somewhat shorter in the liquid and that of the transverse phonons is very much shorter or they are non-existent at almost any frequency.† In a sense, the degrees of freedom associated with phonons thus released by the absence of transverse modes reappear as diffusive modes of motion. With increasing temperature the low-frequency diffusive modes increase in relative importance (we recall that $Z(0) = D/\pi$ and $D$ increases with temperature). The loss of the transverse modes in a liquid is clearly illustrated in Fig. 6.12.

Attempts have been made to obtain expressions for $G_d(\mathbf{r}, t)$ in terms of various models. An early one due to Vineyard (1958) is based on the argument that, given the motion of any one atom, i.e. $G_s(\mathbf{r}, t)$, we can get the motion of two atoms relative to one another by allowing them both to move according to $G_s(\mathbf{r}, t)$ and weighting the motion to allow for the varying probability of finding an atom at a distance $\mathbf{r}'$, which is $g(\mathbf{r}')$. Thus

$$G_d(\mathbf{r}, t) = \rho \int g(\mathbf{r}') G_s(\mathbf{r} - \mathbf{r}', t) \, d\mathbf{r}', \qquad (6.41)$$

giving

$$S(Q, \omega) \simeq S_s(Q, \omega) S(Q). \qquad (6.42)$$

There are various drawbacks to this model, both physical and mathematical, but it is useful as a first approximation. It does show how, for large $Q$, $S(Q, \omega)$ will be large over a range of $\omega$ as is $S_s(Q, \omega)$ (see §6.2), (For small $Q$ it is quite wrong: it does not give the Brillouin lines). Physically it is clear that the motion of each atom of a pair must affect the time variation of the distance between them and so, subject to interference of the two scattered wavelets, the motion contributes to $S(Q, \omega)$. We shall not discuss more sophisticated theories here (see Sjölander 1965).

Further evidence for not including transverse phonons in the liquid is given by the better fit on ignoring them, as illustrated by the cross-section measurements for liquid argon (Sköld and Larsson 1967). The scattered intensity as a function of $Q$ at fixed $\omega$ and for high $\omega$ shows a dip at $Q \simeq Q_0$ which would not appear if transverse phonons were included. It is interesting that quite good fits to experimental results are obtained (Sköld 1967) using the modified diffusion model as developed for $S_s(Q, \omega)$ and also for $S_d(Q, \omega)$, but 'scaled' by $S(Q)$—e.g. for simple diffusion replace $D$ by $D/S(Q)$—which forces $S(Q, \omega)$ to have

† The reader will be aware that macroscopically transverse waves cannot exist in a fluid because it has zero shear modulus. However, there is evidence that at high frequencies a shear modulus may exist since the molecules do not have time to relax (the relevant time is the Maxwell relaxation time $\eta/G$ which is of order $10^{-11}$ s) and transverse waves become possible. On the other hand, for $\omega \gg G/\eta$, their wavelength is comparable with atomic spacings and so their existence is still doubtful.

certain known very general properties (the moment relations). It is a matter for contemplation that this latter entirely non-physical device gives a good fit to experiment and probably proves nothing.

It may be advisable to remind the reader again that, just as for $S_s(Q, \omega)$ (see eqn (6.11) *et seq.*), the deduction of $G(\mathbf{r}, t)$ from $S(Q, \omega)$ involves a Fourier transformation and that one seldom has a satisfactory range of $Q$ and $\omega$ to ensure an adequate accuracy in $G(\mathbf{r}, t)$. Clearly, we still await both entirely convincing experimental results for coherent scattering and a convincing theoretical explanation of such experimental results as are available.

## 6.5. Molecular liquids

Molecular liquids present a more complex atomic motion problem than atomic liquids. We leave aside the possibility of excitation of internal vibrations for which rather more energetic neutrons are usually required and which is a study of the molecules themselves (see Chapter 3). For molecular liquids, and regarding the molecules as rigid bodies, we have the possibility of reorientational as well as translational motions. The question then arises as to the nature and the rate of molecular reorientation. We must also ask if, and in what way, the reorientational and the translational motions are correlated. Thus, does a molecule reorient on the spot and then jump, or does it turn as it translates, and so on . . .? These questions have been attacked by many methods, particularly by nuclear magnetic resonance, but are still largely unanswered. It seems possible that for small (or rather low moment-of-inertia) molecules of pseudo-spherical shape the reorientation is faster than translation—in the sense that a molecule can reorient by a radian in a shorter time than it translates by a molecular diameter. Presumably then the reorientational and translational motions can be regarded as, in large measure, independent. It seems likely that this may be true for methane, for instance, and to a lesser extent for benzene (Powles and Figgins 1966). On the other hand, for a molecule with an awkward shape it seems probable that the reorientational and translational motions are closely linked—e.g. bromobenzene (Powles and Figgins 1967).

It must be remembered that scattered neutrons see *all* motions—neutron scattering is essentially a Doppler effect—and they do not distinguish between that part of the movement of a nucleus which is due to molecular reorientation from that due to translation. This separation can be done experimentally by non-scattering methods, e.g. nuclear magnetic resonance, but in scattering studies it can only be done by analysis of the results. Thus if we divide the actual motion of a nucleus, $\mathbf{r}$, into two parts, that due to translation of the centre of the molecule, $\mathbf{R}$, and that due to reorientation about the centre, $\mathbf{b}$, we have

$$\mathbf{r} = \mathbf{R} + \mathbf{b}. \tag{6.43}$$

But the nucleus sees only, say, $\langle r^2(t) \rangle$ which for the atom is $\langle R^2(t) \rangle$ (see §6.2). There is no way of finding $\langle R^2(t) \rangle$ or $\langle b^2(t) \rangle$ (or more usefully $\langle \mathbf{b}(t_0) \cdot \mathbf{b}(t_0 + t) \rangle$)

ATOMIC AND MOLECULAR MOTION IN LIQUIDS

separately. Of course, if reorientation and translation are statistically independent, then

$$\langle r^2 \rangle = \langle R^2 \rangle + \langle b^2 \rangle, \tag{6.44}$$

and we can try various models for $\mathbf{R}(t)$ and $\mathbf{b}(t)$, but the situation is at present virtually insoluble if we have to know $\langle \mathbf{R}(t)\mathbf{b}(t) \rangle$. The same applies to more complex quantities than $\langle r^2 \rangle$ and so far, *faute de mieux*, $R$ and $b$ are usually assumed to be statistically independent. Just as a first approximation for $\langle R^2 \rangle$ was $6Dt$ (eqn 6.5), so an approximation for $\langle b^2 \rangle$ is $6D_r t$, where $D_r$ is a rotational diffusion constant defined by the 'macroscopic' rotational diffusion equation

$$\frac{\partial p(\Omega, t)}{\partial t} = \frac{D_r}{a^2} \nabla_r^2 p(\Omega, t), \tag{6.45}$$

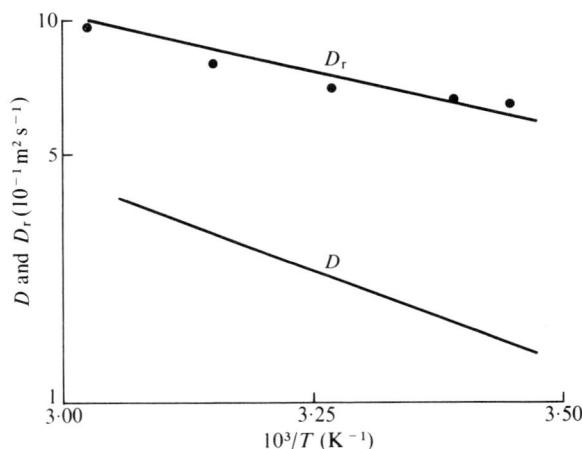

FIG. 6.18. Apparent diffusion constant, $D_r$, for liquid benzene from quasi-elastic neutron scattering, compared with $D$, the macroscopic diffusion constant. (After Holmryd and Nelin 1968).

with $\Omega$ the angular motion and $a$ the radius of the molecule. Unfortunately, there is no truly macroscopic way of getting $D_r$ and it may not even exist, and so we are denied the comforting macroscopic limit at $Q \to 0$ which we found so useful for the analysis of translational motion. Nevertheless, if we had reorientation and no translation we might expect a quasi-elastic peak (cf. eqn (6.16)) of width

$$\Delta\omega = 2D_r Q^2 \tag{6.46}$$

for $Q$ neither too large nor too small and provided macroscopic diffusion is meaningful for angles less than $2\pi$. It is possible that an approximation to this situation has been observed in liquid benzene where the width of the incoherent quasi-elastic peak, analysed as in §6.2, yields an apparent diffusion constant *larger* than the macroscopic translational constant,† as shown in Fig. 6.18 (Holmryd and Nelin 1968). This situation may be contrasted with that in water (Fig. 6.7) where the correct value of $D$ is observed. For water the reorientation is surely very closely related to translation. It is known from n.m.r. work (Powles and Figgins 1966) that in liquid benzene the molecular reorientation is fast compared with translation (in the sense mentioned above) and that its activation energy is lower, so the two experiments are consistent. For macroscopic reorientational diffusion, we have for the reorientation correlation time $\tau_2$ (actually of $Y_2(\theta)$, i.e. of $(\cos^2\theta - \tfrac{1}{3})$) at 25°C,

$$\tau_2 = a^2/6D_r = \frac{(2\cdot 5 \times 10^{-8})^2}{6 \times 9 \times 10^{-5}} = 1\cdot 1 \times 10^{-12} \text{ s}, \tag{6.47}$$

where we have used $D_r$ from Fig. 6.18. From n.m.r. (Powles and Figgins 1966) the value of $\tau_2$ is known to be $1\cdot 4 \times 10^{-12}$ s which is again consistent. If the reorientation is independent of translation, the quasi-elastic width is actually determined by the correlation time of $Y_1(\theta)$, i.e. of $\cos\theta$, since we can write

$$\langle \Delta \mathbf{b}^2 \rangle \langle [\mathbf{b}(t_0 + t) - \mathbf{b}(t_0)]^2 \rangle = 2a^2 [1 - a^{-2}\langle \mathbf{b}(t_0) . \mathbf{b}(t_0 + t)\rangle]$$
$$= 2a^2 [1 - \langle \cos\theta(t_0) \cos\theta(t_0 + t)\rangle]. \tag{6.48}$$

More generally, the whole spectrum involves all spherical harmonics, as mentioned below. It is more difficult to observe the reorientational motion in the way we did for translation. The analogy is similar in that for short times the diffusion equation is surely not valid and we have a gas-like behaviour,

$$\langle \Delta \mathbf{b}^2(t) \rangle = a^2 \langle \Delta\theta^2 \rangle = (a^2 k_B T/I) t^2, \tag{6.49}$$

where $I$ is the moment of inertia. Thus there is again a changeover from $t^2$ to $t$ dependence as for the translational case, although this point does not appear to have been investigated. Moreover, if there are torsional oscillations of the molecules, these will result in inelastic scattering because of the vibrational motion of nuclei which are not at the centre of the molecules. The period of such oscillations is expected to be in the region of $10^{-13}$ s so that such contributions to the $S_s(Q, \omega)$ spectrum are not readily distinguishable from the translational ones.

The quantity for reorientation which most nearly corresponds to $\langle \mathbf{v}(t_0) . \mathbf{v}(t_0 + t)\rangle$ for translation is the angular velocity correlation function $\langle \Omega(t_0)\Omega(t_0 + t)\rangle$, but this is inherently more complex since the reorientation is not about a fixed axis. The correlation time for this quantity, analogous to $\tau_0$ for $v(t)$, is called $\tau_{sr}$ in n.m.r. It is usually very short, of order $10^{-13}$ s (Powles 1966), and it is probably this which is the 'changeover time' seen by slow neutrons scattered by rapidly reorienting molecules. It is a considerably shorter changeover time than for translation (see §6.2).

† It will be recalled that $\Delta\omega$ is always too small in atomic liquids.

Of course, translational diffusional motion still occurs in a molecular liquid and can be investigated in principle. But it must be borne in mind that the relative number of diffusional degrees of freedom is reduced in a molecular liquid since some degrees of freedom are tied up in reorientation and these cannot become diffusive modes as long as the molecule exists. So even apart from the complication due to reorientational diffusion, the translational diffusion is more difficult to see. In general, the proportion of diffusional modes of motion in a complex molecular liquid may be quite low, say 1 per cent, in contrast with an atomic liquid where it might be 10 per cent or more. However, even a small proportion of diffusional degrees of freedom corresponds to a large

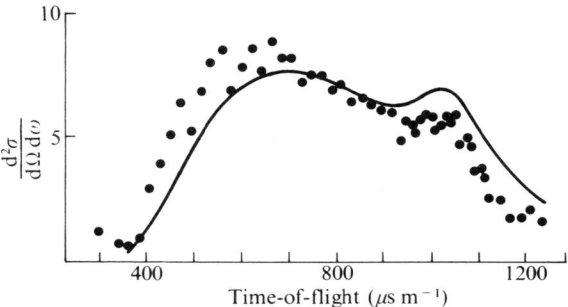

FIG. 6.19. Time-of-flight spectra of neutrons scattered by liquid methane at 98 K. Experimental points are given as closed circles and theoretical result is shown by solid line. Note that the time of flight $\propto$ (frequency)$^{-1/2}$. (After Agrawal *et al.* 1968).

contribution to the scattering because they are at low $\omega$ in $Z(\omega)$ and so are amplified by the factor $1/\omega^2$ in eqn (6.31). That the reorientation is important, if not dominant for many molecular liquids, is illustrated by the case of liquid methane (Agrawal, Desai, and Yip 1968), in Fig. 6.19. The theoretical result (full line) includes both translational and reorientational contributions, the latter motion being derived from the infra-red line shape which gives the Fourier transform of $\langle \cos \theta (t_0) \cos \theta (t_0 + t) \rangle$ (Gordon 1965). Only if the reorientational contribution is included, is the correspondence with experiment reasonable.

For many molecules one has the additional complication of internal flexibility. One also has the possibility of investigating the motion of different parts of the same molecule by isotopic substitution, the most valuable and dramatic being the substitution of deuterons for protons which replaces a strong incoherent by a weak coherent scatterer. Both these effects are illustrated (see Fig. 3.13) by the difference in the spectra of the two methyl alcohols $CD_3OH$ and $CH_3OD$ (Aldred, Eden, and White 1967), for which the molecular motion is probably not very different. The self-diffusion constant (*via* eqn (6.16)) is the same for the methyl or the hydroxyl protons and so presumably one is seeing whole-molecule motions. The large inelastic peak for $CH_3OD$ in Fig. 3.13(c) is undoubtedly associated with methyl-group reorientation. The fall with increasing angle (i.e.

$Q$) of the intensity of the quasi-elastic peak is much greater for $CH_3OD$ than for $CD_3OH$ indicating a larger Debye–Waller effect (eqn (6.13)), as might be expected. For comparison with optical spectroscopic data, one should derive the density of states (eqn (6.31)) from $S_s(Q, \omega)$.

A detailed analysis of purely reorientational motion has been given by Sears (1966) who shows that the corresponding $G_s(\mathbf{r}, t)$ can be expressed in terms of a series involving the correlation functions of all the spherical harmonics $F_l(t) = \langle Y_l[\Omega(t_0 + t)] Y_l[\Omega(t_0)] \rangle$. The result for $S_s(Q, \omega)$ is

$$S_s(Q, \omega) = j_0^2(Qb)\delta(\omega) + \sum_{l=1}^{\infty} (2l+1)j_l^2(Qb)F_l(\omega), \qquad (6.50)$$

where $j_l$ is a spherical Bessel function and $F_l(\omega)$ is the Fourier transform of $F_l(t)$. In principle, by varying $Q$ every spherical harmonic could be found using eqn (6.50) and then could be compared with say $F_1(\omega)$ from dielectric or infra-red measurements and $F_2(\omega)$ from n.m.r., Raman scattering and depolarized Rayleigh scattering. However, this cannot be achieved in practice, both because it is not established that molecular reorientation is independent of translation in any liquid and also because the experimental accuracy required is excessive. This is another example of the embarrassing richness of the information contained in slow-neutron scattering spectra which makes them so difficult to interpret. Comparison with the results of methods which see only parts of, or simpler functions of the motion (because of effectively more severe selection rules) is a valuable aid in the interpretation of neutron scattering data. However, this has so far been little exploited.

It would appear that the most common case, where the reorientational and translational motions are correlated, has not yet been analysed at all for neutron scattering, and a wide field of investigation, both theoretical and experimental, lies open here. Since even the correlation of orientation in molecular liquids is presently not well understood (Egelstaff, Page, and Powles 1971), it would seem likely that progress in understanding the scattering of slow neutrons by molecular liquids will be slow to develop and will be very dependent on other techniques which give less but simpler information more directly.

Finally, we should point out that in most molecular liquids we have more than one species of scattering nucleus whose motions are in general different. The cases we have discussed, and the liquids which have mostly been measured, contain protons which generally dominate the scattering, making it almost completely incoherent. The protons usually have the same motion as, for instance, in a $CH_3$ group. If the protons do not all have the same motion, as in $CH_3OH$, the incoherent scattering intensity is just the sum of that for each type. For coherent scatterers the interpretation of the observed intensity is more difficult with several types of nucleus and little work, either theoretical or experimental, has been done for this very complex situation. Here again slow

ATOMIC AND MOLECULAR MOTION IN LIQUIDS

neutron scattering must be, eventually, a very rich source of information about molecular motion in liquids.

## 6.6. References

AGRAWAL, A. K., DESAI, R. C. *and* YIP, S. (1968). In *Neutron inelastic scattering*, vol. I. IAEA, Vienna. p. 545.
ALDRED, B. K., EDEN, R. C. *and* WHITE, J. W. (1967). *Discuss. of Faraday soc.* **43**, 169.
CHUDLEY, G. T. *and* ELLIOTT, R. J. (1961). *Proc. Phys. Soc.*, **77**, 353.
COCKING, S. J., RANDOLPH, P. D., (1965). Quoted by EGELSTAFF, P. A. (1965) in *Rep. Prog. Phys.* **29**, 333. See also pp. 449 and 463 of *Neutron inelastic scattering*, vol. I. IAEA, Vienna, (1968).
DASANNACHARYA, B. A. *and* RAO, K. R. (1965). *Phys. Rev.* **137**, A417.
EGELSTAFF, P. A. (1961). In *Inelastic scattering of neutrons in solids and liquids*. IAEA, Vienna. p. 25.
EGELSTAFF, P. A. (1966). *Rep. Prog. Phys.* **29**, 333.
EGELSTAFF, P. A., PAGE, D. I. *and* POWLES, J. G. (1971). *Molec. Phys.* **20**, 881.
GORDON, R. G. (1965). *J. chem. Phys.* **43**, 1307.
HOLMRYD, S. *and* NELIN, G. (1968). In *Neutron inelastic scattering*, vol. I. IAEA, Vienna. p. 475.
LARSSON, K. E., DAHLBORG, U. *and* HOLMRYD, S. (1960). *Ark. Fys.* **17**, 369.
LARSSON, I. E. DAHLBORG, U. *and* JOVIC, D. (1965). In *Inelastic scattering of neutrons*, vol. II. IAEA, Vienna. p. 117.
LARSSON, K. E. (1968). In *Neutron inelastic scattering*, Vol. I. IAEA, Vienna, p. 397.
PASKIN, A. *and* RAHMAN, A. (1966). *Phys. Rev. Lett.* **16**, 300.
POWLES, J. G. (1968). In *Neutron inelastic scattering*, vol. I. IAEA, Vienna. p. 379.
POWLES, J. G. *and* FIGGINS, R. (1966). *Molec. Phys.* **10**, 155.
POWLES, J. G. *and* FIGGINS, R. (1967). *Molec. Phys.* **13**, 253.
POWLES, J. G. (1966). In *Molecular relaxation processes*. (*Ed. M. Davis*). The Chemical Society and Academic Press, London.
RAHMAN, A. (1964). *Phys. Rev.* (A) **136**, 405.
RANDOLPH, P. D. (1968). In *Neutron inelastic scattering*, vol. I. IAEA, Vienna. p. 449.
SAKAMOTO, M., BROCKHOUSE, B. N., JOHNSON, R. H. *and* POPE, N. K. (1962). *J. phys. Soc. Japan.* **17**, (*suppl.* B11) 370.
SEARS, V. F. (1966). *Can. J. Phys.* **44**, 1279.
SKÖLD, K. *and* LARSSON, K. E. (1967). *Phys. Rev.* **161**, 102.
SKÖLD, K. (1967). *Phys. Rev. Lett.* **19**, 1023.
VINEYARD, G. H. (1958). *Phys. Rev.* **110**, 999.
VAN HOVE, L. (1954). *Phys. Rev.* **95**, 249.

*General*

There are many books and reviews on liquids. The following take particular note of neutron scattering methods.

EGELSTAFF, P. A. (1962). Neutron scattering studies of liquid diffusion. *Phil. Mag. Suppl.* **11**, 203.

EGELSTAFF, P. A. (1965). The thermal motion of simple liquids. *Br. J. appl. Phys.* **16**, 1219.

EGELSTAFF, P. A. *and* SCHOFIELD, P. (1965). The structure and thermal motion of simple liquids. *Contemp. Phys.* **6**, 274, 453.

EGELSTAFF, P. A. (1967). *An introduction to the liquid state.* Academic Press, London.

ENDERBY, J. E. (1968). Neutron scattering studies of liquids. Chapter 14 of *Physics of simple liquids (Eds. Témperley, Rowlinson and Rushbrooke).* North Holland, Amsterdam.

LARSSON, K. E. (1965). Experimental results in liquids. Chapter 8 in *Thermal Neutron Scattering (Ed. P. A. Egelstaff).* Academic Press, London.

LARSSON, K. E. (1968). Liquid dynamics from neutron scattering, in *Neutron inelastic scattering,* vol. I. IAEA, Vienna.

MARCH, N. H. (1968). *Liquid metals,* Pergamon.

MARCH, N. H. (1968). The liquid state, in *Theory of condensed matter.* IAEA, Vienna.

SJÖLANDER, A. (1965). Theory of neutron scattering by liquids. Chapter 7 in *Thermal neutron scattering (Ed. P. A. Egelstaff).* Academic Press, London.

# 7 Structure and Atomic Motion in Glasses

By A. J. LEADBETTER

School of Chemistry, Bristol University

## 7.1. The glassy state

A glass may be defined simply as a non-crystalline solid. It is non-crystalline because it lacks the periodicity or long-range order of a crystal and it is macroscopically a solid because its viscosity is greater than about $10^{13}$ kg/ms ($10^{14}$ poise). This definition includes a wide variety of materials, the most common and well-known of which are the inorganic glasses formed by supercooling the melt. In this category are the network structure, single-constituent materials such as $SiO_2$, $GeO_2$, $BeF_2$ and $As_2S_3$, plus a wide range of more complex compositions such as alkali and alkaline earth silicates, borosilicates and aluminosilicates. Only Se of the elements readily forms a stable glass from the melt, but this is more akin to the polymeric glasses than the other supercooled inorganic glasses. Many inorganic compositions which cannot be supercooled to give vitreous materials may be prepared as glasses by other techniques such as vapour deposition, sputtering or electrodeposition, e.g. amorphous Ge and Si. Organic polymers generally form an amorphous phase as a result of chain entanglement (see Chapter 5) but glasses are also obtained with small organic molecules. This is often the result of chain or network formation by hydrogen bonding, but it may also arise because of difficulty in packing the molecules in an ordered array. For further reading on glasses and glass formation, see Jones (1956), Rawson (1967), and Mackenzie (1960, 1962, 1964).

The specification of a viscosity of about $10^{13}$ kg/ms as an approximate dividing line between solid and liquid is somewhat arbitrary but is a useful and usual definition. This viscosity corresponds to a stress relaxation time of the order of minutes, so that the criterion specifies when the material will behave as a rigid solid for experiments carried out on a macroscopic time scale, e.g. volume and entropy measurements. However, by waiting long enough (perhaps years) it is possible to observe relaxation to a new configurational equilibrium state of the supercooled liquid at viscosities up to several orders of magnitude higher than $10^{13}$ kg/ms. Furthermore, since the viscosity is strongly temperature dependent, the above criterion also defines a *temperature* at which the relaxation times are of the order of minutes: this is the glass transformation temperature $T_g$.

What is important in an experiment is the relative magnitude of the experimental time scale and an appropriate relaxation time for the glass. For neutron scattering the relevant time scale is around $10^{-12}$ s, and so it is obvious that a glass will show properly solid-like behaviour in a neutron scattering experiment to temperatures considerably in excess of $T_g$. This may be seen more explicitly by considering the diffusion coefficients of the various atomic species in inorganic glasses in terms of the Einstein relation $\langle r^2(t)\rangle = 6Dt$ (eqn 6.5), so as to deduce the time for movement of an atom through a distance of, say, 1 Å. For network ions such as oxygen in silica or germania, values of $D$ near $T_g$ are typically smaller than $10^{-16}$ m$^2$ s$^{-1}$, corresponding to a relaxation time $\tau_0$ for moving 1 Å of greater than $10^{-4}$ s. Small cations located in interstitial sites in the glass network, such as Na$^+$ in sodium silicate glasses, may have much higher mobilities, typically $D \sim 10^{-11}$ m$^2$ s$^{-1}$ and $\tau_0 \sim 10^{-9}$ s near $T_g$. Hence diffusive motions in these materials are too slow to be studied by neutron scattering techniques unless the temperature is considerably higher than $T_g$. Small cations present in low concentration in network glasses like vitreous silica, may have even higher mobilities (e.g. $D \sim 10^{-6}$ m$^2$ s$^{-1}$ and $\tau_0 \sim 10^{-11}$ s for Na$^+$ in SiO$_2$ at $T \sim 1000°$C), but here the concentrations will generally be too low to permit a useful neutron study of diffusive motions. In general, therefore, for $T < T_g$ the neutron scattering from a glass will be clearly separable into elastic and inelastic components which will give information respectively about the average atomic positions and the vibrations of the atoms about these positions, exactly as for a crystal.

The study of glasses is important for two main reasons. Firstly, the vitreous state is a state of matter intermediate between crystal and liquid. The structure is characteristic of the instantaneous structure of a liquid but the atoms are constrained to vibrate, more or less harmonically, about their disordered, quasi-equilibrium sites, so that the dynamics are characteristic of a solid. The structure may be studied by either elastic scattering experiments, as for a crystal, or by total scattering in the same way that the instantaneous structure of liquids is determined. Study of the dynamics gives a means of examining the effects of static geometrical disorder (lack of periodicity) on both the collective and individual particle motions. It is worth noting here that the atomic dynamics of liquids will be closely similar to those of glasses at frequencies $\nu$ high enough so that $\nu^{-1} \ll \tau$ where $\tau$ is some characteristic time for significant structural change in the liquid ($\tau \sim 10^{-12}$ s; see Chapter 6). In the second place, the study of glasses is of interest because of their special (sometimes unique) properties, often imperfectly understood from a fundamental point of view, and their immense technological importance. An old example is the unusual thermal and elastic properties of vitreous silica, and in particular its extraordinarily low thermal expansivity (Anderson and Dienes 1960, Gaskell 1966, Leadbetter 1968); a new example is the electrical switching behaviour of various chalcogenide glasses (see *Amorphous and liquid semiconductors*. 1970, in General References).

In this chapter we shall consider first the basic problem of the determination of the structure of polyatomic inorganic glasses by scattering techniques and the particular contributions which neutron scattering can make. We shall then describe how information about the frequency distribution of the glass modes may be determined using incoherent inelastic neutron scattering, assuming the validity of the so-called incoherent approximation for coherent scatterers. Finally we shall discuss coherent inelastic scattering phenomena and the information they give about the nature of collective motions in inorganic glasses.

## 7.2. Structure

The elastic component in both X-ray and neutron scattering experiments from crystals is separable from the inelastic component because of its occurrence as sharp Bragg peaks. For a glass this separation is only possible experimentally with neutron scattering, and not with X-ray scattering, because only for the former is sufficient energy discrimination available. For liquids no such rigorous distinction exists between elastic and inelastic scattering, although an approximate separation of a quasi-elastic component (including diffusive motion contributions) is often possible and useful (see Chapters 6 and 8).

The elastic differential cross-section may be obtained from eqn (1.24a) of Chapter 1 by writing the correlation function $G(\mathbf{r}, t)$ as the sum of two parts:

$$G(\mathbf{r}, t) = G(\mathbf{r}, \infty) + G'(\mathbf{r}, t)$$

where $G(\mathbf{r}, \infty)$ is the asymptotic part as $t \to \infty$, and $G'(\mathbf{r}, t)$ is the time-dependent part which tends to zero as $t \to \infty$. Substituting in (1.24a)

$$\frac{d^2\sigma}{d\Omega\, dE'} = N \frac{k'}{k} \frac{\langle b \rangle^2}{2\pi\hbar} \iint d\mathbf{r}\, dt\, e^{i(\mathbf{Q}\cdot\mathbf{r} - \omega t)} G(\mathbf{r}, \infty)$$
$$+ N \frac{k'}{k} \frac{\langle b \rangle^2}{2\pi\hbar} \iint d\mathbf{r}\, dt\, e^{i(\mathbf{Q}\cdot\mathbf{r} - \omega t)} G'(\mathbf{r}, t). \quad (7.1)$$

$G(\mathbf{r}, \infty)$ is independent of time, so that the integration over $t$ in the first term of eqn (7.1) is the $\delta$-function

$$\delta(\omega) = \frac{1}{2\pi} \int_{-\infty}^{\infty} e^{-i\omega t}\, dt$$

(cf. eqn (1.18b)). Clearly, this first term gives the elastic scattering, with $\omega = 0$ and $k' = k$. Integrating eqn (7.1) over $E' = \hbar \omega$ gives

$$\left(\frac{d\sigma}{d\Omega}\right)_{\text{elastic}} = N \frac{\langle b \rangle^2}{2\pi} \iint 2\pi\delta(\omega)\, e^{i\mathbf{Q}\cdot\mathbf{r}}\, G(\mathbf{r}, \infty)\, d\mathbf{r}\, d\omega$$

$$= N\langle b \rangle^2 \int e^{i\mathbf{Q}\cdot\mathbf{r}}\, G(\mathbf{r}, \infty)\, d\mathbf{r}. \quad (7.2)$$

The time-averaged positions $G(\mathbf{r}, \infty)$ may be described in terms of static equilibrium positions $G^e(\mathbf{r})$ and thermal (Gaussian) displacements about these sites. This leads in the usual way (see §1.5), to the appearance of a Debye-Waller factor $\exp(-2W)$ whose exponent is

$$W = \tfrac{1}{6} \langle u^2 \rangle Q^2,$$

for isotropic mean square displacements $\langle u^2 \rangle$. $G(\mathbf{r}, \infty)$ is given by

$$G(\mathbf{r}, \infty) = G^e(\mathbf{r}) \exp(-2W),$$

and the elastic scattering cross section becomes

$$\left(\frac{d\sigma}{d\Omega}\right)_{\text{elastic}} = N\langle b \rangle^2 \int e^{i\mathbf{Q}\cdot\mathbf{r}} G^e(\mathbf{r}) \exp(-2W)\, d\mathbf{r}. \tag{7.3}$$

The total correlation function $G^e(\mathbf{r})$ may be separated further into its self and its distinct parts, as follows,

$$\begin{aligned} G^e(\mathbf{r}) &= G_s^e(\mathbf{r}) + G_d^e(\mathbf{r}) \\ &= \delta(\mathbf{r}) + \rho g_e(\mathbf{r}), \end{aligned} \tag{7.4}$$

where $g_e(r)$ is the pair distribution function for the equilibrium atomic positions and $\rho$ is the number density of atoms. (It is customary to normalize $g(r)$ to unity at large $r$).

Thus finally we have for the monatomic case, by substituting eqn (7.4) into eqn (7.3):

$$\left(\frac{d\sigma}{d\Omega}\right)_{\text{elastic}} = N\langle b \rangle^2 \left[1 + \int \rho g_e(r)\, e^{i\mathbf{Q}\cdot\mathbf{r}}\, d\mathbf{r}\right] e^{-2W}. \tag{7.5}$$

Glass structure can also be examined by *total* scattering experiments, as for the determination of liquid structures (Chapter 8). The appropriate differential cross-section is again derived from eqn (1.24a) by integrating with respect to energy,

$$\left(\frac{d\sigma}{d\Omega}\right)_{\text{total}} = \int_{-\infty}^{\infty} dE'\, N \frac{k'}{k} \frac{\langle b \rangle^2}{2\pi\hbar} \iint d\mathbf{r}\, dt\, e^{i(\mathbf{Q}\cdot\mathbf{r} - \omega t)}\, G(\mathbf{r}, t). \tag{7.6}$$

Unfortunately, this equation contains the factor $k'/k$ which depends on the energy transfer, and to proceed further it is necessary to assume that $k' \simeq k$. This is the 'static approximation', involving the physical assumption that energy transfers are small compared with the incident energy. It is an excellent approximation for X-ray scattering but is less good for neutrons, although adequate corrections may usually be made (see Enderby 1968, and §8.2.3.5 of Chapter 8). Putting $k' = k$ in eqn (7.6), and recalling the properties of the $\delta$-function given in eqns (A.1) and (A.3) of Appendix II (p. 300), we have,

$$\begin{aligned} \left(\frac{d\sigma}{d\Omega}\right)_{\text{total}} &= \int_{-\infty}^{\infty} \hbar\, d\omega\, N \frac{\langle b \rangle^2}{2\pi\hbar} \iint d\mathbf{r}\, dt\, e^{i(\mathbf{Q}\cdot\mathbf{r} - \omega t)}\, G(\mathbf{r}, t) \\ &= N\langle b \rangle^2 \iint d\mathbf{r}\, dt\, e^{i\mathbf{Q}\cdot\mathbf{r}} \delta(t) G(\mathbf{r}, t) \\ &= N\langle b \rangle^2 \int G(\mathbf{r}, 0)\, e^{i\mathbf{Q}\cdot\mathbf{r}}\, d\mathbf{r}. \end{aligned}$$

## STRUCTURE AND ATOMIC MOTION IN GLASSES

Separating the total correlation function, $G(\mathbf{r}, 0)$, into its self and distinct parts, as before, leads for the monatomic case to

$$\left(\frac{d\sigma}{d\Omega}\right)_{total} = N\langle b\rangle^2 [1 + \int \rho g(r) e^{i\mathbf{Q}\cdot\mathbf{r}} d\mathbf{r}] \tag{7.7}$$

where $g(r)$ is the usual pair distribution function describing the probability of an atom being at a distance $r$ from the origin atom, as determined at the same instant of time for both atoms.

The difference between $g_e(r)$ in eqn (7.5) and $g(r)$ in eqn (7.7) is simply that the effects of the thermal cloud are included in the atom positions described by the latter. Only from elastic scattering experiments, however, through the determination of (average) Debye-Waller factors, may information about thermal displacements be obtained. From comparison of $g_e(r)$ and $g(r)$ it may also be possible to determine distortions in $g(r)$ arising from the effects of the static approximation. Although elastic scattering experiments are in principle straightforward, they require the use of a triple-axis spectrometer and only one reasonably extensive experiment (on vitreous $SiO_2$) has been reported (Lorch 1970) so far.

In studying real glasses the main difficulty lies not with the static approximation, for which quite good corrections may be applied, but arises from the fact that nearly all glasses are at least diatomic in composition. This means that at least three pair distribution functions are required fully to describe the average structure: $g_{12}(r), g_{11}(r)$ and $g_{22}(r)$.

Generalizing eqn (7.7) to a polyatomic system and averaging over all orientations of $\mathbf{Q}$ with respect to $\mathbf{r}$ gives

$$\frac{1}{N}\left[\frac{d\sigma_{coh}}{d\Omega}\right]_{total} = I(Q) = \sum_i \langle b_i\rangle^2 + \sum_i \sum_j \langle b_i\rangle\langle b_j\rangle$$

$$\times \rho_j \int_0^\infty 4\pi r^2 g_{ij}(r) \frac{\sin Qr}{Qr} dr$$

where $i$ labels the atoms in a unit of composition (e.g. $SiO_2$ in vitreous silica) and $j$ labels the chemically distinct atom types. Defining

$$i(Q) = I(Q) - \sum_i \langle b_i\rangle^2 - I_0(Q)$$

where $I_0(Q)$ is the unobservable scattering near $Q = 0$ arising from the average density of the system, then

$$Qi(Q) = \sum_i \sum_j \langle b_i\rangle\langle b_j\rangle \int_0^\infty d_{ij}(r) \sin Qr\, dr \tag{7.8}$$

where

$$d_{ij}(r) = 4\pi r \rho_j [g_{ij}(r) - 1].$$

Eqn (7.8) may be Fourier transformed to give the total correlation function

$$D(r) = \sum_i \sum_j \langle b_i \rangle \langle b_j \rangle d_{ij}(r) = \frac{2}{\pi} \int_0^\infty Qi(Q) \sin Qr \, dQ. \quad (7.9)$$

Thus a single scattering experiment gives only a weighted sum of the component correlation functions $d_{ij}(r)$, all of which are required to describe the structure fully (see also Chapter 8).

The presence of more than one type of atom introduces an additional difficulty in the elastic scattering experiment because each atom type will, in general, have a different Debye–Waller factor which will not be obtainable from the single experiment. The fact that experimental data terminate at finite values of $Q$ must also be taken into account (for further details, see Leadbetter and Wright 1972; Lorch 1970).

In order to obtain the separate correlation functions, at least the same number of distinct diffraction experiments as functions is required. This has been achieved for liquid CuCl, as described in Chapter 8, using isotopic substitution to alter the neutron scattering lengths. Unfortunately, most simple glasses do not have suitable isotopes. It might be possible to use a combination of X-ray, neutron and electron diffraction, but there are considerable experimental difficulties with electron diffraction studies of bulk glass structure and, in any case, the scattering lengths of atoms for X-rays and electrons are closely related. Both neutron and X-ray diffraction, however, have now been applied to a number of glasses. The problem remains of attempting to determine the structure with inadequate data and it is therefore necessary to use some kind of model. The availability of two separate sets of scattering data will considerably narrow the range of possible models without, of course, permitting a unique solution. For most simple stoichiometric glasses it is possible to draw extensively on a knowledge of the crystal chemistry of the material (e.g. the prevalence of $SiO_4$ tetrahedra in the various silica polymorphs). We shall not discuss the various model approaches which might be used, but shall deal briefly with one particular model which will exemplify the problem and also has relevance to the interpretation of inelastic scattering data.

The principal role which any model must fulfil is to account for the lack of long-range order, that is, to reproduce the more or less gradual approach of the correlation functions $d_{ij}(r)$ to zero with increasing $r$. A simple device for achieving this is to use, in eqn (7.8), the correlation functions for a crystalline modification of the glass in question and to modify this with a function $F(r)$ which is unity at low $r$ and zero above some value $r = L_0$, defined as the maximum correlation length for the structure. A clear physical interpretation of this quasi-crystalline model is possible: it is the scattering (per unit of composition) from a spherical crystallite of diameter $L_0$ set in a structureless matrix of the same

STRUCTURE AND ATOMIC MOTION IN GLASSES

average density and averaged over all orientations of the crystal with respect to **Q**. $F(r)$ then has the analytical form (Leadbetter and Wright 1972)

$$F(r) = (r/L_0 - 1)^2 (r/L_0 + 2)/2.$$

This extremely simple type of model has been successfully used for liquid metals (Kaplow, Strong, and Averbach 1965) and its application to glasses has been discussed in more detail by Leadbetter and Wright (1972). It is not expected

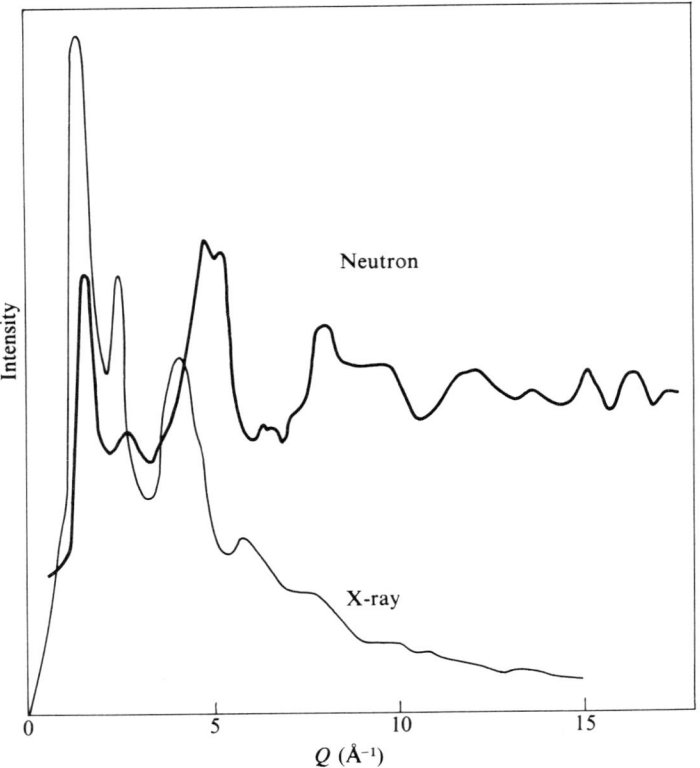

FIG. 7.1. X-ray and total neutron scattering intensities from vitreous $GeO_2$.

to give a complete description of the structure of even a simple glass, but it should, however, provide a simple conceptual basis from which to discuss the structure. Thus the component $Qi_{ij}(Q)$ functions and their corresponding correlation functions are obtainable separately. Furthermore, by first calculating the intensity, then Fourier transforming this under precisely the same conditions as for the experiment, convolution effects due to the finite upper limit of $Q$ or to the $Q$-dependence of the X-ray atomic form factors may readily be accounted for.

As an example of a structural investigation using both X-ray and neutron diffraction and the above simple model, we shall consider the case of vitreous $GeO_2$. In Fig. 7.1 are shown the X-ray and neutron intensities (total scattering) and in Fig. 7.2 the total correlation functions $D(r)$ derived from these data (eqn (7.9)). These correlation functions are consistent with a structure based on $GeO_4$ tetrahedra with a Ge–O distance of 1·74 Å. Also included in Fig. 7.2 are the component correlation functions $d_{ij}(r)$ calculated from a quasi-crystalline mode based on the quartz structure with a maximum correlation length $L_0 = 10·5$ Å. The effect is shown of increasing the separation of three of

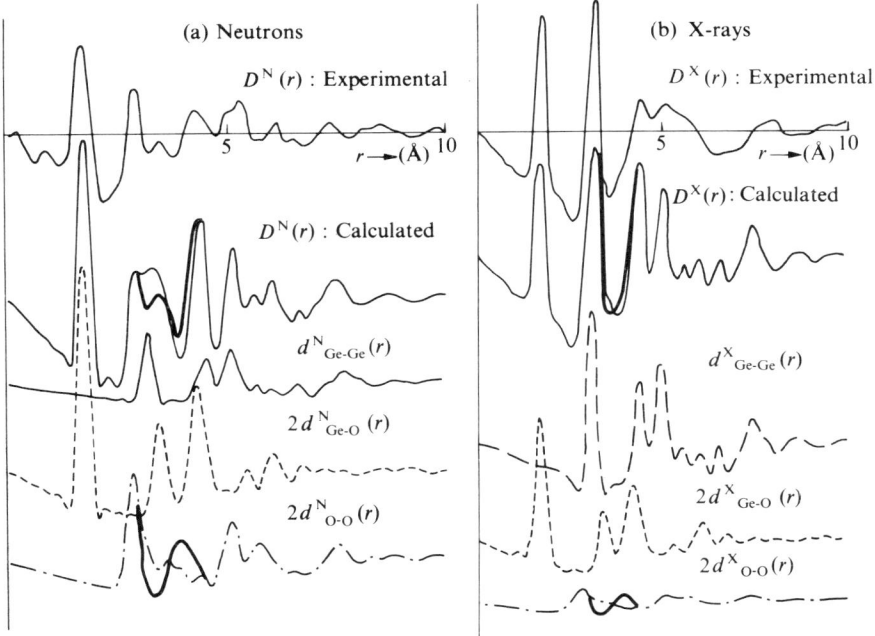

FIG. 7.2. Neutron (a) and X-ray. (b) correlation functions for vitreous $GeO_2$. Neutron experimental data are from Lorch (1969) (see also Ferguson and Haas 1970) and X-ray data from Leadbetter and Wright (1972). Model calculations are based on the quartz structure with $L_0 = 10·5$ Å. The effect on the $d_{O-O}(r)$ function and on the total correlation function $D(r)$ of a slight rearrangement of O atoms is indicated by heavy lines; this significantly improves the agreement with experiment (see text). Note that the neutron $D(r)$ functions for model and experiment are drawn on different scales.

the eight O—O pairs between 3 and 4 Å by about 0·7 Å: this significantly improves the agreement with experiment of the resultant total correlation function. This comparison confirms that vitreous $GeO_2$ is composed of nearly regular tetrahedra, and that their short range organization differs slightly from that of the quartz crystal modification in a direction towards a more isotropic structure such as the cubic (cristobalite) modification. The simple model still retains too much order at higher distances; the additional disorder in the real glass arises mainly from a randomization in orientation of tetrahedra, due partly to a distribution of Ge-O-Ge angles about an average value of $\sim 133°$ and more particularly to relative rotation of tetrahedra with this intertetrahedral angle.

The dangers of too simple an interpretation of a single scattering experiment are evident from Fig. 7.2. The neutron correlation function peak at 3·45 Å was originally interpreted as arising from Ge-Ge atom pairs, giving a Ge-O-Ge bond angle of 180° compared with about 133° suggested by a similar interpretation of the X-ray data. This simple interpretation happens to be more nearly correct for the X-ray case: for the neutron experiment the first Ge-Ge peak is not in fact seen and the peak in $D(r)$ at 3·45 Å is a composite dominated by Ge-O and O-O pairs.

## 7.3. Atomic motions

### 7.3.1. General remarks

The dynamics of the atomic motions in a glass as in a liquid or a crystal, may be described in terms of the space-time correlation functions $G_s(\mathbf{r}, t)$ and $G_d(\mathbf{r}, t)$. For the glass these will correspond at all times to the short-time liquid behaviour as shown in Fig. 6.15 of Chapter 6, reflecting the fact that the atoms are constrained to vibrate, more or less harmonically, about their quasi-equilibrium sites. Within the limits of the harmonic approximation, therefore, it is possible rigorously to define proper normal modes of vibration, and a frequency distribution $Z(\omega)$ for these modes. $Z(\omega)$ will in fact be temperature and volume dependent (because of the anharmonicity of the vibrations) and will determine the thermal properties of the glass exactly as for a crystal.

The *character* of the modes in a glass is a separate and more difficult problem. In the low frequency, long wave limit, they will be elastic plane waves and the excitations will be properly describable as phonons just as for crystals, except that the waves will be more heavily damped (the phonon lifetime will be shorter) than for the crystal. At higher frequencies and shorter wavelengths, the modes will in general no longer be plane waves and they will be more localized than the corresponding crystal modes (Bell and Dean 1970, 1971). The important question is whether the collective motions, or vibrational excitations, may be described in any simplified (e.g. quasi-phonon) manner rather than by knowing the individual displacements of each atom for every mode. A quasi-phonon description of the high-frequency (acoustic) excitations in liquids is useful (Cocking 1967, Randolph and Singwi 1966), so that it might be expected to be even more so for a glass, at least for the acoustic modes. The physical idea here is that the collective motions can still be resolved into a superposition of plane wave contributions but these will no longer be pure normal modes. Each wave will contain contributions from a number of modes, so that a given wave vector is associated with a band of frequencies, and vice versa. These mode contributions will change as the wave propagates, corresponding to intrinsic scattering by the disorder in the glass.

Such a description will be useful if the phonon lifetimes determined by this disorder are longer than about a vibration period and this question must be answered by experiment. There is finally the question of how important in glasses are highly localized and resonance vibrations.† These might be expected to be of some importance because the disorder in a glass will not be homogeneous, as envisaged in the simple random network theory of glass structure, but will be characterized by regions of locally greater disorder corresponding, for example, to the vacancies, interstitials or even dislocations of crystal lattices.

### 7.3.2. *The frequency distribution–incoherent scattering and the incoherent approximation*

If a glass contains enough hydrogen, then the observed scattering will be effectively completely incoherent. In this case it will be possible to determine the frequency distribution of the modes of vibration, but not the character of the individual modes. The experiments may be interpreted quite generally in terms of eqn (1.24b) of Chapter 1 but more usefully using the phonon formulation of eqn (1.34), which may be modified for the general solid by labelling the modes individually rather than by branch ($j$) and wave vector (q) which are explicitly crystal language. The main point is that in the case of purely incoherent scattering there is little in the formalism for interpreting experimental data which might distinguish glass from crystal. Both are related in a similar way to the frequency distribution, although this might differ for the different phases. However, no comparative experiments of crystalline and glassy phases of organic materials (e.g. glycerol) have been reported, but only work on molecular crystals or liquids. While it seems unlikely that there will be significant differences in $Z(\omega)$ at higher frequencies this is not necessarily true at low frequencies where the acoustic branches of the spectrum are all-important.

There have of course been many studies of incoherent inelastic scattering on polymers, but these have been concerned with the dynamics of the polymer chain or side groups themselves, rather than the effect of long range organization of the chains which will only be of secondary importance (see Chapter 5). Most materials of interest as glasses are almost purely coherent scatterers, but under certain circumstances eqn (1.34) may still be used to interpret the results. Thus the coherent inelastic scattering cross-section is related to the total correlation function $G(\mathbf{r}, t)$ which may be separated into self and distinct parts according to eqn (1.27):

$$G(\mathbf{r}, t) = G_s(\mathbf{r}, t) + G_d(\mathbf{r}, t).$$

For experiments at high $Q$, the important part of $G_d(\mathbf{r}, t)$ is that at low $r$. Here, $G_d(\mathbf{r}, t)$ tends to zero because only the origin atom can be found at later times close to its position at $t = 0$, as all atoms are constrained to vibrate about their quasi-equilibrium sites and do not approach closer than the normal interatomic spacings ($r_0$). Hence, for $Q \gtrsim 2\pi/r_0$, $G(\mathbf{r}, t)$ approaches $G_s(\mathbf{r}, t)$, $S(\mathbf{Q}, \omega)$ approaches

† For a description of these motions see Maradudin (1968).

## STRUCTURE AND ATOMIC MOTION IN GLASSES

$S_s(\mathbf{Q}, \omega)$ (eqns (1.25a) and (b)), and $d^2\sigma_{coh}/d\Omega\, dE'$ approaches the form of $d^2\sigma_{incoh}/d\Omega\, dE'$ but with $(\langle b^2 \rangle - \langle b \rangle^2)$ replaced by $\langle b^2 \rangle$, which is identical with $\langle b \rangle^2$ for a purely coherent scatterer. Labelling the individual modes $j$, as indicated above, eqn (1.34) may then be written

$$\frac{d^2\sigma_{incoh}}{d\Omega\, dE'} = \frac{k'}{k} \sum_j \delta(\omega \mp \omega_j) \frac{(n_s + \tfrac{1}{2} \pm \tfrac{1}{2})}{2\omega_j}$$

$$\times \sum_\rho \frac{\langle b_\rho^2 \rangle}{M_\rho} |\mathbf{Q} \cdot \mathbf{U}_\rho(j)|^2 \, e^{-2W_\rho}. \qquad (7.10)$$

The term in the cross-section containing $G_d(\mathbf{r}, t)$ describes the effect of interference of neutron waves scattered by different atoms; our assumption that this term is zero is known as the 'incoherent approximation'.

Eqn (7.10) may be simplified by assuming that chemically equivalent atoms are dynamically equivalent. This will be a good approximation for a glass like $SiO_2$ or $GeO_2$ but not for a complex molecule where, for example, the dynamics of H atoms bonded to different groups may be very different. The averaging over all orientations of $\mathbf{Q}$ with respect to $\mathbf{U}$ may be done by writing $\langle [\mathbf{Q} \cdot \mathbf{U}_\rho(j)]^2 \rangle = \tfrac{1}{3} Q^2 \langle U_\rho^2(\omega) \rangle$ where $\langle U_\rho^2(\omega) \rangle$ is an average taken at constant frequency over all atoms of a given species and all orientations. In general $\langle U^2 \rangle$ will be different for different atomic species and the relative amplitudes will be frequency-dependent. Finally, the summation over the modes can be transformed to an integration over the frequency distribution (cf. §1.5.4) to give

$$\frac{d^2\sigma_{incoh}}{d\Omega\, dE'} = \frac{k'}{k} \frac{Q^2 (n_s + \tfrac{1}{2} \pm \tfrac{1}{2})}{2\omega} Z(\omega) N \sum_{\rho=1}^{p} \frac{\langle b_\rho^2 \rangle}{M_\rho} \langle U_\rho^2(\omega) \rangle e^{-2W_\rho} \qquad (7.11)$$

where the summation is over the $p$ atoms in a unit of composition ($GeO_2$, $SiO_2$, $BeF_2$, etc.) The lower sign $(-)$ in $(n_s + \tfrac{1}{2} \pm \tfrac{1}{2})$ applies to the case of neutron energy gain and the upper sign $(+)$ to neutron energy loss, where $n_s = [\exp(\hbar\omega_j/k_B T) - 1]^{-1}$.

The experiment thus gives $Z(\omega)$ weighted by a frequency-dependent term $\dfrac{\langle b^2 \rangle}{M} \langle U^2(\omega) \rangle e^{-2W}$ for each atom in the composition unit. It is therefore not possible to determine $Z(\omega)$, even given the incoherent approximation, unless the relative amplitudes $\langle U^2(\omega) \rangle$ of the different atoms are known.

Two examples of experimental results are given below: these are the data on vitreous $GeO_2$ obtained by neutron energy-gain (phonon destruction) scattering of cold neutrons (Leadbetter and Litchinsky 1970), and the measurements on vitreous $SiO_2$ by neutron energy-loss. We may define, quite generally, from eqn (7.11) for incoherent scattering or from eqn (1.33a) for coherent scattering, a function $g(Q, \omega)$ in terms of experimentally known quantities:

$$g(Q, \omega) = \frac{d^2\sigma}{d\Omega\, dE'} \frac{k}{k'} \frac{2\omega}{Nn_s Q^2}. \qquad (7.12)$$

Coherence effects show up as a $Q$ dependence of $g(Q, \omega)$ (see §7.3.3) and if the incoherent approximation is valid then

$$g(Q, \omega) = Z(\omega) \left[ \left\{ \frac{\langle b \rangle^2}{M} \langle U^2(\omega) \rangle e^{-2W} \right\}_{Ge} + 2 \left\{ \frac{\langle b \rangle^2}{M} \langle U^2(\omega) \rangle e^{-2W} \right\}_0 \right] .$$

The results of experiments with neutrons of incident energy ~4meV, energy-analysed by time-of-flight techniques is shown in Fig. 7.3 for three scattering angles. Different scattering angles correspond to different $Q$ values for the same energy transfer $\hbar\omega$, so that the differences in $g(Q, \omega)$ at low $\omega$ show that the

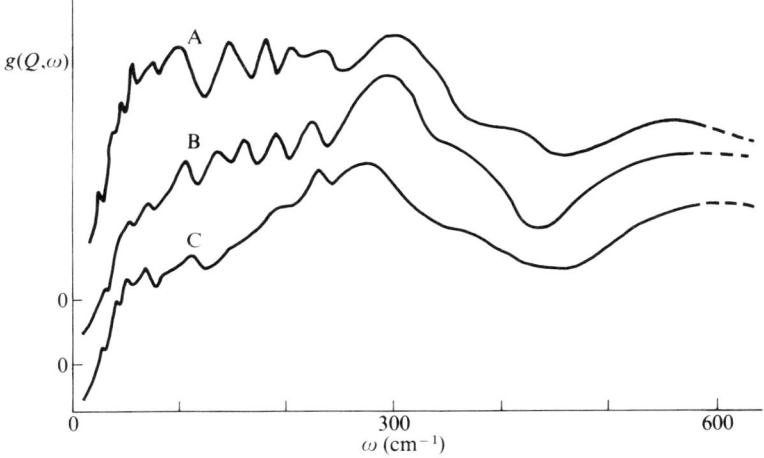

FIG. 7.3. $g(Q, \omega)$ versus $\omega$ for vitreous $GeO_2$ from cold neutron time-of-flight experiments at scattering angles of 30, 60 and 90° (curves A, B and C respectively). The curves are displaced vertically for clarity.

incoherent approximation is here inadequate. With increasing $\omega$, which also means higher $Q$, the results for different scattering angles tend towards equality, showing that the incoherent approximation becomes more reasonable. Thus the main features of $Z(\omega)$ are obtained, at least for $200 \lesssim \omega \lesssim 600$ cm$^{-1}$, although without knowledge of the $\langle U_\rho^2(\omega) \rangle$ the detailed shape of $Z(\omega)$ is not determined. The amplitude values, however, may probably be obtained sufficiently accurately from model calculations such as those of Bell and co-workers. Unfortunately, because of the influence of the phonon occupation number $n_s$ (which varies as $1/\omega$ at high frequencies) and the decreasing energy resolution with increasing energy transfer (decreasing time-of-flight), the scattered intensities in these energy-gain experiments were inadequate to give $g(Q, \omega)$ for $\omega \gtrsim 600$ cm$^{-1}$.

Data may be obtained at high frequencies by an energy-loss technique in which relatively high energy neutrons, selected by a crystal monochromator, are scattered from the sample through a Be filter, so that only neutrons of energy less than ~5 meV (~40 cm$^{-1}$) are detected (Chapter 2). Results on vitreous silica obtained in this way (Leadbetter and Stringfellow 1972) are shown in Fig. 7.4. The cross-sections were converted to $g(Q, \omega)$ using eqn (7.12) with $n_s$ replaced by $n_s + 1$, and they clearly show the three main peaks in $Z(\omega)$ near

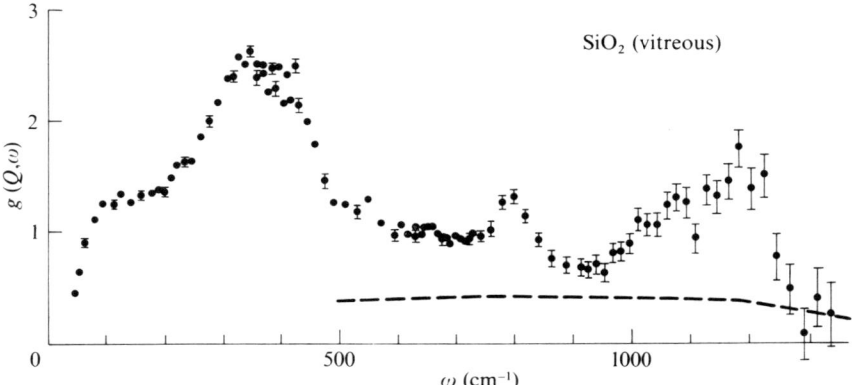

FIG. 7.4. $g(Q, \omega)$ versus $\omega$ for vitreous SiO$_2$ using the Be-filter, down-scattering technique. The dashed line shows the estimated two-phonon contribution.

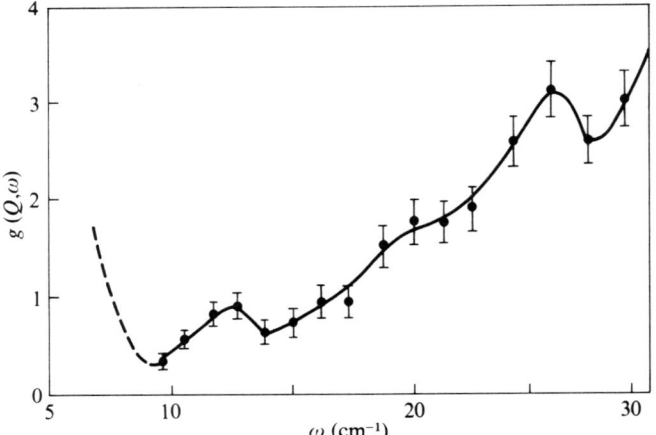

FIG. 7.5. $g(Q, \omega)$ versus $\omega$ for vitreous GeO$_2$ at very low frequencies.

350, 800 and 1100 cm$^{-1}$ which are expected from theoretical and optical spectroscopic work.

At the other extreme of very low frequencies the cold neutron scattering results on vitreous GeO$_2$ suggest the possibility of a band of resonance modes near 13 cm$^{-1}$. The peaks in the cross-section are very small and the results shown in Fig. 7.5 were obtained by summing the data over all angles to improve the statistical precision.

The features in the cross-section near 20 and 26 cm$^{-1}$ are coherent peaks arising from the lowest-energy acoustic branch of the spectrum, but the 13 cm$^{-1}$ peak cannot be explained in this way. The probable existence of a band of modes at about 13 cm$^{-1}$ is also indicated by low-temperature heat capacity data as was shown in research by Wycherley (1969), and the most likely origin of these modes is resonance-type vibrations associated with defect regions in the structure.

### 7.3.3. Coherent scattering

The problem of interpreting coherence effects in the inelastic scattering of neutrons from glasses is very similar to that for liquids at high enough energy transfer (short time behaviour). A substantial amount of experimental work has been done on coherently scattering simple liquids but relatively little so far on the much more complex glassy systems. Coherent scattering is related to the relative motions of atoms and, since no general theory of the collective motions in liquids or glasses exists there is no theory of coherent scattering for these materials comparable with that for crystals. For glasses, however, the nature of the collective motions in the long wavelength limit is clear: here there exist transverse and longitudinal sound waves, just as for a crystal, and these have been observed for example in Brillouin light scattering experiments (Flubacher et al.(1959). The microscopic disorder in the glass will cause the waves to be damped and this damping will increase as the wavelength decreases. The situation is somewhat analogous to that in a polycrystal of small crystallite size in which a propagating wave will be damped as it traverses regions of differing elastic properties. This damping, or decrease of phonon lifetime, will become more serious as the wave length of the excitations approaches the size of the interatomic spacing, when the nature of the excitation will be very directly influenced by the atomic disorder characteristic of the glass. The question of the most useful description of the collective excitations away from the long wave limit is unlikely to have any single answer as it will depend on the particular glass and on the frequency (or wavelength) of the excitation. The answer must be determined ultimately by experiment.

Consideration of the coherent scattering from a polycrystal provides a useful starting point both for understanding the nature of the problem and for extrapolation to the glass or liquid. The one-phonon scattering from a single crystal is governed by the conditions of conservation of energy and wave vector,

$$\hbar\omega = \pm \hbar\omega_q^s \quad \text{and} \quad Q \pm q = \tau$$

the latter of which is illustrated in Fig. 7.6(a) for a particular phonon, propagating in the [010] direction. Fig. 7.6(b) is a typical phonon dispersion curve for the same direction of propagation, illustrating the results obtained from single-crystal experiments, reduced to the first Brillouin zone. For a polycrystal, $Q$ can now take all orientations relative to the crystal, as illustrated

# STRUCTURE AND ATOMIC MOTION IN GLASSES

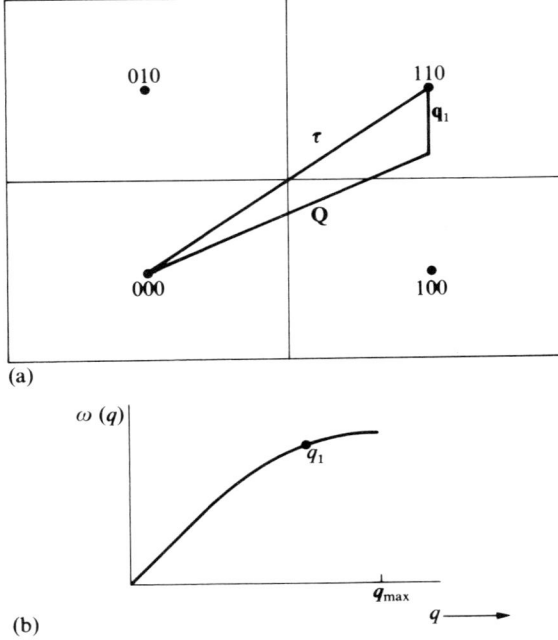

(a)

(b)

FIG. 7.6. (a) Illustration of one-phonon scattering in a single crystal. The scattering is in the (001) plane and is associated with a phonon of wavevector $\mathbf{q}_1$ in the [010] direction. The rectangles represent a cross-section through the first Brillouin zone centred on each reciprocal lattice point. (b) Dispersion curve for modes propagating along [010].

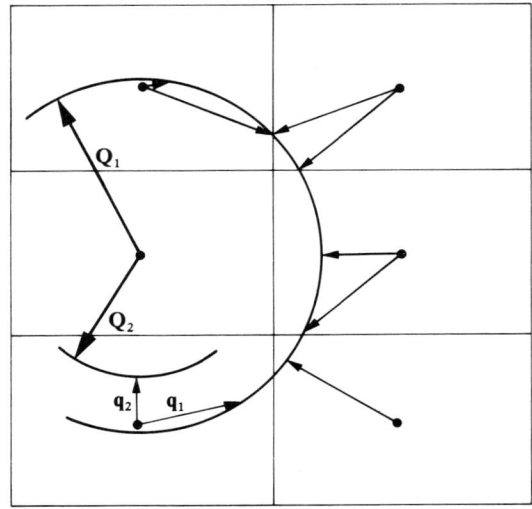

FIG. 7.7. Representation of one-phonon scattering from a polycrystal. The diagram shows a plane of the reciprocal lattice and the wavevectors of some of the phonons which may be observed with a given scattering vector $\mathbf{Q}$.

160

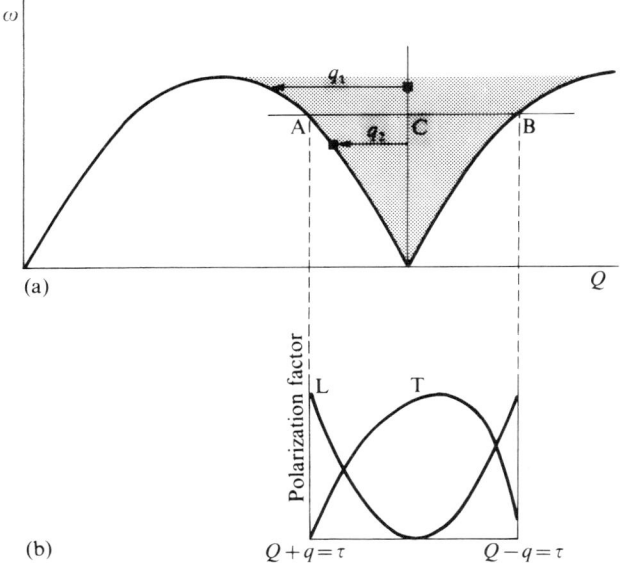

FIG. 7.8. (a) Schematic $\omega(Q)$ diagram for a polycrystal showing a single, isotropic, acoustic branch and only one reciprocal lattice vector. In a real polycrystal the $\omega(Q)$ relations will be bands due to anisotropy, and dispersion curves based on more than one reciprocal lattice point will be superimposed. The area of allowed scattering is denoted by shading (see eqn (7.13)) and the wavevectors of two particular phonons corresponding to those labelled $q_1$ and $q_2$ in Fig. 7.7 are shown. For these cases scattering peaks will be observed at $Q_1 = \tau$ and $Q_2 = \tau - q_2$. (b) The polarization factors for the scattering intensities from transverse (T) and longitudinal (L) components of the mode polarization vectors.

in Fig. 7.7, and only its magnitude is important. Thus, the wavevector conservation condition is modified because scattering involving a given reciprocal lattice vector $\tau$ is now allowed for all magnitudes of **q** (and corresponding directions, see Fig. 7.7) consistent with

$$Q + q \geqslant |\tau| \geqslant Q - q. \tag{7.13}$$

Here $q \leqslant q_{\max}$, where $q_{\max}$ is determined by the size of the first Brillouin zone. (The meaning of eqn (7.13) may be seen most easily by considering what phonon wavevectors in Fig. 7.7 are allowed as **Q** increases in magnitude.) The region of the $\omega$-**Q** plane where scattering is allowed by eqn (7.13) is indicated in Fig. 7.8(a). This widening of the allowed values of $q$, together with that resulting from the anisotropy of the dispersion relations which will make the $\omega(Q)$ lines of Fig. 7.8 into bands, plus the possibility that scattering at a given $Q$ may involve more than one reciprocal lattice point, obviously make the scattered spectrum much more complex than for the single crystal.

For not too high $Q$, where only one or two reciprocal lattice points contribute to the scattering, it may still be possible to determine dispersion relations averaged over all crystal directions. This, in fact, is much easier for the longitudinal than for the transverse branches. Thus, the intensity of scattering is determined by the dynamic structure factor (see eqn (1.33b)) of which the important part here is the polarization term $\mathbf{Q} \cdot \mathbf{U}_\rho^j(\mathbf{q})$. For pure transverse and longitudinal components this takes the form $QU \sin \phi$ and $QU \cos \phi$ respectively where $\phi$ is the angle between $\mathbf{Q}$ and the direction of polarization. Hence, the intensity of scattering will be proportional to $\sin^2 \phi$ for the transverse and to $\cos^2 \phi$ for the longitudinal components. The detailed intensity profile for an experimental scattering locus such as AB in Fig. 7.8(a) will depend on the relationship between $Q$ and $\phi$ for the allowed scattering, and this will be determined by the geometry of the reciprocal lattice and the shape of the constant energy surfaces in reciprocal space. However, the profile will be of the general form indicated in Fig. 7.8(b) which means that pure longitudinal modes will give maximum intensity where $Q = \tau \pm q$ (e.g. points A and B in Fig. 7.8(a)) and pure transverse modes will give maximum intensity at $Q \approx \tau$. In the former case peaks in the experimental cross-section (e.g. $q_2$ in Fig. 7.8(a)) will give an average dispersion relation while in the latter case (e.g. $q_1$ in Fig. 7.8(a)) they will only give the position of the maximum in $Z(\omega)$ corresponding to the top of the TA branch. On the other hand, the positions of the intensity cut-offs (where there is a sharp change in intensity) as functions of $Q$ and $\omega$ might, in favourable cases, give an indication of the shape of the TA branch (cf. Cocking 1967). The polycrystalline scattering will be further complicated because modes will not in general be purely transverse or longitudinal and furthermore the resolved components of the polarization vector parallel and perpendicular to $\mathbf{q}$ will depend on $|\mathbf{q}|$. These effects will be particularly important for the relatively complex crystals, such as quartz, which are the crystalline modifications of the simple glasses and hence, by implication, for the glasses themselves. In such cases the intensity profile along a locus such as AB will be some weighted sum of functions like the pure T and L factors shown in Fig. 7.8(b).

The above general ideas may be applied directly to liquids and glasses. Thus in the experimental cross-sections, or in the corresponding $S(Q, \omega)$ functions, coherence effects will result in the positions of peaks and cut-offs being functions of $\omega$ and $Q$. The data may then be examined to see if any pattern in the $\omega$-$Q$ plane emerges which can be interpreted in terms of the general ideas derived from the consideration of polycrystals. Cocking (1967) has reviewed work of this sort on simple liquids and its application to glasses will be discussed below. Before doing this it will be useful to consider briefly attempts at a more quantitative treatment of coherent scattering from simple liquids, as these provide further insight into the nature of the coherence effects to be expected in glasses.

Egelstaff (1962, 1963) derived a formula for the scattering law for a liquid by direct extrapolation from the polycrystal formulation. He first obtained an approximate expression for the polycrystalline cross-section (for a Bravais lattice) by carrying out the averaging over all orientations of $\mathbf{q}$ and $\boldsymbol{\tau}$ about $\mathbf{Q}$, assuming an isotropic Debye-type frequency spectrum ($\omega = v_s q$).† The expression is

$$S(Q, \omega) = S_s(Q, \omega) Z(Q, \omega), \tag{7.14}$$

or, more fully,

$$S(Q, \omega) = [S_s(Q, \omega) Z]_L + [S_s(Q, \omega) Z]_T, \tag{7.15}$$

where

$$Z_{L,T} = \sum_\tau \frac{\pi F_\tau}{2BQq\tau} (\cos^2 \phi, \sin^2 \phi), \tag{7.16}$$

with the subscripts L, T referring to longitudinal and transverse polarizations. $F_\tau$ is the structure factor for the planes $\tau$, $B$ is the volume per atom, separate $Z$ factors are used for the two polarizations and $S_s(Q, \omega)_{L,T}$ is the incoherent scattering law for modes of given polarization. Even for a simple Bravais lattice this result is only approximate (de Wette and Rahman 1968) and no attempt has been made to extend it to more complex crystals. Nevertheless, it may usefully be extended to simple liquids by replacing the summation over $\tau$ by an integration over the continuous structure factor $S(Q)$ (which is simply the total diffraction pattern less the independent scattering). The physical assumption involved here is that the peaks in $S(Q)$ act in the scattering process like smeared-out reciprocal lattice points. The result is identical in form to eqn (7.14), with

$$Z_{\text{liquid}} = \frac{1}{2Qq} \int_{\tau_{\min} = Q-q}^{\tau_{\max} = Q+q} QS(Q) \, dQ \, (\cos^2 \phi, \sin^2 \phi). \tag{7.17}$$

This result is formally similar to that resulting from Vineyard's convolution approximation discussed in Chapter 6 (§ 6.4), namely,

$$S(Q, \omega) = S_s(Q, \omega) S(Q), \tag{7.18}$$

with $Z_{\text{liquid}}$ replacing $S(Q)$.

Singwi (1964/5) has modified the convolution approximation to obtain a result very similar to Egelstaff's. The basic assumption of the convolution approximation is that the motion of two atoms is independent. Singwi's modification is to assume that this is true only for atoms separated by distances greater than some correlation range $R$. Within this correlation range the relative motion is assumed to be described by crystal-like modes. The final result may be written

$$S(Q, \omega) = [S(Q) + S'(Q, \omega)] S_s(Q, \omega)$$

† $v_s$ is the sound velocity.

## STRUCTURE AND ATOMIC MOTION IN GLASSES

where $S'(Q, \omega)$ is a complicated function, even when evaluated for a Bravais lattice and assuming an isotropic Debye-type frequency spectrum, with no account of mode polarization. The physical basis of this approach is attractive because there indeed exists in a liquid or glass a 'correlation length' $L_0$ beyond which no correlation in atom positions is detectable in scattering experiments (see §7.2). It is therefore to be expected that coherence effects in inelastic scattering will be unimportant for atoms separated by distances greater than about $L_0$, so that it is only necessary to know the relative motions of atoms separated by distances less than around 10-20 Å.

For $R \to \infty$ it may be shown (Cocking 1968) that $S(Q) + S'(Q, \omega) = Z_{\text{liquid}}$, if polarization terms are omitted. The Egelstaff and Singwi results are thus essentially similar but the latter involves no assumption about an effective reciprocal lattice in the liquid (or glass). Conversely, this agreement may be taken to show that coherence effects in liquids and glasses are indeed describable in terms of the structure factor $S(Q)$ acting as a continuum of reciprocal lattice vectors in the scattering process (eqn (7.13)).

The detailed models discussed above have been applied to a number of simple liquids (Cocking 1968; Randolph and Singwi 1966; Dahlborg and Larsson 1965; Sköld and Larsson 1967). Although the models contain too many approximations or unknown parameters to determine useful quantities such as the correlation range $R$, with any degree of confidence, a number of general conclusions emerge which are directly applicable also to glasses. These are:

(a) Only relatively short-range correlations ($R \gtrsim 20$ Å) in the motions of the atoms are important and these may usefully be described in terms of 'lattice' waves (hence phonons).

(b) Polarization of the modes is important, and although transverse modes sometimes contribute only slightly to the scattering in liquids, they should be much more important for rigid glasses.

(c) The broadening in the experimentally observed dispersion ($\omega$ versus $Q$) relations arising from the disorder appears to be primarily accounted for by using a continuum of $Q$-values (weighted by the structure factor $S(Q)$) in place of the reciprocal lattice vectors in the wavevector conservation condition (7.13). The size of the correlation range plays some part in determining these widths, or phonon lifetimes, and the disorder within the correlation sphere itself does not appear to be an important factor. Note, however, that the use of a correlation range is itself only a simple means of accounting for the disorder in the structure (§7.2).

For glasses the basic structures are considerably more complex than for the simple liquids discussed above and it is likely, by analogy with the crystal, that anisotropy will play a major role in broadening the average dispersion relations. Furthermore, this complexity of structure will imply that most of the modes in the glass will be analogous to the modes in the closely spaced optical branches of the crystal (e.g. 24 optical branches for the quartz structure). It seems highly unlikely that coherence effects in the scattering from these modes will be interpretable in any simple way and we should only expect this to be at all possible for the acoustic branches, that is at relatively low frequencies.

The experimental consequences of the above discussion of coherence effects in glasses is illustrated schematically in Fig. 7.9. This shows the shape and width of a single average acoustic branch in a disordered solid, together with the experimental scattering law expected for various $\omega(Q)$ loci. Curves 1–4 in (b) represent

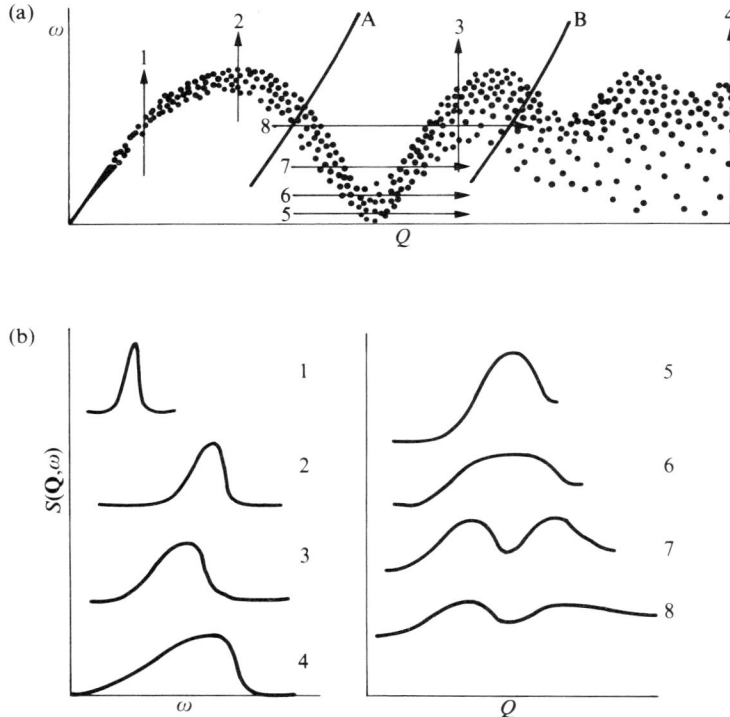

FIG. 7.9. (a) Schematic representation of a dispersion $\omega(Q)$ relation for an acoustic branch in a liquid or glass. (b) The form of the scattering law expected for various experimental loci (1 to 8) through this curve.

the results of experimental scans at constant $Q$, showing the increasing broadening of $S(Q, \omega)$ with increasing $Q$, and curve 4 represents the situation where the incoherent approximation is good. Curves 5–8 are constant $\omega$ scans; 5 and 6 represent increasingly broadened versions of the first peak in $S(Q)$, while 7 and 8 show the dispersion curves on each side of this peak. Experimentally, one determines $S(Q, \omega)$ for certain scattering loci in the $\omega$–$Q$ plane and investigates to what extent they correspond to a pattern like that of Fig. 7.9(a), remembering of course that this represents only one branch of the spectrum.

## STRUCTURE AND ATOMIC MOTION IN GLASSES

The most extensively investigated disordered 'solid' is liquid lead (we noted earlier that at sufficiently high frequencies a liquid behaves much like a solid). The results of a number of investigations have been summarized by Randolph and Singwi (1966), from whose work Fig. 7.10 is taken. Here the longitudinal branch is dominant and the data clearly resemble the schematic picture of Fig. 7.9(a).

Four inorganic glass-forming materials have now been investigated by inelastic neutron scattering, all of them by cold neutron time-of-flight techniques, which give experimental loci like the lines labelled A and B in Fig. 7.9(a).† The simplest glass studied is Se, for which no investigation of coherence effects was reported (Axmann *et al.* 1970), but a high scattering cross-section at very low energy-transfers strongly suggests the presence of an appreciable number of 'soft' modes in the glass which are not observed in the crystal. The other glasses studied

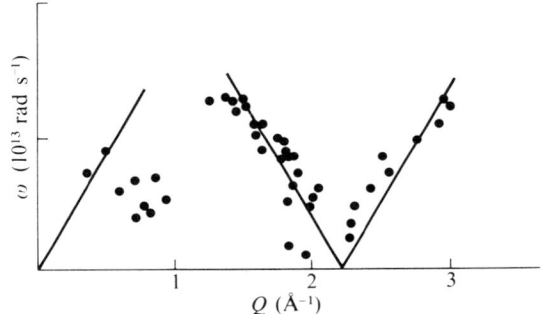

FIG. 7.10. Experimental $\omega(Q)$ relation for liquid lead (after Randolph and Singwi 1966). Data also from Cocking and Egelstaff (1965) and Dorner *et al.* (1965). Sound velocity lines are shown.

are $SiO_2$ (Egelstaff 1965; Leadbetter 1969), $BeF_2$ (Leadbetter and Wright 1970) and $GeO_2$ (Leadbetter and Litchinsky 1970). These are all of the three-dimensional network type and in each case results have also been obtained for one or more (poly)crystalline modifications. Typical spectra for $GeO_2$ and $BeF_2$ are shown in Fig. 7.11. For both polycrystal and glass specimens, coherence effects give patterns in the cross-sections which may be simply (if incompletely) interpreted in terms of the discussion of this section, particularly remembering that the polarizations will in general not be purely transverse or longitudinal, so that peaks might be expected in $S(Q, \omega)$ at positions corresponding to points A, B and C of Fig. 7.8. Thus, low frequency peaks in the cross-sections (Fig. 7.11), together with the known sound velocities and crystal structure, enable $\omega(Q)$

---

† For the time-of-flight technique the experimental loci occur in directions like those shown for the lines A and B in Fig. 7.9(a). Constant $Q$ scans, lines 1 to 4 in Fig. 7.9(a), or constant $\omega$ scans, lines 5 to 8, are possible with the triple-axis method (see § 2.4).

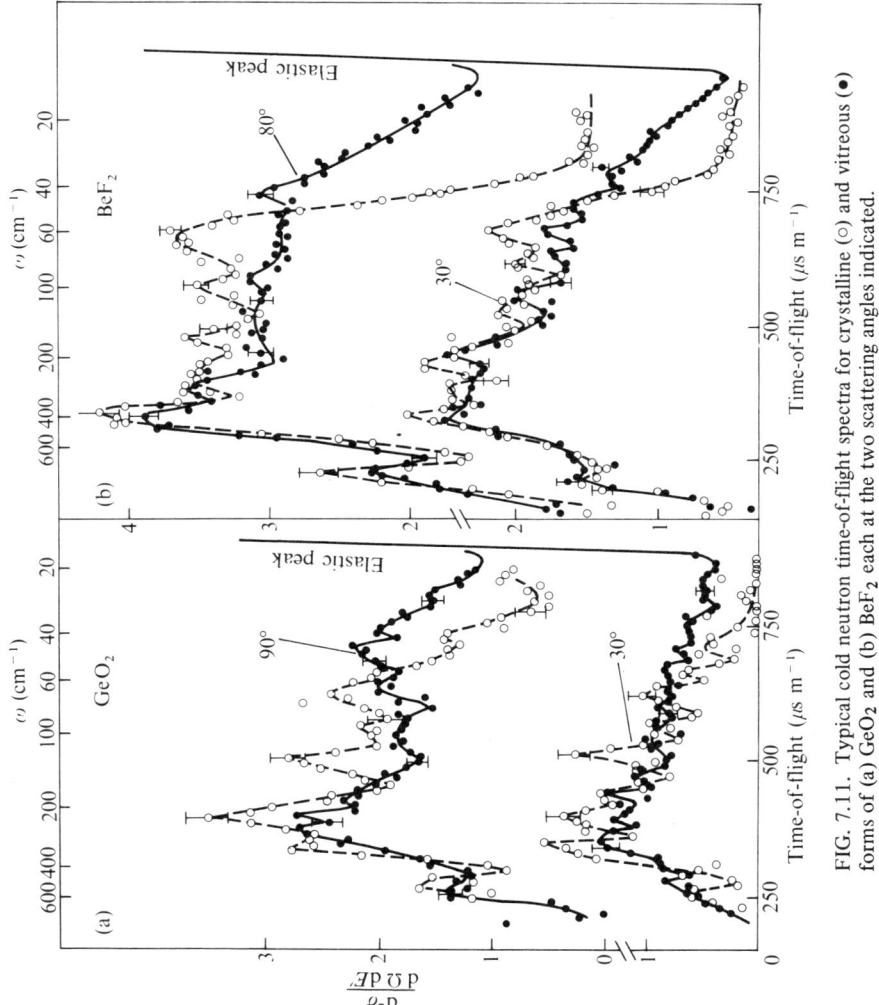

FIG. 7.11. Typical cold neutron time-of-flight spectra for crystalline (○) and vitreous (●) forms of (a) $GeO_2$ and (b) $BeF_2$ each at the two scattering angles indicated.

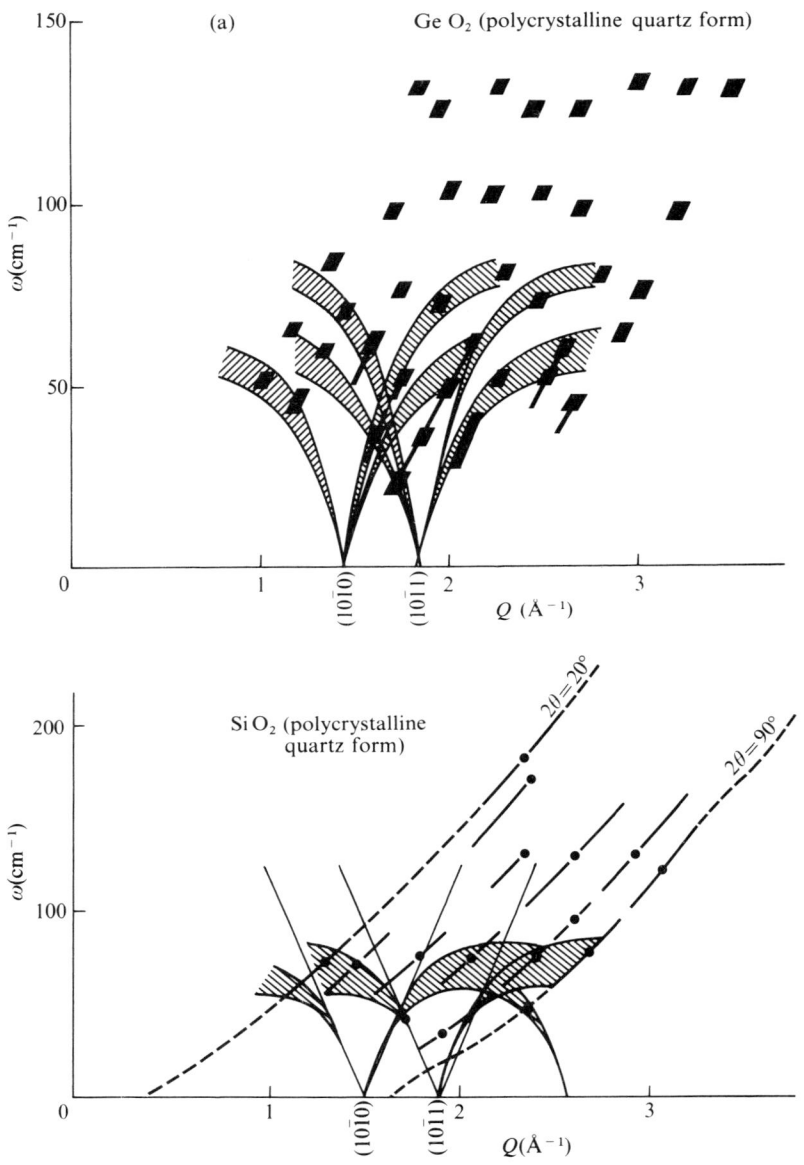

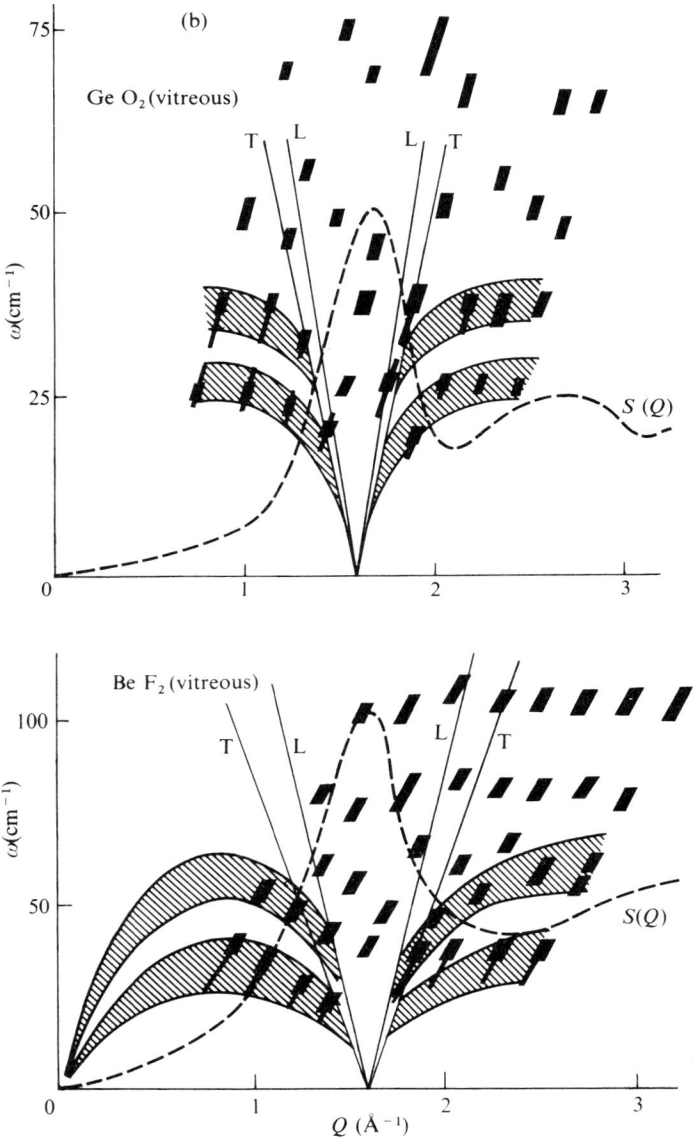

FIG. 7.12. $\omega(Q)$ data for (a) polycrystalline quartz forms of $SiO_2$ and $GeO_2$ and (b) vitreous forms of $GeO_2$ and $BeF_2$. Points and bars denote peak positions in spectra like those of Fig. 7.11, and bars extend to the half-height positions. The sound velocities are represented by lines, which radiate from the positions of the reciprocal lattice vectors for the crystals, or structure factor maxima for the glasses. (T is the transverse sound velocity and L the longitudinal velocity). Estimates of average dispersion curves are shown by the hatched areas. The structure factors $S(Q)$ are also shown in (b), drawn on an arbitrary scale.

diagrams for the polycrystal to be constructed. These are shown in Fig. 7.12 for the quartz modifications of $SiO_2$ and $GeO_2$. The average dispersion curves or 'bands' have simply been drawn in, to represent the data in accordance with the ideas discussed above. The data for $SiO_2$ quartz show this procedure to be reasonable, as here the directionally-averaged lowest-energy acoustic branch estimated in this way is in good agreement with the single-crystal work and detailed spectral computations by Elcombe (1967). For $GeO_2$ quartz, the approximate form of two average branches has been determined in a similar way, and for both crystals the effects of crystal anisotropy in broadening these branches is striking.

For vitreous $SiO_2$, little information about the shape of any average dispersion relations was directly obtainable, as the low frequency peaks in $S(Q, \omega)$ were very diffuse. This was due mainly to the anisotropy and intrinsic width of the dispersion curves rather than to poor experimental resolution. Nevertheless, coherence effects similar to those for the polycrystalline cristobalite modification were observed and the scattering from the vitreous form may be crudely interpreted as a smeared-out version of that from cristobalite. For vitreous $BeF_2$ and $GeO_2$, however, it has been possible to construct $\omega(Q)$ diagrams and to estimate the form of two average acoustic branches using the experimental $S(Q, \omega)$ data, the sound velocities and the structure factor $S(Q)$. The shape of the dispersion curves was determined using the results for $Q$ less than the value $Q_0$, corresponding to the first maximum in the structure factor, and assuming $S(Q)$ at $Q_0$ to act, in the scattering process, as a sharp reciprocal lattice sphere. The curves for $Q > Q_0$ are essentially reflexions through $Q_0$ of those for $Q < Q_0$. This is a much oversimplified interpretation because the structure factor is, of course, continuous (see Fig. 7.12(b)) and this will lead to a considerable broadening with increasing $Q$ of 'phonon' peaks in the cross-section. The detailed behaviour of $\omega(Q)$ near $Q_0$ is of considerable theoretical interest but this will not be further discussed here. For $Q < Q_0$ the widths of the dispersion curves for glasses do not appear to be very much wider than for the polycrystal, although there is definitely a low energy tail in $S(Q, \omega)$ (or $(d^2\sigma/d\Omega\, dE')$, see Fig. 7.11). This may arise from the real presence of very low frequency modes (e.g. resonance modes) or from the relaxation of the wavevector conservation condition implicit in the use of the continuous $S(Q)$ instead of $\tau$. It is worth noting that the peaks in $Z(\omega)$ corresponding to the tops of the dispersion curves of Fig. 7.12 are in all cases in excellent agreement with the low-temperature heat capacity data, which is a necessary test of their validity. As expected, no simple interpretation of coherence effects other than those arising from the acoustic branches was possible.

To summarize, coherence effects in a few simple glasses have been observed and these effects are consistent with average acoustic-phonon dispersion curves, which are very similar in general character to those observed for polycrystalline specimens but somewhat more broadened by the disorder. Further progress must await higher resolution experiments and a more detailed and extensive experimental coverage of the $\omega-Q$ plane, particularly the investigation of the first Brillouin Zone for the glasses.

## 7.4. References

ANDERSON, O. L. *and* DIENES, G. J. (1960). *Non-crystalline solids* (*Ed. V. D. Fréchette*). Wiley, New York. p. 449.
AXMANN, A., GISSLER, W., KOLLMAR, A. *and* SPRINGER, T. (1970). *Discuss. Faraday soc.* **50**, 74.
BELL, R. J. BIRD, N. F. *and* DEAN, P. (1968). *J. phys. Chem.* (2) **1**, 299.
BELL, R. J. *and* DEAN, P. (1970). *Discuss. Faraday Soc.* **50**, 55.
BELL, R. J. *and* DEAN, P. (1972). *Amorphous Materials.* (*Eds. R. W. Douglas and B. Ellis*). Wiley, London.
COCKING, S. J. (1967). *Adv. Phys.* **16**, 189.
COCKING, S. J. (1968). *Rep. U.K. atom. Energy Auth.* (*London*).
COCKING, S. J. *and* EGELSTAFF, P. A. (1965). *Phys. Lett.* **16**, 130.
DAHLBORG, U. *and* LARSSON, K. E. (1965). Ark. Fys. **33**, 271.
DE WETTE, F. W. *and* RAHMAN, A. (1968). *Phys. Rev.* **176**, 784.
DORNER, B., PLESSER, T. *and* STILLER, H. (1965). *Physica's Grav.* **31**, 1537.
EGELSTAFF, P. A. (1962). *Adv. Phys.* **11**, 203.
EGELSTAFF, P. A. (1963). *Rep. U.K. atom. Energy Auth.* (*London*) R4101.
EGELSTAFF, P. A. (1965). *Physics of Non-Crystalline Solids.* (*Ed. Prins, J. A.*). North Holland, Amsterdam. p. 127.
ELCOMBE, M. M. (1967). *Proc. phys. Soc.* **91**, 947.
ENDERBY, J. E. (1968). *Physics of simple liquids* (*Eds. Temperley, H. N. V., Rowlinson, J. S. and Rushbrooke, G. S.*) North Holland, Amsterdam. Chapter 14.
FERGUSON, C. A. *and* HAAS, M. (1970). *J. Am. Ceram. Soc.* **53**, 109.
FLUBACHER, P., LEADBETTER, A. J., MORRISON, J. A. *and* STOICHEFF. B. P. (1959). *J. physics Chem. Solids.* **12**, 53.
GASKELL, P. H. (1966). *Trans. Faraday Soc.* **62**, 1505.
KAPLOW, R., STRONG, S. L. *and* AVERBACH, B. L. (1965). *Phys. Rev.* A**138**, 1336.
LEADBETTER, A. J. (1968). *Phys. Chem. Glasses.* **9**, 1.
LEADBETTER, A. J. (1969). *J. chem. Phys.* **51**, 779.
LEADBETTER, A. J. *and* LITCHINSKY, D. (1970). *Discuss. Faraday Soc.* **50**.
LEADBETTER, A. J. *and* STRINGFELLOW, M. W. (1972). *Neutron inelastic scattering.* I.A.E.A., Vienna.
LEADBETTER, A. J. *and* WRIGHT, A. C. (1970). *J. non-cryst. Solids.* **3**, 239. North Holland, Amsterdam.
LEADBETTER, A. J. *and* WRIGHT, A. C. (1972). *J. non-cryst. Solids.* **7**, 23, 37, 141 and 156.
LORCH, E. A. (1969). *J. phys. C.* **2**, 229. © Am. Chem. Soc.
LORCH, E. A. (1970). *J. phys. C.* **3**, 1314.
MARADUDIN, A. A. (1968). *Proc. Int. Conf. Localised Excitations in Solids*, California 1967, (*Ed. Wallis, R. F.*). Plenum Press, New York. p. 1.
RANDOLPH, P. D. *and* SINGWI, K. S. (1966). *Phys. Rev.* **152**, 99.
SINGWI, K. S. (1964). *Phys. Rev.* (A) **136**, 969.

SINGWI, K. S. (1965). *Physica.* **31**, 1257.
SKÖLD, K. *and* LARSSON, K. E. (1967). *Phys. Rev.* **161**, 102.
WYCHERLEY, K. E. (1969). *Ph.D. Thesis*, University of Bristol.

*General*

(i) Books on the glassy state:

JONES, G. O. (1956). *Glass.* Methuen, London.
MACKENZIE, J. D. (1960, 1962 and 1964). *Modern aspects of the Vitreous State*, Vol. I, II and III. Butterworths London.
RAWSON, H. (1967). Inorganic glass-forming systems. Academic Press, London.

(ii) Conference proceedings on glasses:

*Non-crystalline solids.* (*Ed. V. D. Fréchette*). (1960). Wiley, New York.
*Physics of non-crystalline solids* (*Ed. J. A. Prins*). (1965). N. Holland, Amsterdam.
*Amorphous materials* (1972). (*Eds. R. W. Douglas and B. Ellis*). Wiley, London.
The Vitreous State (1970). *Discuss. Faraday Soc.* **50**.
Amorphous and Liquid Semiconductors (1970), Proc. Int. Conf. Cambridge (Ed. N. F. Mott). (1969). *J. non-cryst. Solids*, **4**.

# 8 The Structure of Liquids by Neutron Scattering

*By* D. I. PAGE

*Atomic Energy Research Establishment, Harwell*

## 8.1. Introduction

The term 'structure' as applied to a physical sample usually means the spatial distribution of the atoms or molecules in the sample. In a solid the concept is simple because one talks about the distribution in space of the equilibrium sites of the component atoms and this distribution is largely time invariant. In a liquid, however, the atoms have no equilibrium positions and one assumes an instantaneous picture ('snapshot') of the sample and works out an appropriate distribution

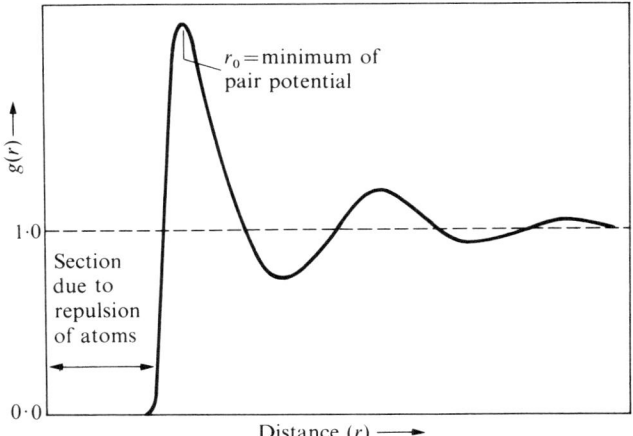

FIG. 8.1. Typical form for the pair distribution $g(r)$: for a dense liquid the principal peak of $g(r)$ occurs at a position close to the principal minimum of $u(r)$. (After Egelstaff 1967).

function. The structure is usually defined in terms of the two-particle or pair distribution function, $g(r)$, which describes the probability of two atoms being a distance $r$ apart in our 'snapshot'.

Clearly, this distribution depends on the force developed between the atoms, and Fig. 8.1 shows a typical $g(r)$ and its relationship to the interatomic pair potential, $u(r)$. Theories of the liquid state relate the pair distribution function to the pair potential. It can be shown (Egelstaff 1967) that almost all the properties of a classical liquid may be calculated from the hypothesis that the interatomic forces and energies are dominated by the sum of pair interactions. Thus 3-body or higher-order forces are neglected in the simplest theories of the liquid state; more sophisticated theories (see Rice and Gray 1965) assume a suitable approximation for the 3-body distribution functions. In principle, the pair potential can be regarded as a basic quantity from which all other liquid properties could be calculated, and it is the aim of the experimentalist to ultimately determine this quantity. Neutron scattering affords a method of obtaining $g(r)$ and hence, by use of a suitable theory, $u(r)$. This chapter will discuss the acquisition from scattering experiments of reliable data which can be used to determine $g(r)$.

## 8.2. The structure factor $S(Q)$

### 8.2.1. Theory

In §1.4.3 we saw that the scattering function $S(\mathbf{Q}, \omega)$ is the Fourier transform of the Van Hove space-time correlation function $G(\mathbf{r}, t)$ which describes the spatial and dynamical behaviour of all the atoms in the scattering system. The 'structure' of a liquid will show up in the spatial part of the correlation function, i.e. in the momentum-transfer part of the scattering function. To illustrate this, consider the simple case of radiation being Bragg reflected through an angle $2\theta$ by a crystal. Bragg's law relates the wavelength of the radiation ($\lambda$) to the spacing ($d$) of the lattice planes in the crystal:

$$\lambda = 2d \sin \theta. \tag{8.1}$$

From Fig. 1.3,

$$Q^2 = k'^2 + k^2 + 2k'k \cos(2\theta) \tag{8.2}$$

and therefore, for elastic scattering ($k' = k$),

$$Q = 2k \sin \theta = (4\pi/\lambda) \sin \theta. \tag{8.3}$$

Comparing eqns (8.1) and (8.3) we see that $Q$ behaves as ($2\pi$/spacing) and note intuitively that Fourier transforming with respect to $Q$ will provide information about the spacing between atoms in the scattering system.

The pair distribution function $g(r)$ denotes the relative positions of a pair of atoms at the same instant of time, i.e. it refers to $G_d(r, 0)$. We define the structure factor $S(Q)$ by

$$S(Q) = \int_{-\infty}^{\infty} S(\mathbf{Q}, \omega) \, d\omega, \tag{8.4}$$

and this can be related to $g(r)$ as follows. From eqn (1.25a)

$$S(\mathbf{Q}, \omega) = \frac{1}{2\pi} \iint G(\mathbf{r}, t) \, e^{i(\mathbf{Q} \cdot \mathbf{r} - \omega t)} \, d\mathbf{r} \, dt,$$

so that,

$$S(Q) = \int S(\mathbf{Q}, \omega) \, d\omega = \frac{1}{2\pi} \iint G(\mathbf{r}, t) \, e^{i\mathbf{Q}\cdot\mathbf{r}} \, d\mathbf{r} \, dt \int e^{-i\omega t} \, d\omega$$

$$= \iint G(\mathbf{r}, t) \, e^{i\mathbf{Q}\cdot\mathbf{r}} \, d\mathbf{r} \, dt \, \delta(t)$$

$$= \int G(\mathbf{r}, 0) \, e^{i\mathbf{Q}\cdot\mathbf{r}} \, d\mathbf{r}.$$

Putting

$$G(\mathbf{r}, 0) = \delta(r) + \rho g(r),$$

where $\rho$ is the number density of atoms, then gives

$$S(Q) = 1 + \rho \int g(r) \, e^{i\mathbf{Q}\cdot\mathbf{r}} \, d\mathbf{r}. \tag{8.5}$$

Starting from eqn (1.25b), the corresponding expression for incoherent scattering is

$$S_s(\mathbf{Q}, \omega) \, d\omega = \int G_s(\mathbf{r}, 0) \, e^{i\mathbf{Q}\cdot\mathbf{r}} \, d\mathbf{r}$$

$$= \int \delta(r) \, e^{i\mathbf{Q}\cdot\mathbf{r}} \, d\mathbf{r}$$

$$= 1. \tag{8.6}$$

Thus, in principle, to measure the structure factor, $S(Q)$, the specimen is placed in a beam of radiation and the scattered radiation measured as a function of angle (i.e. $Q$), taking care to include all the scattered radiation at each $Q$ value, irrespective of any energy transfers involved. For X-ray scattering this integration over energy transfers is done automatically by the detector, as the energy transfers are small compared with the incident energy, and the scattering process appears to be perfectly elastic; for neutron scattering this is not usually so, and the conversion of the scattered intensities to $S(Q)$ is less straightforward (see below).

Fig. 8.2 shows the qualitative behaviour of $S(Q)$ and $S(\mathbf{Q}, \omega)$ for a simple liquid. The uppermost curve shows $S(Q)$ and the remaining curves the behaviour of $S(\mathbf{Q}, \omega)$ at several values of $Q$. At low values of $Q$, $S(\mathbf{Q}, \omega)$ splits into three components where the displaced lines correspond to sound-wave modes and their size and shape may be described by simple hydrodynamic theory; the central peak corresponds to the thermal damping of the sound waves or entropy fluctuations in the system. At high values of $Q$, $S(\mathbf{Q}, \omega)$ has a shape appropriate to a perfect gas because one is considering atomic movement on a very small time scale. This movement will be gas-like as the atom will travel in a straight line for a very small distance until it encounters the field of force generated by

# STRUCTURE OF LIQUIDS BY NEUTRON SCATTERING

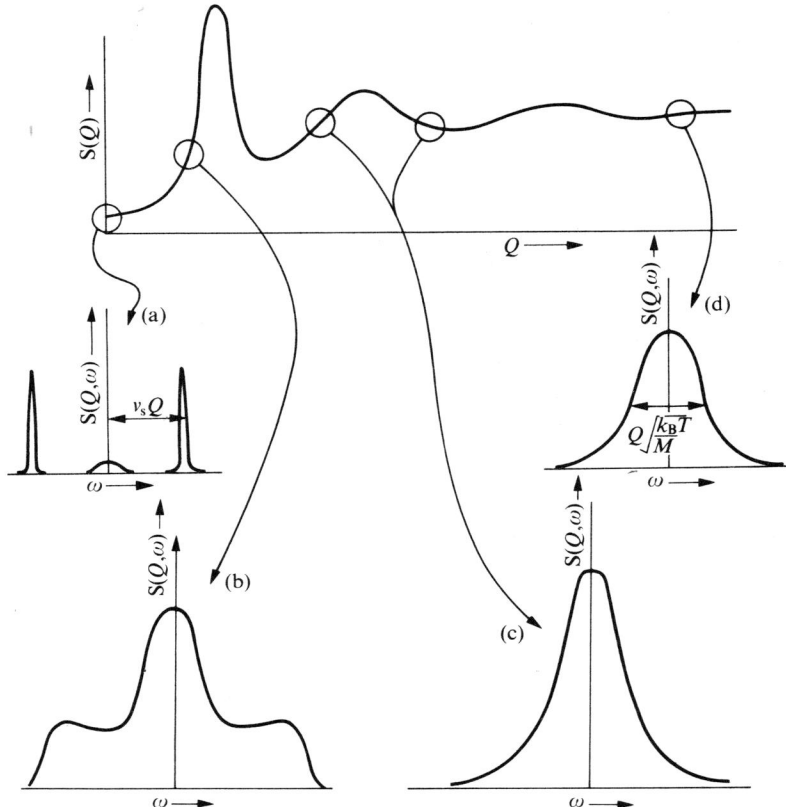

FIG. 8.2. Qualitative behaviour of $S(\mathbf{Q}, \omega)$ for several values of $Q$. The upper curve is the function $S(Q)$, and the remaining curves, (a) to (d), show the spectral shape at fixed values of $Q$ marked by the circles on $S(Q)$. At low $Q$, the Brillouin lines in (a) are separated by $2v_s Q$ where $v_s$ is the sound velocity. In curve (d), at high $Q$, classically the full-width at half-height is $Q(k_B T/M)^{1/2}$ where $M$ is the mass of the scattering atom. The width in $\omega$ cannot be calculated from simple considerations for the intermediate cases, (b) and (c). (After Egelstaff 1967).

its neighbours. At intermediate values of $Q$, the cooperative modes merge into the single-particle modes.

### 8.2.2. Application to neutron scattering

The double differential cross-section $(\mathrm{d}^2 \sigma / \mathrm{d}\Omega\,\mathrm{d}\omega)$ for the coherent scattering of neutrons from a specimen is given by eqn (1.25c):

$$\frac{\mathrm{d}^2 \sigma_{\mathrm{coh}}}{\mathrm{d}\Omega\,\mathrm{d}\omega} = N \langle b \rangle^2 \frac{k'}{k} S(\mathbf{Q}, \omega) \tag{8.7}$$

where $N$ is the number of nuclei in the specimen and $b$ is their scattering length. Integrating over all energy transfers at a constant value of $Q$ gives the angular differential cross-section:

$$\frac{d\sigma_{coh}}{d\Omega} = N\langle b\rangle^2 \int_{-\infty}^{\infty} \frac{k'}{k} S(\mathbf{Q}, \omega)\, d\omega. \tag{8.8}$$

Comparison with eqn (8.4) shows that the integral expression is $S(Q)$ except for the factor $k'/k$. If we consider the scattering to be purely elastic (*the static approximation*), this factor becomes unity and the structure factor is immediately accessible from the measured intensity.

The scattering from the sample may be both coherent and incoherent and $S(Q)$ comes only from the coherent part. We must therefore subdivide $d\sigma/d\Omega$, writing

$$\frac{d\sigma}{d\Omega} = \frac{d\sigma_{coh}}{d\Omega} + \frac{d\sigma_{incoh}}{d\Omega} \tag{8.9}$$

where $\sigma_{coh}$ and $\sigma_{incoh}$ are the coherent and incoherent scattering cross-sections respectively. In the static approximation,

$$\left(\frac{d\sigma_{coh}}{d\Omega}\right)_t = Nb_{coh}^2 \int_{-\infty}^{\infty} S(\mathbf{Q}, \omega)\, d\omega$$

$$= Nb_{coh}^2 S(Q) \tag{8.10}$$

from eqn (8.4), and

$$\left(\frac{d\sigma_{incoh}}{d\Omega}\right)_t = Nb_{incoh}^2 \int_{-\infty}^{\infty} S_s(\mathbf{Q}, \omega)\, d\omega$$

$$= Nb_{incoh}^2 \tag{8.11}$$

from eqn (8.6). The subscript 't' in eqn (8.10) and eqn (8.11) reminds us that these are 'theoretical' expressions, based on the assumption that $k' = k$. $b_{coh} (= \langle b \rangle)$ and $b_{incoh} [= (\langle b^2 \rangle - \langle b \rangle^2)^{\frac{1}{2}}]$ are the scattering lengths associated with $\sigma_{coh}$ and $\sigma_{incoh}$.

In a conventional diffraction experiment we measure an integrated intensity ($I$) of scattered neutrons at a given angle of scattering, the integration over the energy transfers being performed by the detector. Three effects must be taken into account in interpreting $I$:

(a) the static approximation does not hold, as energy transfers do occur;
(b) the detector efficiency is a function of the neutron energy and has a cut-off when the energy loss, $\hbar\omega$, equals the incident neutron energy, $E$;
(c) the measurements are made at constant scattering angle $2\theta$ and not constant $Q$. This is particularly important when large energy transfers are involved.

Thus the measured differential cross-section $(d\sigma/d\Omega)_m$ is given by (c.f. eqn. (8.8))

$$I(\theta) \propto \left(\frac{d\sigma}{d\Omega}\right)_m = \left|Nb^2 \int_{-E/\hbar}^{\infty} \frac{k'}{k} S(\mathbf{Q}, \omega)\, d\omega\right|_{\text{const } \theta} \quad (8.12)$$

so that $S(Q)$ is not easily accessible. Furthermore, the scattered intensity is obtained as the difference between the scattering from the specimen-plus-container $(I_{\text{tot}})$ and the empty container $(I_c)$; each intensity must be corrected for absorption in the scatterer and for multiple scattering, as $S(Q)$ only relates to single scattering events. The corrected scattered intensity must then be put on an absolute cross-section scale. Thus

$$I_{\text{tot}} - I_c \equiv I(\theta) = \frac{\alpha(\theta)}{\beta(\theta)}\left[\left(\frac{d\sigma_{\text{coh}}}{d\Omega}\right)_m + \left(\frac{d\sigma_{\text{incoh}}}{d\Omega}\right)_m + \delta_{\text{coh}} + \delta_{\text{incoh}}\right] \quad (8.13)$$

where $\alpha(\theta)$ is the absorption correction, $\beta(\theta)$ is the normalization correction and $\delta_{\text{coh}}$ and $\delta_{\text{incoh}}$ are the coherent and incoherent multiple scattering corrections, respectively. From eqns (8.10) and (8.11), eqn (8.13) may be written

$$I(\theta) = \frac{\alpha(\theta)}{\beta(\theta)}\left\{[Nb_{\text{coh}}^2 S(Q)]_m + [Nb_{\text{incoh}}^2]_m + \delta_{\text{coh}} + \delta_{\text{incoh}}\right\}, \quad (8.14)$$

where the subscript 'm' reminds us that these terms must be corrected for the effect of the static approximation.

To extract $S(Q)$ from the intensity data, $\alpha, \beta$, and $\delta$ must be either measured or calculated and the relationship between the theoretical cross-section and the measured cross-section determined. We shall now consider the individual corrections in more detail.

### 8.2.3. Corrections to neutron diffraction data

The corrections to be applied to the intensity data can rarely be calculated in a simple, generalized way. They depend on the details of the experiment—for example, on the geometry of both the diffractometer and the specimen, which vary for individual experiments. For this reason detailed calculation of the correction is not discussed here; instead we refer to the methods of correction which may then be adapted for each particular experiment. Anyone carrying out a scattering experiment should give considerable thought to the geometry of his equipment before doing any measurements; in particular, the size and shape of his sample container will determine the degree of complexity of the correction to be applied.

#### 8.2.3.1. Specimen container.
The size of the specimen to be placed in the neutron beam is determined by two factors: the desirability of using as much of the beam area as possible, and the need to choose the thickness of the sample as a com-

promise between having a lot of scattering material present, to shorten the experimental running time, and having little material present so that absorption and multiple scattering are small (see below). A generally acceptable criterion is that 5-10 per cent of the main beam should be scattered. Thus large, plane, thin containers are often used but these are difficult to manufacture with the accurately spaced, parallel walls which are needed to calculate the number of scattering nuclei in the beam. A good design is reported by Striffler and Carpenter (1963). Care is also needed in using plane samples in that measuring over a large angular range usually means that some of the measurements are in transmission and some in reflexion. The angular setting of the container must be altered to measure at least part of the range. Cylindrical specimen containers are much more convenient to use (and are essential for high pressure and temperature work), but are inefficient in their use of the total available flux as the diameter of the cylinder is fixed by the '10 per cent scattering' criterion. If this diameter is small, the amount of material in the walls of the container becomes comparable to the amount in the sample, and produces a high background. For thick-walled specimens (i.e. either a small bore cylinder or a truly thick-walled pressure vessel), one can improve the signal-to-noise ratio by limiting the beam width to about one half of the cylinder diameter. In this way, about two-thirds of the scattering from the cylinder walls is lost but only two-fifths of that from the specimen. This gain is at the expense of increased counting times and more complex corrections. Cylindrical containers should be as thin-walled as practicable and preferably not of glass. Glass itself has a liquid-like structure factor (see Chapter 7) and it is usually easier to allow for several Bragg peaks in the background than to subtract two similar spectra. A tube of vanadium or a null-matrix alloy is easily the best container; this gives a flat background and the scattering from the empty container can often be used for normalization (see §8.2.3.3.).

*8.2.3.2. Absorption corrections.* These are usually calculated either numerically or graphically, depending on the specimen geometry. Paalman and Pings (1962) have evaluated the cases of cylindrical and annular geometries for X-rays and their methods can be applied directly to neutrons. The corrections must be applied to both the full and empty containers before subtracting the background, as the presence of the specimen in the can affects the number of detected counts from the can.

*8.2.3.3. Normalizing correction.* Scattering data are put on an absolute cross-section scale by comparing them with the intensity from a specimen of known scattering cross-section. Vanadium is the usual standard as it scatters almost totally incoherently ($\sigma_s$ = 5·13b, $\sigma_{coh}$ < 0·03b), giving an isotropic distribution (except for the Placzek correction, see §8.2.3.5.). It should be noted that vanadium has an absorption cross-section of 5b. Matching the geometry of the vanadium sample to that of the scattering sample (for example, by using tube instead of solid rod to match diameters) allows for inhomogeneities in the incident beam and probably makes the relative corrections more reliable. A simple point sometimes overlooked is that, for correct normalization, the total number of scattering atoms in both the sample and calibration material must be known accurately. This can be calculated by masking the specimens to a known area, but is better done by stopping down the beam (so that the same height and/or width is used in both cases) and using the relative thicknesses and number densities.

Another point is that the scattering length for the specimen needs to be known at least as accurately as the required accuracy in $S(Q)$ (eqn 8.10). In many cases scattering lengths are not known to better than 5-10 per cent and in some cases $b_{coh}$ is not known at all. This is particularly a cause for concern in using separated isotopes (see §§8.4 and 8.5).

*8.2.3.4. Multiple scattering corrections.* Multiple scattering (M-S) corrections to diffraction (i.e. total scattering) data are simpler than to inelastic data. The full calculation of the M-S contribution demands a knowledge of the complete scattering law $S(\mathbf{Q}, \omega)$ for the material and of the detailed geometry of the experiment. The subsequent computer calculations are very complex, but less rigorous procedures are much easier and are probably adequate. Calculations have been published using various approximations (Vineyard 1954; Blech and Averbach 1965; Cocking and Heard 1965; Slaggie 1967; Rao et al. 1971) and suggest that for total scattering the M-S contribution is isotropic. This isotropic assumption suggests that for incoherent scattering there is no M-S correction, because in any given direction as many neutrons are lost by multiple scattering as are gained by multiple scattering from other directions. Carrying this idea further, there is the implication that in any region where the total scattering has a flat distribution (e.g. the high $Q$-value region of simple liquids) the M-S correction is also very small. Undoubtedly, however, M-S tends to decrease peak heights in a spectrum and fill in the minima. Failing a complete calculation, a practical method of correction is to calculate the percentage of multiple scattering using one of the approximations referred to previously, multiply each point on the curve by this factor, and then subtract the mean number of M-S counts assuming an isotropic distribution of the multiply-scattered neutrons. Note that when data are normalized by dividing the specimen counts by the calibration counts, the M-S contributions are also divided and can cancel each other out to a large extent, if similar scattering powers and geometries are used.

*8.2.3.5. Placzek corrections.* Eqn (8.14) was written on the assumption that the static approximation applies, i.e. $k' = k$ in eqn (8.12). The effect of this assumption was first considered by Placzek (1952), who expanded the integrand of eqn (8.12) in energy moments of $S(\mathbf{Q}, \omega)$. The first few terms in the resultant expansion, involving the usual conversion from centre-of-mass to laboratory co-ordinates, include the recoil term, but higher terms contain the derivatives of $S(Q)$. Attempts to apply more than the recoil term have been made by

Ascarelli and Caglioti (1966) using a model, but they appear to over-correct their data. The best discussion of this correction is given by Egelstaff and Poole (1969). For neutrons of high incident energy and for medium or heavy scattering atoms, the correction is usually only a few per cent and the first few terms of the Placzek expansion appear to be sufficient. Fig. 8.3 shows the (truncated) Placzek correction for vanadium. For $Q \to 0$ it is possible that the Placzek corrections are of the order of $S(Q)$. For light elements, especially in molecules with high energy modes of vibration (e.g. $D_2O$), the correction is large and cannot yet be properly applied.

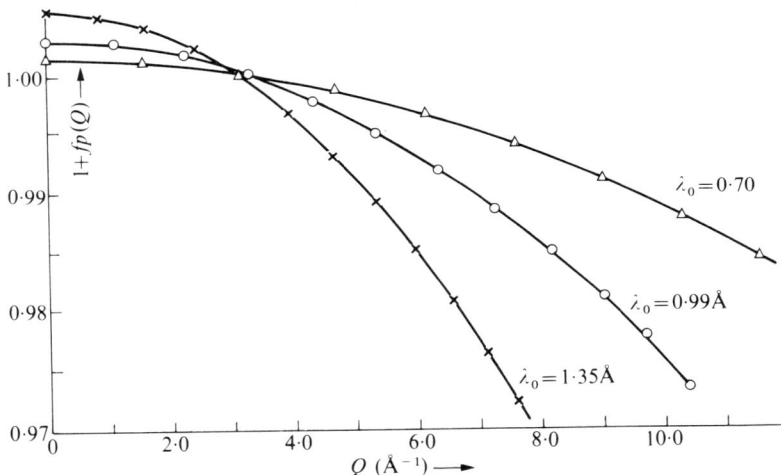

FIG. 8.3. Placzek corrections for vanadium at 293 K using a $1/v$ detector. $f(Q)$ is the correction expressed as a fraction of the total scattering and $\lambda_0$ is the incident neutron wavelength. (After North 1966.)

## 8.3. Monatomic liquids

There are many reported measurements of the structure factors for the rare gases and liquid metals, the so-called simple liquids, and a recent compilation is given by Enderby (1968). Few measurements are complete in that not all of the corrections to the data have been applied, in particular an absolute cross-section scale is rarely given. Fig. 8.4 shows a typical measured spectrum (natural argon at 85 K) after the corrections have been applied.

Only recently has any degree of consistency been achieved between measurements from different laboratories: Fig. 8.5 shows three independent sets of data on liquid zinc obtained by Egelstaff in 1970, which are consistent to within 5 per cent. All simple liquids have similar structure factors, consisting of a main peak followed by an oscillatory curve which is rapidly dampened to a straight line at $Q \gtrsim 10$ Å$^{-1}$. The asymptotic limit of $S(Q)$ is unity and this is often used to normalize the observed curves. This provides no check as to whether or not the corrections have been done properly. The main peak occurs at a $Q$-value

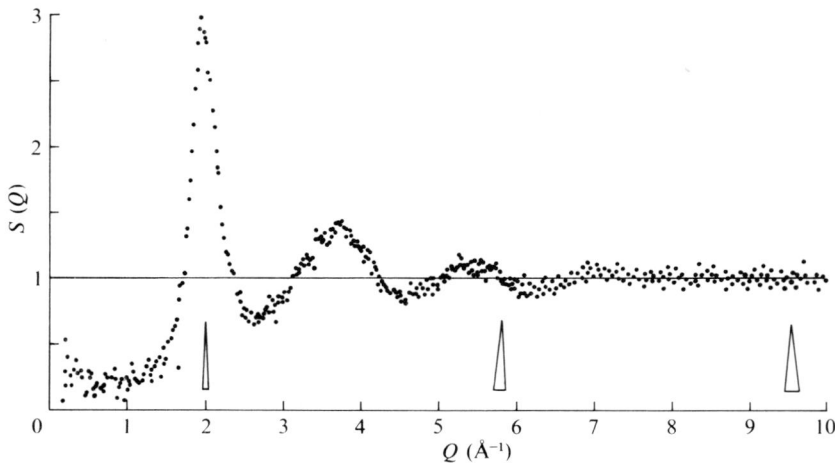

FIG. 8.4. Neutron determination of the structure factor for natural argon at 85 K. The triangles indicate the experimental resolution function. (After Page 1972).

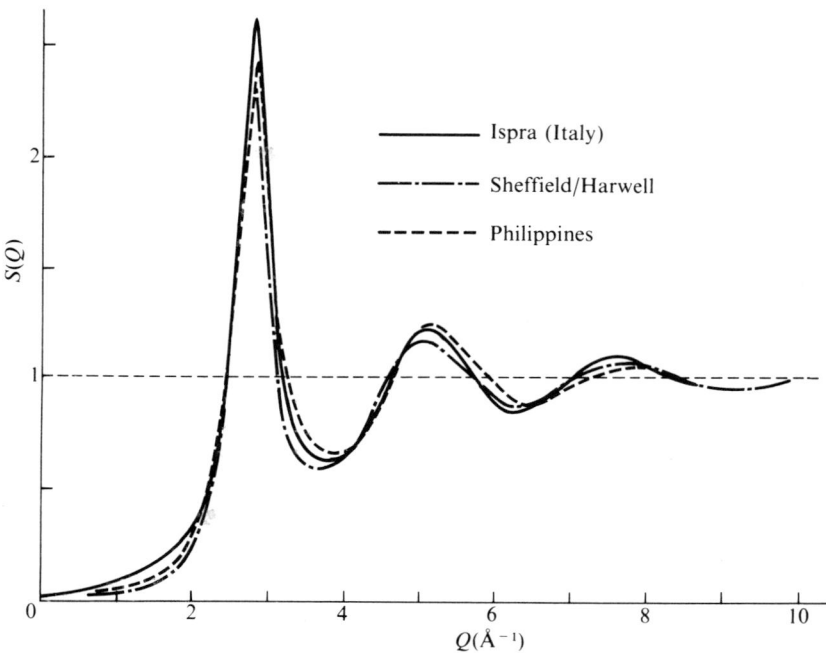

FIG. 8.5. Comparison of neutron diffraction data on zinc from three independent laboratories. (After Egelstaff 1970).

equal to ($2\pi$/mean interatomic distance), as in Bragg scattering, and the intercept at $Q = 0$ is given by

$$S(0) = \chi_T \rho k_B T$$

where $\chi_T$ is the isothermal compressibility. This value is achieved (at least for liquid metals, see Egelstaff et al. (1966)) at $Q \gtrsim \frac{1}{4} Q_0$ where $Q_0$ corresponds to the main peak (often $1 \cdot 5 \to 2$ Å$^{-1}$). As neutron scattering measurements usually have $Q > 0 \cdot 5$ Å$^{-1}$ (corresponding to a 1 Å neutron scattered through 4°), and considerable uncertainty occurs near this scattering angle as the main neutron beam begins to intrude, this low-$Q$ limit is not reached with conventional diffractometers and the data can only be extrapolated back to the theoretical intercept. Smoothing functions (Powell 1970) are often applied to the data at this stage to give a smooth curve more amenable to analysis.

The next step is to Fourier transform $S(Q)$ to give the radial distribution function $g(r)$ (eqn (8.5)). Large errors occur when this is done, due to the truncation of the data at finite $Q$-values (both high and low) and to experimental errors in the measured data. Typically, rapid oscillations occur in $g(r)$ at small values of $r$ (see, e.g. Fig. 8.13(c)). This transformation problem is discussed by Lorch (1969, 1970). The transformed data are usually too unreliable to extract an interatomic potential, although this has been done for argon and sodium (Johnson et al. 1964). Ascarelli (1966) and North et al. (1968) have employed models to analyse $g(r)$.

An alternative approach is to assume a potential and calculate the pair-distribution function. Three such potentials (see, e.g. Hirschfelder et al. 1954) are:

(a) the hard-sphere potential

$$u(r) = +\infty \text{ for } r < \sigma$$
$$= 0 \quad \text{for } r > \sigma.$$

(b) the Lennard-Jones potential

$$u(r) = 4\epsilon [(\sigma/r)^{12} - (\sigma/r)^6].$$

(c) the modified Buckingham potential

$$u(r) = [\epsilon/(1 - 6/\alpha)] \left[ \frac{6}{\alpha} \exp \alpha \left(1 - \frac{r}{r_m}\right) - \left(\frac{r_m}{r}\right)^6 \right]$$

$$\text{for } r > R$$

$$= +\infty \text{ for } r \leq R.$$

These potentials are illustrated and the symbols defined in Fig. 8.6. The calculations always involve an approximation for the non-pair effects in the liquid (see, e.g. Egelstaff 1967).

# STRUCTURE OF LIQUIDS BY NEUTRON SCATTERING

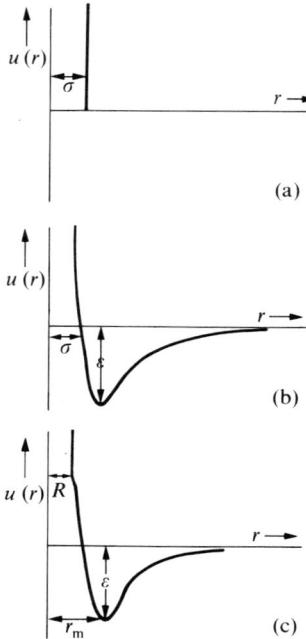

FIG. 8.6. Illustrations of spherically symmetrical potentials: (a) hard-sphere, (b) Lennard–Jones, (c) modified Buckingham.

Ashcroft (1967) has used the hard-sphere potential to calculate the structure factor for randomly packed spheres and this shows reasonable agreement with the measured data on simple liquids. This is because in a real liquid the attractive parts of the potentials overlap to a large extent and the dominant part of $u(r)$ is the repulsive core. Comparison of the structure factors for hard spheres, argon and rubidium (see Fig. 8.7 and Page et al. 1969) shows that the repulsive core for argon must be less steep than that for hard spheres, and that for rubidium must be less steep again. Thus the atomic overlap in collisions is greater for liquid metals than for rare gases. The softer core gives a less steep rising edge to the main peak in $g(r)$ which, on transforming, leads to greater damping of the oscillations in $S(Q)$.

Fig. 8.8 shows calculations by Khan (1964) using a Lennard–Jones potential to fit the early argon data of Henshaw (1957). The agreement is good. An important extension of this approach are the 'molecular dynamics calculations' initiated by Rahman (1964), in which a system of many 'particles' in a computer store are allowed to interact through a suitable potential, the laws of motion applied and, after many collisions, the required distribution functions extracted. This technique is clearly the most fruitful one at present. Computer calculations on an argon-like system using a Lennard–Jones potential show very good agreement with experimental data (Page 1972).

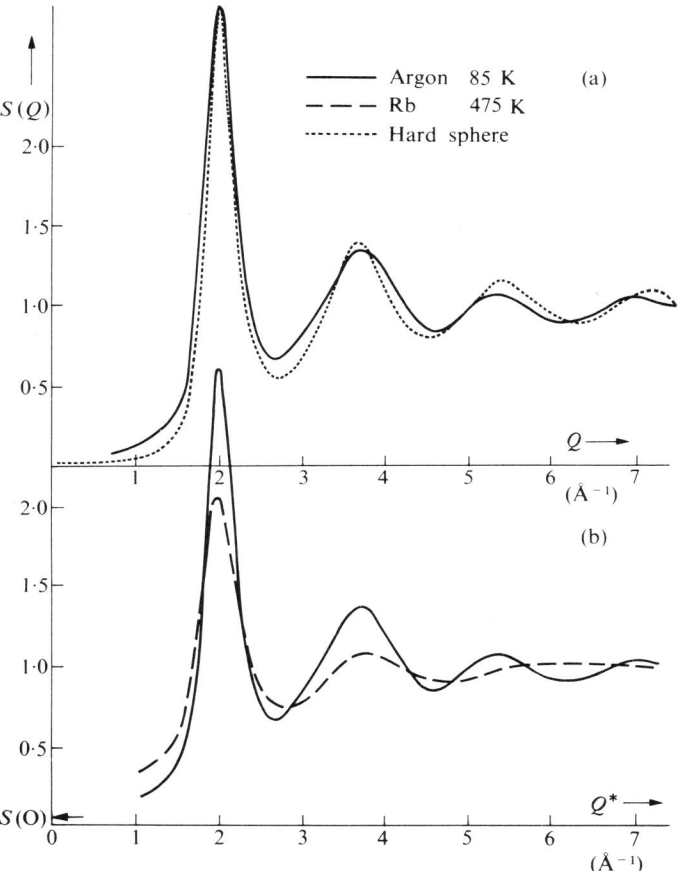

FIG. 8.7. Comparison of liquid structure factors (a) for hard spheres and argon, (b) for argon and rubidium: note the normalized $Q$-scale. (After Page *et al.* 1969).

## 8.4. Binary liquids

In studying liquid structure, the next step in complexity after a simple liquid is a binary liquid. By this we do not mean a diatomic molecular system but rather an 'atomic mixture' of two types of atom. Examples we shall consider are the binary alloys and the fully ionized molten salts. In both cases a reasonable estimate of the potential can be made to calculate the expected distribution functions; the atoms in the alloy interact through an atomic potential while the molten salts act through an ionic potential.

# STRUCTURE OF LIQUIDS BY NEUTRON SCATTERING

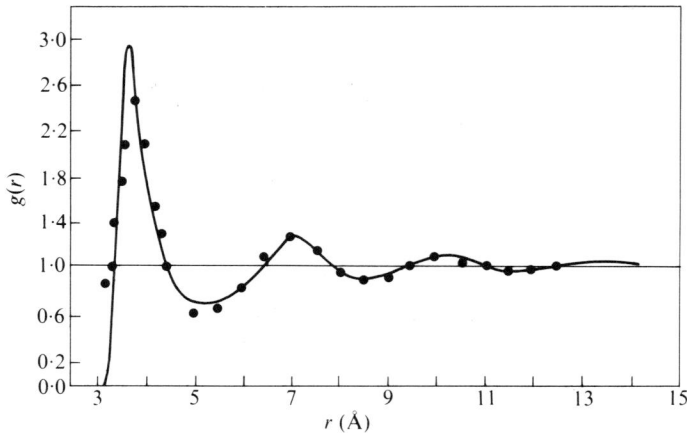

FIG. 8.8. Calculated $g(r)$ for argon at 84·4 K compared with values determined from neutron diffraction measurements. Full line is calculated $g(r)$ using Lennard-Jones potential (from Kahn 1964). Full circles are experimental points of Henshaw (1957). (After Egelstaff 1971).

Experimentally the presence of two different types of atom greatly complicates the situation. The differential scattering cross-section is written (from eqn (1.11)) as

$$\frac{d\sigma}{d\Omega} = \left\langle \sum_p \sum_q b_p b_q \exp[i\mathbf{Q} \cdot (\mathbf{r}_p - \mathbf{r}_q)] \right\rangle \qquad (8.15)$$

where $b_p, b_q$ and $\mathbf{r}_p, \mathbf{r}_q$ are respectively the coherent scattering lengths and position co-ordinates of the $p$th and $q$th atoms and $\langle \rangle$ denotes a time average. For a monatomic liquid $b_p = b_q$ and eqn (8.15) reduces to

$$\frac{d\sigma}{d\Omega} = Nb_p^2 S(Q) \qquad (8.16)$$

as in §8.2. For a binary system, in general $b_p$ does not equal $b_q$ and we define a generalized structure factor $S_{pq}(Q)$ by the equation

$$S_{pq}(\mathbf{Q}) = \delta_{pq} + 2\left(\frac{n_p n_q}{n_p + n_q}\right) \int_0^\infty [g_{pq}(\mathbf{r}) - 1] \exp(i\mathbf{Q} \cdot \mathbf{r}) \, d\mathbf{r} \qquad (8.17)$$

where $p, q$ may take the values 1 or 2 for the two types of atom present and $\delta_{pq}$ is the Kronecker $\delta$-function. Thus there are three independent structure factors, $S_{11}(Q), S_{22}(Q)$ and $S_{12}(Q)$, related to the probability $g_{pq}$ of finding an atom of type $q$ at a distance $r$ if an atom of type $p$ is at the origin. We call the $S_{pq}(Q)$'s the 'partial structure factors' for the system. Eqn (8.15) now reduces to

$$\frac{d\sigma}{d\Omega} = N[c_1 b_1^2 S_{11}(Q) + c_2 b_2^2 S_{22}(Q) + b_1 b_2 S_{12}(Q)], \qquad (8.18)$$

where $c_1, c_2$ are the atomic concentration of the constituents. Eqn (8.18) shows that a single scattering experiment can give only average information on the partial structure factors (see Levy and Danford 1964). At least three independent measurements are required for a complete solution of the equation and this can be most readily achieved by varying the scattering lengths of the two components, either by combining different scattering methods (e.g. X-ray, neutron, and electron) or, and this is more satisfactory, by carrying out neutron scattering experiments with different sets of isotopes.†

Isotopic substitution measurements on the alloy $Cu_6Sn_5$ have been reported by Enderby et al. (1966). They measured the scattering from three specimens consisting of 45 atomic per cent natural Sn alloyed with (a) natural copper, (b) copper enriched with $Cu^{63}$ to 99 per cent and (c) copper enriched with $Cu^{65}$ to 99 per cent. The values of the copper scattering lengths are respectively 0·76, 0·67, and 1·11 f.‡ Because of the small intensity differences, particularly between the scattering patterns of the $Cu^{63}$ and natural Cu alloys, the calculated partial structure factors showed large fluctuations. Using X-ray data as a guide, together with some mathematical constraints on the allowable values of the neutron data, they deduced a final set of partial structure factors which reproduced the total scattering function fairly well. These are shown in Fig. 8.9, where both Cu–Cu and Sn–Sn correlations are clearly defined. The authors pointed out that for more accurate results, there should be a wider variation in the values of the scattering lengths of the different samples. Calculations of structure factors for liquid alloys using a hard-sphere potential have been made by Enderby and North (1967) and show good agreement with experiment for the Na–K system.

An experiment using large variations in the scattering lengths was carried out by Page and Mika (1971). They measured molten CuCl, substituting isotopically for both the copper and the chlorine atoms. Four specimens were prepared consisting of $Cu^{63}Cl^{35}$, $Cu^{nat}Cl^{nat}$, $Cu^{65}Cl^{nat}$, and $Cu^{65}Cl^{37}$ for which the ratio $b_{Cu}/b_{Cl}$ was 0·58, 0·82, 1·16, and 2·41 respectively. The enrichment was 99 per cent for both copper isotopes, 97 per cent for $Cl^{35}$, and 78 per cent for $Cl^{37}$. Fig. 8.10 shows a typical set of scattering data ($Cu^{nat}Cl^{nat}$) while Fig. 8.11 shows the smoothed, corrected data for all the specimens. Note how the first peak in the scattering patterns decreases as the contribution from the chlorine decreases. The smoothed partial structure factors extracted from the data are shown in

† It may be possible to obtain approximate partial structure factors by varying the atomic concentration in cases where $S(Q)$ is nearly independent of concentration. See, e.g. the X-ray investigations on Ag–Mg (Bühner and Steeb 1969) Au–Sn (Wagner et al. 1969), Cu–Ge (Isherwood and Orton 1970) and Cu–Sn (North and Wagner 1970).
‡ f = fermi unit: see p. xv.

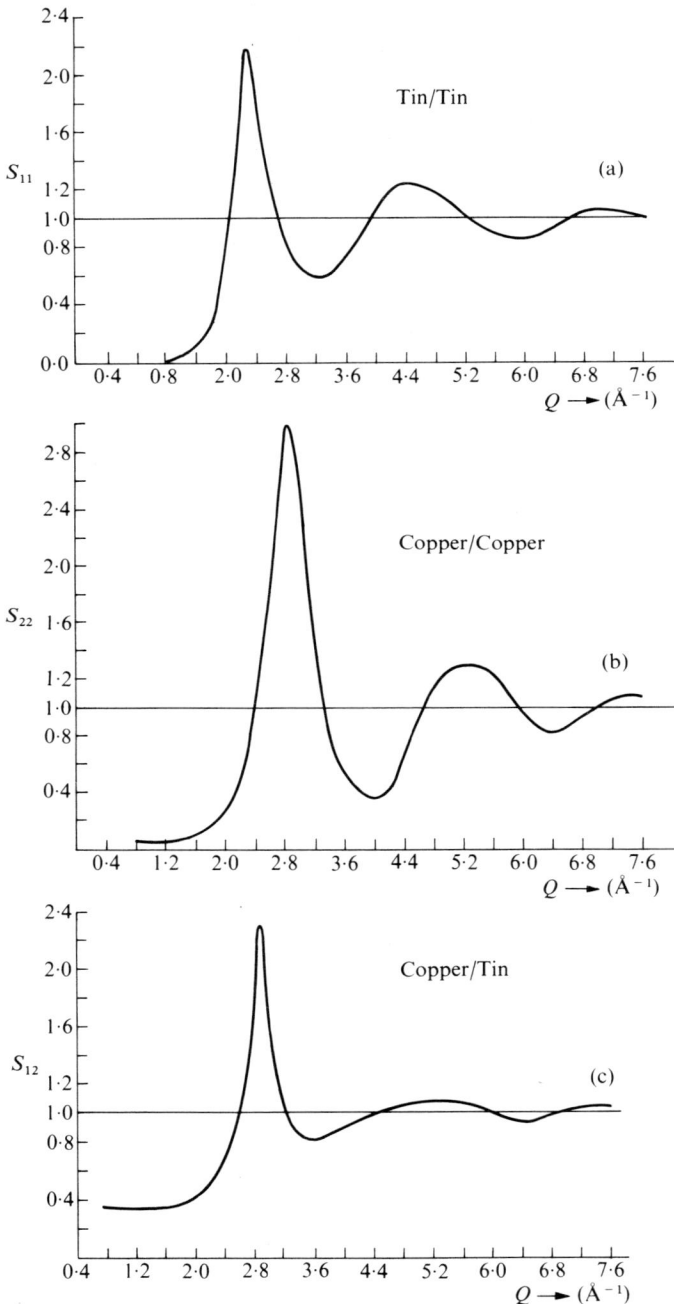

FIG. 8.9. Partial structure factors for: (a) Sn–Sn; (b) Cu–Cu; (c) Cu–Sn. (After Enderby *et al.* 1966).

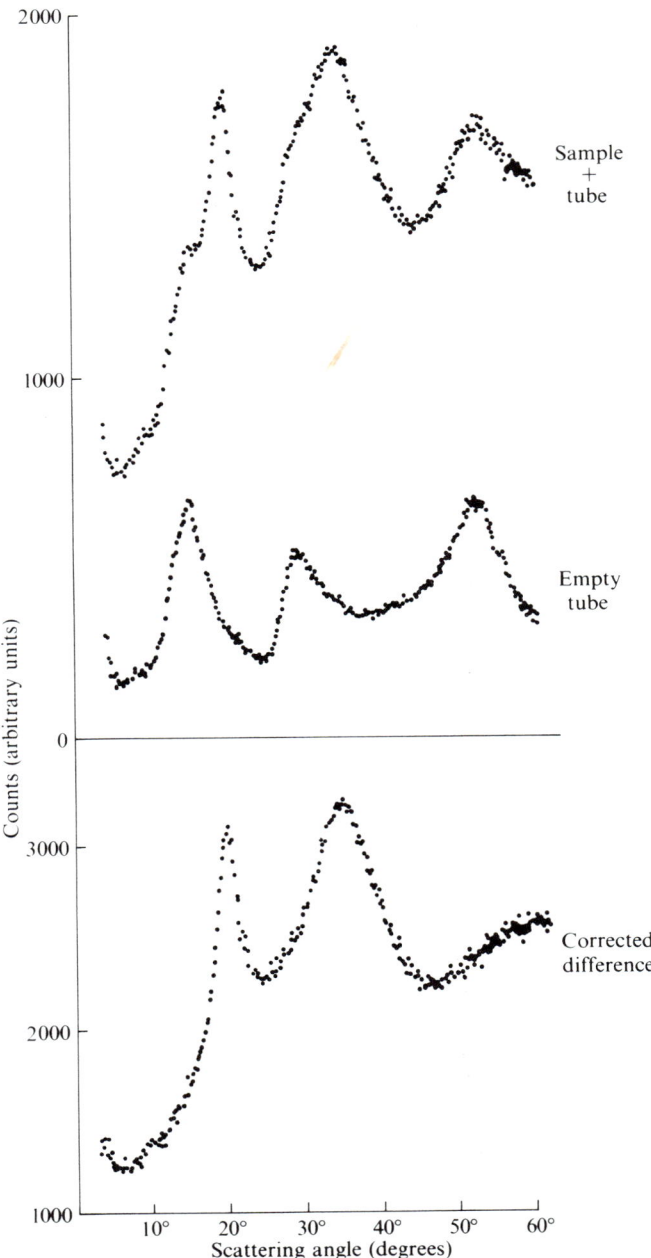

FIG. 8.10. Neutron diffraction measurements on molten CuCl. (After Page and Mika 1971).

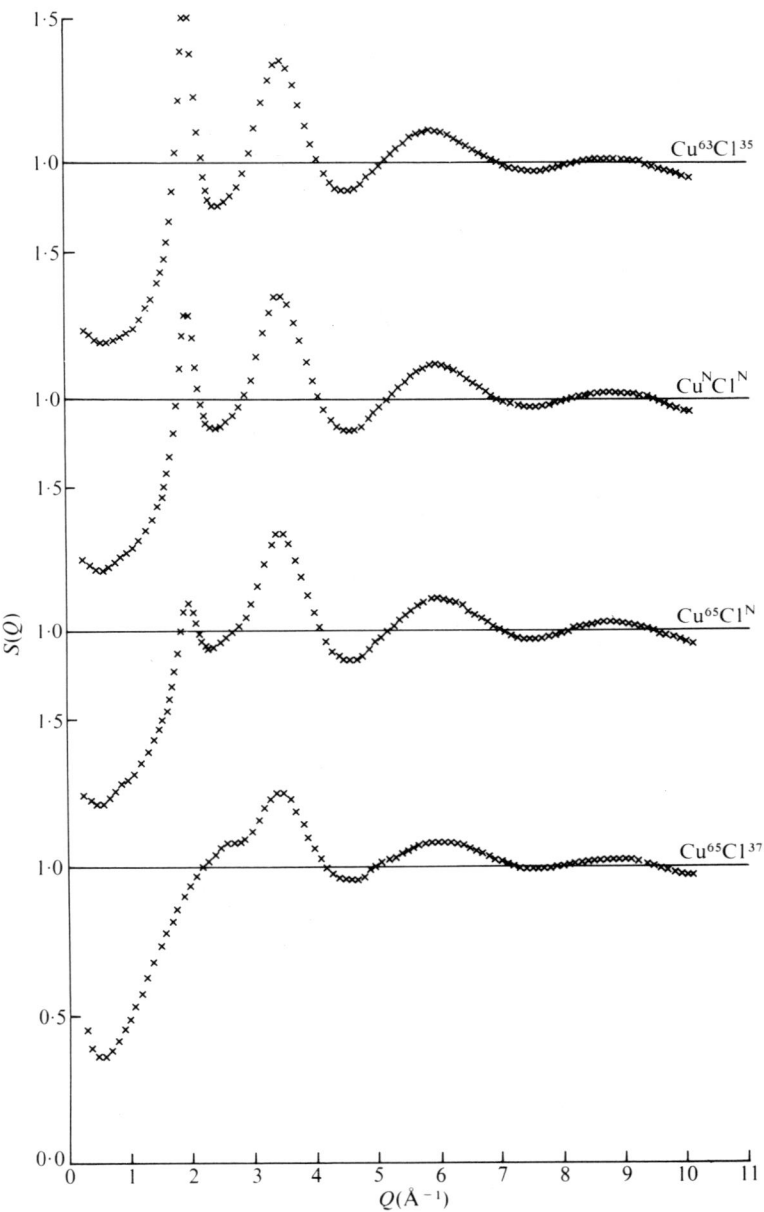

FIG. 8.11. Observed structure factors for molten $Cu^+Cl^-$ (750 K) as a function of isotopic composition. (After Page and Mika 1971).

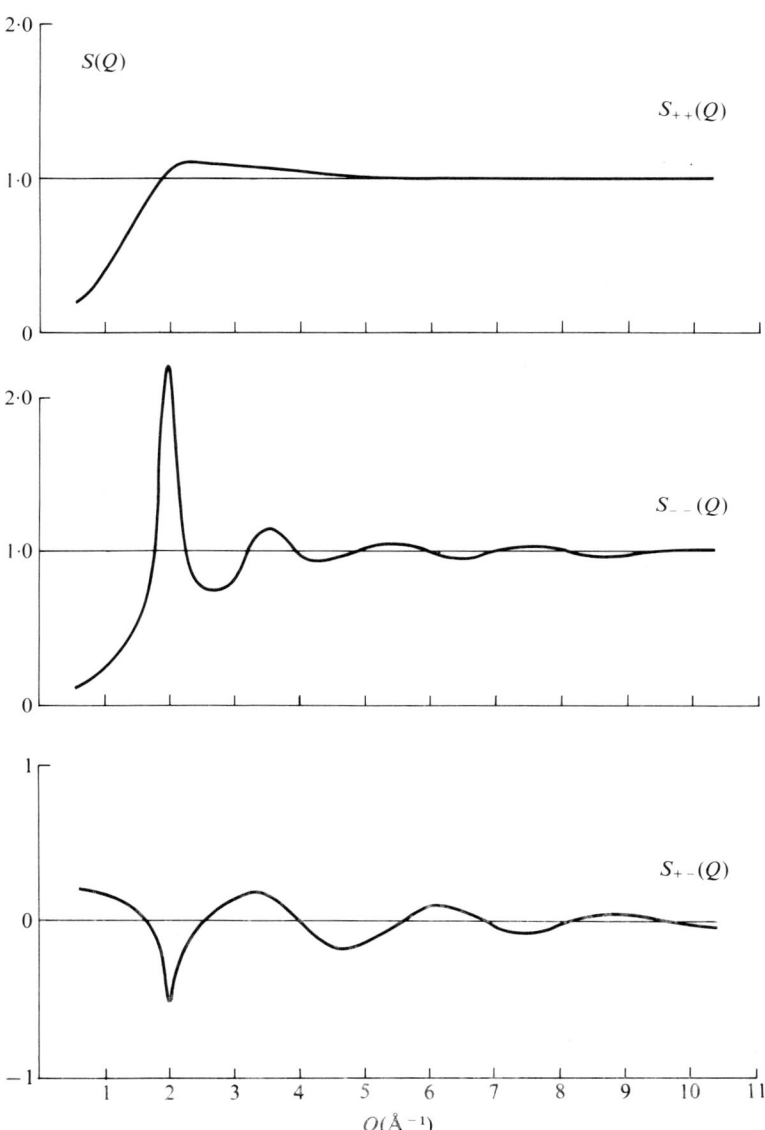

FIG. 8.12. Smoothed partial structure factors for molten $Cu^+Cl^-$. (After Page and Mika 1971).

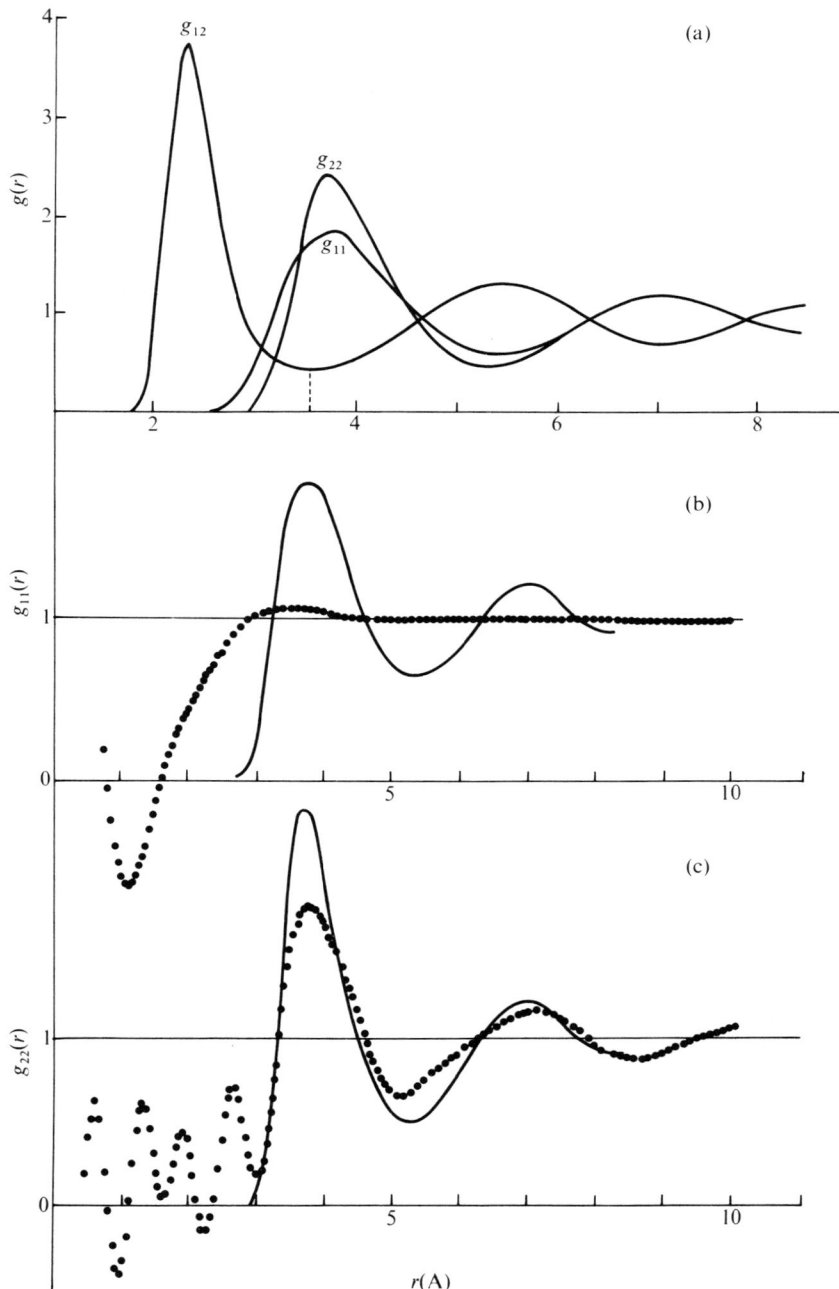

FIG. 8.13. Radial distribution functions for $Cu^+Cl^-$. (a) Monte-Carlo calculations, atom 1 = copper and atom 2 = chlorine; (b) copper–copper distribution; full line Monte-Carlo, closed circles experimental points; (c) chlorine–chlorine distribution; full line Monte-Carlo, closed circles experimental points. (After Page and Mika 1971).

Fig. 8.12; the scattered intensities calculated from these *via* eqn (8.18) agreed with the observed data to within experimental error. The Cl–Cl distribution is similar to that for a simple monatomic liquid discussed in §8.3, and, indeed, agrees well with the $S(Q)$ for argon. The Cu–Cu distribution is markedly different, however, and is more like that associated with a point plasma. Monte-Carlo calculations were made on the $Cu^+Cl^-$ system by Woodcock and Singer in 1970, using a potential of the type suggested by Fumi and Tosi (1964) and assuming complete dissociation of the ions. Their results are compared with the transformed experimental data in Fig. 8.13. The agreement for the Cl–Cl distribution is good, that for the Cu–Cu distribution is bad. Page and Mika suggest that either the CuCl is not fully ionized, having a large number of associated molecules of the form $[Cu^+Cl^{n-}]^{(n-1)-}$ which would give a number of free $Cu^+$ ions, or there is a geometric effect due to the ionic radius of the $Cu^+$ ion being approximately half of that of the $Cl^-$ ion, leading to interference effects in the scattering. The experiment is a good illustration of the type of information now available in the study of liquid mixtures. However, to quote Enderby *et al.* (1966): 'To obtain the partial structure factors for an alloy with the accuracy which is now possible for one-component liquids still represents a major experimental challenge'.

## 8.5. Molecular liquids

### 8.5.1. General

As for a binary system, several partial structure factors are needed to define the structure of a molecular liquid (e.g. three are needed for diatomic molecules) but there are two additional complicating features. The first is in the collection of data; the well-defined interatomic distances in a molecule give rise to sharp peaks in $g(r)$ which, in turn, give long-range oscillations in $S(Q)$. This means that measurements should be taken up to large $Q$-values (say 20–25 $Å^{-1}$) to get accurately transformable data, and also that the oscillations make the normalization more difficult—we cannot use the high-$Q$ limit of $S(Q) \to 1$ as a guide line in processing the data. The second complication concerns the nature of the intermolecular forces. For both simple and binary liquids, the assumption is made that the force between atoms is spherically symmetric (this is a crucial assumption in the theories used), but for irregularly-shaped molecules, orientational correlation may occur between the molecules and the intermolecular forces are then, of course, no longer spherically symmetric. It is sometimes assumed that a symmetric force law holds for 'spherical' molecules, such as tetrahedrally bonded $CCl_4$, but if any angle-dependent forces are present, a more complex theory is needed.

Consequently, we must consider, not only the positions of the molecules, but also their orientational correlations,† and it is not surprising that, as yet, little progress has been made in the neutron and X-ray fields on the structure of

---

† A similar complication arises, for the same reason, in discussing the atomic motion in molecular liquids: see section 5 of Chapter 6.

## STRUCTURE OF LIQUIDS BY NEUTRON SCATTERING

molecular liquids. Many measurements (mainly X-ray) have been made, but little analysis has been done beyond the use of models or the transformation of the data to identify 'co-ordination shells' (see, e.g. Rao 1968; Gruebel and Clayton 1967; Narten et al. 1967). This is chiefly because a single scattering measurement only gives averaged data on the partial structure factors. Clearly, the isotopic substitution method is needed. But here we run into the practical difficulty that suitable isotopes are rarely available in the quantities required for a scattering specimen and so alternative approaches must be used. One such approach is to use 'species' substitution. This means that in the molecule $XY_4$, say, instead of substituting isotopes of the atom X to vary the scattering length, a completely different atom is substituted for X. It is assumed that this does not alter the liquid structure; for the series $XCl_4$, this may be reasonable as chlorine is a large atom which mainly determines the geometric size of the molecule. The scattering lengths may also, of course, be varied by the use of different types of incident radiation. An additional piece of information which is available for a molecular system is the structure of the isolated molecule; this was first exploited by Zachariasen (1935).

We shall now outline recent work on two molecular systems, tetrahedrally-bonded molecules and water, which illustrate the use of these various approaches.

### 8.5.2. Tetrahedrally-bonded molecules

To study orientational effects, we consider first the distribution, $g_c(r)$, of the centres of the molecules (i.e. the positions of the molecules themselves) which gives rise to the molecular-centres structure factor $S_c(Q)$. This is not measurable experimentally, except under certain conditions. The observed scattering is then related to the scattering expected from single molecules whose centres are distributed according to $g_c(r)$ but modified by the orientation of the molecules on these sites. We call the observed distribution $S_m(Q)$. The theory for orientational correlation is given by Egelstaff et al. (1971) and only some of the results will be quoted here.

For certain special cases,

$$S_m(Q) = f_1(Q) + f_2(Q)[S_c(Q) - 1], \tag{8.19}$$

where $f_1(Q)$ is the molecular form factor observed for a single isolated molecule and $f_2(Q)$ is a form factor related to the correlation of molecular orientation. To solve eqn (8.19) still requires three independent pieces of information, but if a knowledge of $f_1(Q)$ is assumed and if $f_2(Q)$ is calculated by the use of models, only one is required. For completely uncorrelated molecular orientation, then $f_2(Q)$ is written $f_{2u}(Q)$ and

$$S_m(Q) = f_1(Q) + f_{2u}(Q)[S_c(Q) - 1], \tag{8.20}$$

or

$$S_c(Q) = 1 + [S_m(Q) - f_1(Q)]/f_{2u}(Q).$$

Consequently, *if* molecular orientations are uncorrelated and the molecular structure is known, then the positional distribution of the molecular centres can be calculated from a single diffraction pattern through eqn (8.20).

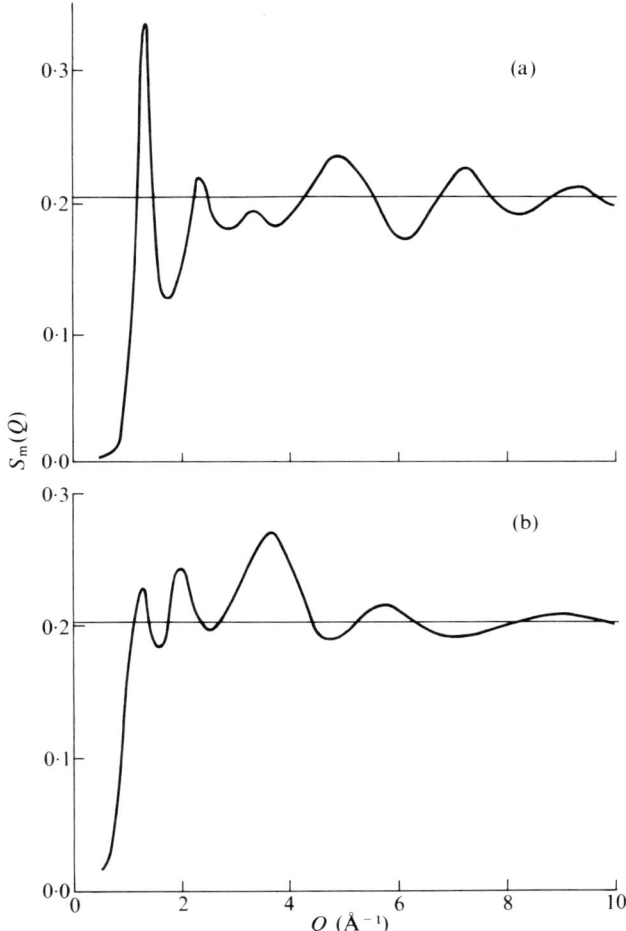

FIG. 8.14. Experimentally-determined absolute neutron structure factors $S_m(Q)$ (a) for liquid carbon tetrachloride at 22°C and (b) for liquid germanium tetrabromide at 60°C. (N.B. $S_m(Q)_{Q \to \infty} \to \Sigma\, b^2 (\Sigma b)^2$). (After Egelstaff *et al.* 1971).

Egelstaff *et al.* used data from carbon tetrachloride to show that orientation effects must exist in the liquid. The observed scattering pattern is shown in Fig. 8.14(a) and the functions $f_1(Q), f_{2u}(Q)$ and $f_{2c}(Q)$ (which allows for complete correlation of the molecules—see later) are shown in Fig. 8.15(a). It is important to notice that $f_{2u}(Q)$ is very small in the ranges $Q \approx 1\cdot9 \to 3\cdot3\,\text{Å}^{-1}$ and $5\cdot7 \to 6\cdot2\,\text{Å}^{-1}$ and indeed goes to zero at $Q = 2\cdot15$ and $2\cdot95\,\text{Å}^{-1}$. This means that, if we use eqn (8.20), extremely large and unacceptable values of $S_c(Q)$ are obtained unless the values of $S_m(Q)$ and $f_1(Q)$ are very close, or equal. However, $S_m(Q)$ and $f_1(Q)$ are substantially different and so the authors concluded that this is clear evidence that correlation of molecular orientation must be taken into account and has an important influence on $S_m(Q)$. Using a model in which a Cl–Cl

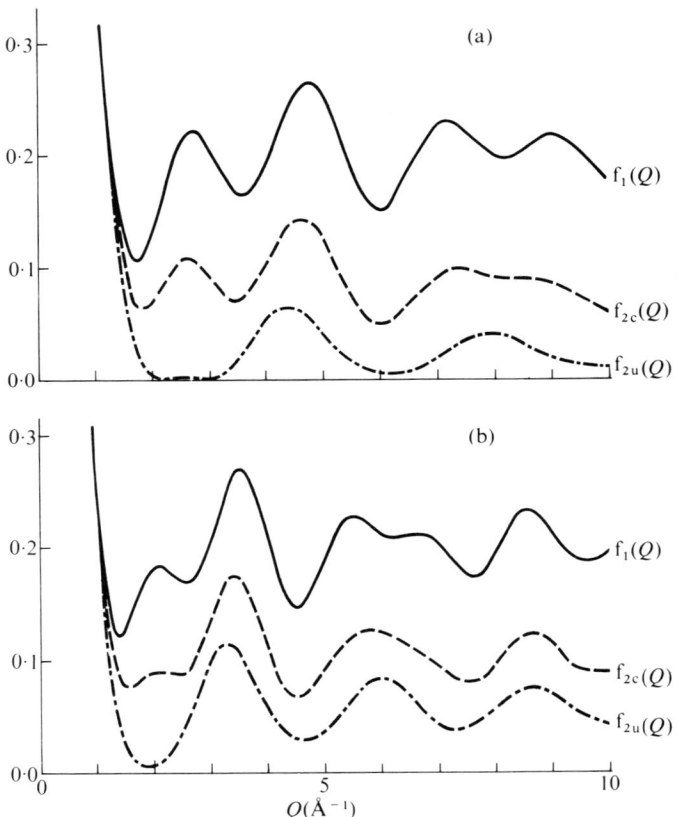

FIG. 8.15. The structure factors $f_1(Q)$, $f_{2u}(Q)$ and $f_{2c}(Q)$ (a) for carbon tetrachloride for $r_{C-Cl} = 1.77$ Å and (b) for germanium tetrabromide for $r_{Ge-Br} = 2.36$ Å (N.B. All the functions approach unity as $Q$ approaches zero). (After Egelstaff et al. 1971).

bond in one molecule is collinear with that of another molecule, so that the chlorine atom lies in the hollow formed by the three off-axis chlorine atoms of the other molecule, and in which the molecules are all completely correlated ('Apollo' model), $S_c(Q)$ was then calculated, see Fig. 8.16(a). For comparison, the dotted curve is $S(Q)$ for argon at 85 K, with the $Q$-scale adjusted to bring the main peaks into coincidence. Species substitution was then tried by repeating the procedure on GeBr$_4$, and the corresponding results are shown in Figs. 8.14(b), 8.15(b), and 8.16(b). The two molecular-centres structure factors are quite similar and much more realistic than for the uncorrelated case. The method, however, depends on model calculation, as species substitution in this case did not allow the simultaneous solution of eqn (8.20) for the two measurements of $S_m(Q)$.

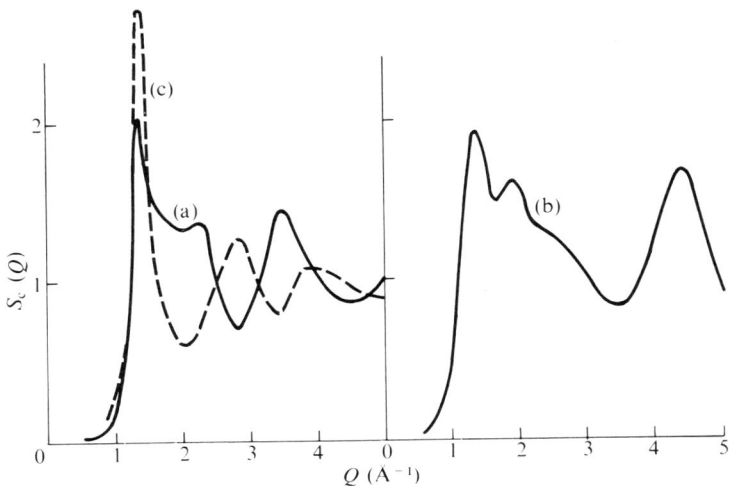

FIG. 8.16. The molecular-centres structure factor $S_c(Q)$ for complete orientational correlation for the Apollo model. (a) Carbon tetrachloride at 22°C and for $r_{C\text{-}Cl}$ = 1·77 Å. (b) Germanium tetrabromide at 60°C and for $r_{Ge\text{-}Br}$ = 2·36 Å. (c) $S(Q)$ as observed for liquid argon, scaled to make the first maximum coincide with liquid carbon tetrachloride. (N.B. $S_c(Q)$ is virtually the same for any $R_c > 3\cdot 5$ Å). (After Egelstaff et al. 1971).

### 8.5.3. Water

The ready availability of the hydrogen isotopes $H^1$ and $D^2$ appears to make water a suitable system for isotopic substitution measurements. Light water, however, scatters neutrons almost totally incoherently, and it is impossible, with present resources, to measure an accurate structure factor. The neutron structure factor of heavy water can readily be measured, with the limitation that the Placzek corrections are difficult to apply to light atoms. The X-ray data for light and heavy water are indistinguishable (Narten et al. 1966), but the effective scattering lengths of the atoms for the two types of radiation are quite different and give us two independent measurements. Page and Powles (1971), following the approach outlined in the previous section, have suggested that the neutron scattering (from both the deuterium and oxygen) gives a value of $S_m(Q)$ while the X-ray scattering, which comes predominantly from the oxygen atoms, gives $S_c(Q)$. This assumes that the scattering centres of the molecules have the same distribution as the oxygen atoms. Knowing these two quantities, together with the molecular structure which gives $f_1(Q)$, should allow eqn (8.19) to be solved for $f_2(Q)$. Unfortunately, the uncertainty of the corrections to the water data makes this determination unreliable. For this reason, Page and Powles adopted the alternative procedure of calculating $f_2(Q)$ using various assumptions for the correlation, and combining this $f_2(Q)$ with $f_1(Q)$ and $S_c(Q)$ to predict the neutron data. The measured X-ray data of Narten *et al.* (1966) are shown in Fig. 8.17(b), while $f_1(Q)$ and $f_{2u}(Q)$ (for completely uncorrelated orientation) are shown in Fig. 8.17(c). Combining these values in eqn (8.20) gives a calculated $S_m(Q)$, which is shown in Fig. 8.17(d). This is to be compared with the measured neutron structure factor in Fig. 8.17(a). The agreement is remarkably good except in the region of $Q = 4$ Å$^{-1}$. Page and Powles concluded that most of the

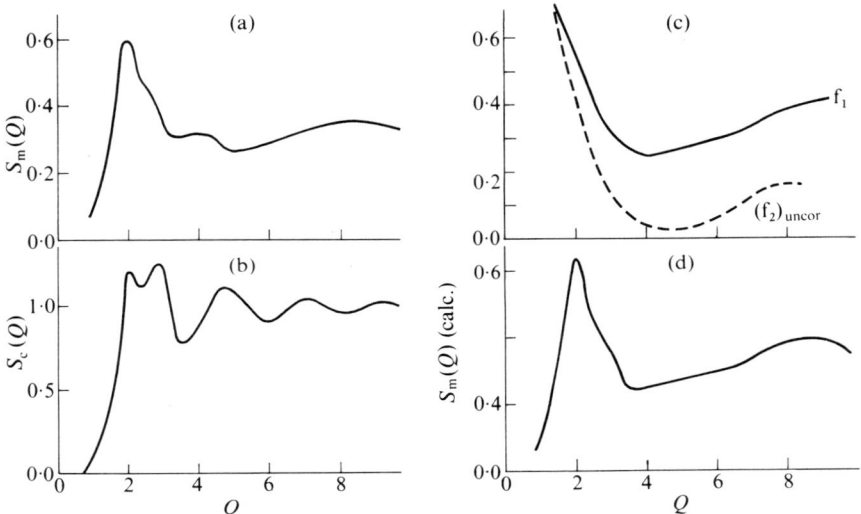

FIG. 8.17. Data on water at 22°C. (a) Experimentally determined neutron structure factor for D$_2$O. (b) Experimentally determined X-ray structure factor for H$_2$O. (c) Molecular form factor $f_1$ and uncorrelated form factor $f_{2u}$. (d) Calculated neutron structure factor. (After Page and Powles 1971).

information regarding orientational correlation is contained in this region (e.g. $f_{2c}(Q)$ is negative here), but were unable to formulate a correlation which predicts the neutron data over the whole range of $Q$.

We can see then that the determination of the structure of molecular liquids is only just beginning to be feasible. Wider application of the isotopic substitution method (which is unique to neutron scattering) and the combination of data from different experimental techniques are the most promising lines for future work.

## 8.6 References

ASCARELLI, P. *and* CAGLIOTI, G. (1966). *Nuovo Cim.* **X43**, 375.
ASCARELLI, P. (1966). *Phys. Rev.* **143**, 36.

ASHCROFT, N. W. (1967). *Physica.* **35**, 148.
BLECH, I. A. *and* AVERBACH, B. L. (1965). *Phys. Rev.* (A) **137**, 1113.
BÜHLER, H. F. *and* STEEB, S. (1969). *Z. Naturf.* (A) **24**, 428.
COCKING, S. J. *and* HEARD, C. R. T. (1965). *RQ.U.K. Atom. Energy Auth. (London)* R. 5016.
EGELSTAFF, P. A., DUFFILL, C., RAINEY, V., ENDERBY, J. E. *and* NORTH, D. M. (1966). *Phys. Lett.* **21**, 286.
EGELSTAFF, P. A. (1967). *An introduction to the liquid state.* Academic Press, London.
EGELSTAFF, P. A. (1969). *Experimental neutron thermalization.* Pergamon Press, Oxford. p. 153.
EGELSTAFF, P. A. (1970). *Current Problems in Neutron Scattering* (C.N.E.N, Rome). p. 56.
EGELSTAFF, P. A., PAGE, D. I. *and* POWLES, J. G. (1971). *Molec. Phys.* **20**, 881.
ENDERBY, J. E., NORTH, D. M. *and* EGELSTAFF, P. A. (1966). *Phil. Mag.* **14**, 961.
ENDERBY, J. E. *and* NORTH, D. M. (1967). *Phys. and Chem. Liq.* **1**, 1.
ENDERBY, J. E. (1968). *Physics of simple liquids.* N. Holland, Amsterdam. p. 620.
FUMI, F. G. *and* TOSI, M. P. (1964). *J. Physics Chem. Solids* **25**, 31.
GRUEBEL, R. W. *and* CLAYTON, G. T. (1967). *J. chem. Phys.* **46**, 4875.
HENSHAW, G. D. (1957). *Phys. Rev.* **105**, 976.
HIRSCHFELDER, J. O., CURTISS, C. F. *and* BIRD, R. B. (1954). *Molecular theory of gases and liquids.* Wiley, N.Y.
ISHERWOOD, S. P. *and* ORTON, B. R. (1970). *Phys. Lett.* (A) **31**, 164.
JOHNSON, M. D., HUTCHINSON, P. *and* MARCH, N. H. (1964). *Proc. R. Soc.* (A) **282**, 283.
KHAN, A. (1964). *Phys. Rev.* (A) **134**, 367.
LEVY, H. A. *and* DANFORD, M. D. (1964). *Molten salts chemistry.* Interscience, N.Y.
LORCH, E. (1969). *J. phys. C.* **2**, 229; (1970). *J. phys. C.* **3**, 1314.
NARTEN, A. H., DANFORD, M. D. *and* LEVY, H. A. (1966). Rep. Oak Ridge Nat. Lab. Number 3997.
NARTEN, A. H., DANFORD, M. D. *and* LEVY, H. A. (1967). *J. chem. Phys.* **46**, 4875.
NORTH, D. M. (1966). *Ph.D. Thesis.* University of Sheffield.
NORTH, D. M., ENDERBY, J. E. *and* EGELSTAFF, P. A. (1968). *J. phys.* C **1**, 1075.
NORTH, D. M. *and* WAGNER, C. N. J. (1970). *Phys. and Chem. Liq.* **2**, 87.
PAALMAN, N. H. *and* PINGS, C. J. (1962). *J. appl. Phys.* **33**, 2635.
PAGE, D. I., EGELSTAFF, P. A., ENDERBY, J. E. *and* WINGFIELD, B. R. (1969). *Phys. Lett.* (A) **29**, 296.
PAGE, D. I. (1972). Rep. U.K. atom. Energy Auth. (London). R. 6828.
PAGE, D. I. *and* MIKA, K. (1971). *J. Phys. C.* **4**, 3034.
PAGE, D. I. *and* POWLES, J. G. (1971). *Molec. Phys.* **21**, 901.
PLACZEK, G. (1952). *Phys. Rev.* **86**, 377.

POWELL, M. J. D. (1970). *Numerical approximations to functions and data.* Athlone Press, London. p. 65.
RAO, K. R. (1968). *J. Chem. Phys.* **48**, 2395.
RAO, K. R., DASANNACHARYA, B. A. *and* YIP, S. (1971). *J. Phys.* C **4**, 2725.
RICE, S. A. *and* GRAY, P. (1965). *The statistical mechanics of simple liquids.* Wiley, N.Y.
SLAGGIE, E. L. (1967). *Nucl. Sci. and Engng.* **30**, 199.
STRIFFLER, C. D. *and* CARPENTER, J. M. (1963). *Univ. of Michigan Report* 03712-3-T.
VINEYARD, G. H. (1954). *Phys. Rev.* **96**, 93.
WAGNER, C. N. J., HALDER, N. C. *and* NORTH, D. M. (1969). *Z. Naturf.* (A) **24**, 432.

# 9 Hydrogen Bonding, and Some Results of its Study by Neutron Diffraction

By J. C. SPEAKMAN
Chemistry Department, Glasgow University.

## 9.1. Introduction

Though the name *hydrogen bond* dates back only about forty years, its history began much earlier when it was first realized that water is a most unorthodox liquid. The abnormality is shown in particular by its physical properties. An example, amongst many, is the boiling point which is 260°C higher than that of methane, a compound with a very similar molecular weight.

With the preparation and characterization of a variety of simple compounds, it became clear that a similar, if less marked, abnormality occurred in all compounds whose molecules were polar; notably those molecules containing HO-groups, but also those with $H_2N$-groups and some others. The abnormality diminished when the hydrogen atoms of the HO- or $H_2N$-groups were replaced by methyl, or other similar radicals. The situation is typified in the following sample of boiling points (Table 9.1).

TABLE 9.1

|            | $H_3C-CH_3$ | $H_3C-OH$ | $H_3C-O-CH_3$ | $H_3C-CH_2-OH$ |
|------------|-------------|-----------|---------------|----------------|
| Mol. Wt.   | 30          | 32        | 46            | 46             |
| B.Pt. (°C) | −172        | 64·5      | −24           | 78·3           |

Numerous differences of this sort between the two classes of 'normal' and 'abnormal' liquids suggested some fundamental cause of difference. This gradually became attributed to *molecular association* of the latter. In the liquid state, water was more appropriately represented by the formula $(H_2O)_n$, where $n$ was some number larger than unity, though not specifically known; alcohols might be $(ROH)_n$; ammonia $(NH_3)_n$. No methods were then available for providing direct evidence of such association in liquids. Though some liquids— but not water—gave vapours in which density measurements implied association, this was not a satisfactory proof of association in the liquid state. Nevertheless, the concept proved to be a valid one.

To explain how molecules, ordinarily regarded as stable and 'saturated' systems, might link themselves together was more difficult. It was believed that nitrogen might increase its valency from 3 to 5, and oxygen from 2 to 4; granted this,

## HYDROGEN BONDING

such formulae as (9.1) for associated water could be written. But this took no account of the fact that abnormal behaviour ceases when the hydrogen attached to the oxygen is replaced by methyl. It ignores the essential role of the hydrogen.†

$$\begin{array}{c} H \\ \diagdown \\ H \diagup \end{array} O=O \begin{array}{c} H \\ \diagup \\ \diagdown H \end{array} \qquad (9.1)$$

In the early days of valency theory, the valency of hydrogen was unity, almost by definition. During the first decade of this century, several chemists began to write formulae such as (9.2), (9.3), and (9.4). The dotted lines represented some sort of rather weak bonding between a hydrogen atom and fluorine, or oxygen. These formulae implied, even if this was not explicitly stated, that hydrogen might become bivalent—that it might form a second, and weaker, bond in addition to its primary valency, represented by the unbroken line which stood for what we should now call a covalent bond. Thus two saturated molecules could be held together; if they were of the same kind, this would be molecular association; if not, it would be the formation of a 'molecular compound'.

$$\text{H–F} \cdots \text{H–F} \qquad (9.2)$$

$$R\text{–O–H} \cdots O \begin{array}{c} \diagup H \\ \diagdown R \end{array} \qquad (9.3)$$

$$H_2 N\text{–H} \cdots O \begin{array}{c} \diagup H \\ \diagdown H \end{array} \qquad (9.4)$$

This is what is now known as hydrogen bonding. It has turned out to be a widely occurring phenomenon, extending far beyond the case of abnormal liquids. We may instance its vital part in organizing the secondary structure of many proteins, and in the structure and replication of the nucleic acids.

A hydrogen bond can be represented generally by the scheme

$$A\text{–H} \cdots B.$$

The atoms $A$ and $B$ must be highly electronegative. Strong hydrogen bonding is confined to cases where $A$ and $B$ are fluorine, or oxygen, or nitrogen, the three most electronegative elements. In an alternative, though related, view, $AH$ must be reasonably acidic; it must be a *proton donor*. And $B$ must be reasonably basic; it must be a *proton acceptor*. In this article we are mainly concerned with systems in which $A$ and $B$ are both oxygen atoms.

† An authoritative monograph on 'Molecular Association', published in 1915, does not even contemplate the possibility that the hydrogen atom may be involved.

## 9.2. Crystallographic evidence for hydrogen bonding

Until the 1930s belief in the validity of hydrogen bonding as an explanation of molecular association was necessarily based on inference. More direct evidence became accessible when the development of X-ray crystallography reached a stage when it could be applied to a variety of simple molecular crystals. However, as we shall describe in more detail later, X-ray analysis has difficulty in locating hydrogen atoms even today. Forty years ago they could not be located at all. To that extent therefore detection of hydrogen bonding was still inferential, though it was convincing enough.

When two oxygen atoms are directly linked by a covalent bond, as in the hydrogen peroxide molecule, their interatomic distance is about 1·48 Å. But when, on the other hand, the two atoms are not bonded at all, their centres may not come closer together than the sum of their van der Waals radii. This radius is about 1·4 Å for oxygen. So, on the face of it, hydrogen bonding might be suspected whenever two oxygen atoms in a crystal are separated by a distance between 1·5 and 2·8 Å. In fact, these limits need reconsideration. No hydrogen bonds are known with the O · · · O distance less than 2·4 Å. The upper limit also needs augmentation. In an O—H · · · O bond, one of the oxygen atoms at least must carry a hydrogen atom. Should this hydrogen atom lie close to the O · · · O line— and it must, for effective bonding—then we must take it into account in assessing our upper limit; to the 1·40 Å van der Waals radius of the one oxygen we must add about 1·0 Å for the O-H distance at the other oxygen, as well as a further 0·9 Å for the van der Waals radius of the hydrogen atom. We arrive at a total distance of about 3·3 Å. To summarize, when two oxygen atoms are separated by a distance between 2·4 and 2·8 Å, hydrogen bonding can be confidently accepted; and, when the chemical geometry of the system makes it likely that a hydrogen atom lies near the O · · · O line, bonding may be suspected up to distances of 3·2 Å. From evidence of this sort a great number of O—H · · · O bonds were recognized. Most carboxylic acids, for instance, crystallize in such a way that their molecules are linked together by a pair of hydrogen bonds between their carboxyl groups. With mono-carboxylic acids this leads to the formation of dimers, as in formula (9.5), and with simple dicarboxylic acids to infinite chains of molecules (9.6). The experimental finding was that each

$$R-C\begin{matrix}OH\cdots O\\ \\ O\cdots HO\end{matrix}C-R \qquad (9.5)$$

$$\begin{matrix}\cdots O\\ \\ \cdots HO\end{matrix}C-C\begin{matrix}OH\cdots O\\ \\ O\cdots HO\end{matrix}C-C\begin{matrix}OH\cdots\\ \\ O\cdots\end{matrix} \qquad (9.6)$$

oxygen atom of one carboxyl group is 2·64—2·66 Å from the opposite oxygen of the other. Fig. 9.1(a) shows part of the electron-density map obtained by Robertson and his co-workers in their study of β-succinic acid (1959). It reveals

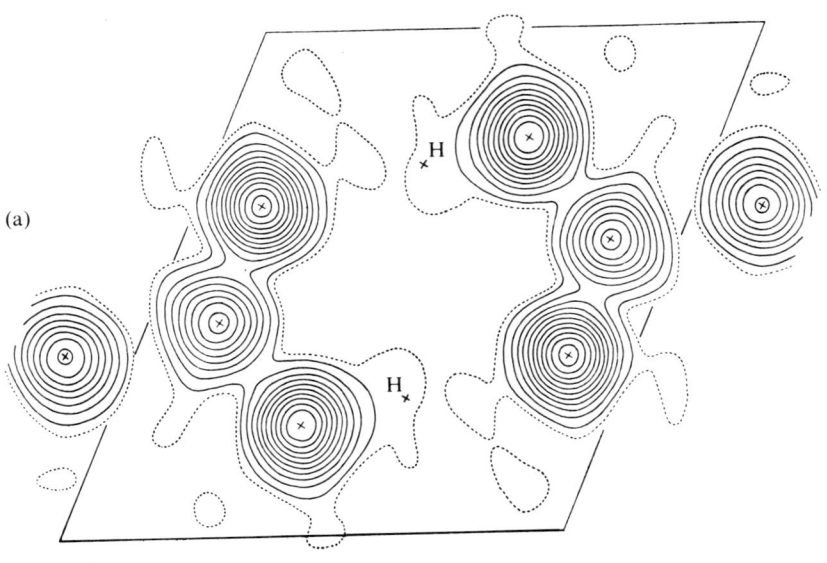

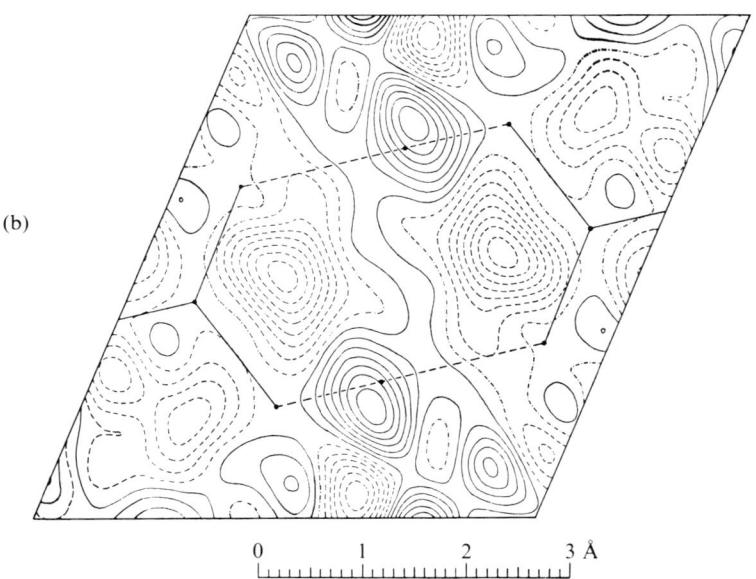

FIG. 9.1. Hydrogen bonding in $\beta$-succinic acid, from the X-ray study by J. M. Robertson and co-workers. (a) Electron density in the plane of two contiguous carboxyl groups. Contour lines are drawn at intervals of 1 electron/$\text{Å}^3$, except for the half-electron line which is broken. Crosses indicate atomic centres. (b) 'Difference' map in the same plane. (The contour-lines here are more closely spaced than in (a); negative contours are represented by broken lines.) Dots indicate positions of oxygen atoms and expected positions of hydrogen atoms, assuming them to lie on O $\cdots$ O lines. (After Broadley *et al.* 1959)

contiguous carboxyls, facing one another across a centre of symmetry. In primitive X-ray studies, the hydrogen atoms are not located though their presence is chemically certain. However, this analysis of succinic acid was a remarkably careful one for its period. Vestiges of the hydrogens can indeed be seen in Fig. 9.1(a) as it stands. They can be made more explicit by the tactic of the 'difference synthesis', the result of which is in Fig. 9.1(b). Whereas, in the calculation of the map in Fig. 9.1(a), the coefficients of the Fourier series were the observed structure amplitudes, $|F_o|$, with calculated signs, the coefficients for Fig. 9.1(b) were the algebraic differences $F_o - F'_c$, where $F_o$ is the appropriately signed observed structure factor, and $F'_c$ is the structure factor calculated for all the heavier atoms (C and O in this case), but with the smaller effects of the hydrogens omitted. Provided always that the observational data are good enough, the 'difference map' shows up what had been omitted. It is as if the main mountain range had been demolished, so that the foothills, previously unimpressive, now stand out prominently. Some, at least, of the minor detail in Fig. 9.1(b) is 'noise'. It is due to small errors in the observed intensity data or in the model used in calculating $F'_c$, but the largest positive peaks are unquestionably genuine, and they represent the hydrogen atoms, lying slightly off the direct O $\cdots$ O line. This map also shows regions of negative density—some of them comparable in prominence with the positive peaks. These 'hollows' are due to the use of a non-infinite Fourier series, to errors in the X-ray intensity measurements, to errors in scaling, and to inadequacies in the model upon which refinement had been based. A consideration of the effects of such errors leaves no doubt that the large positive peaks really do correspond to hydrogen atoms, when the X-ray analysis has been carefully done.

The accumulated evidence is that the actual positions found for hydrogen atoms in this way from X-ray data are a little in error. Measurement of Fig. 9.1(b), for instance, shows that the peak due to hydrogen is 0·95 Å from its oxygen atom. There are compelling reasons for supposing that the true centre of the hydrogen atom—i.e. its proton—is more than 1·0 Å away. The centre of the 'difference' peak does not exactly correspond to the true centre. The apparent position of the hydrogen is rather too close to its electronegative meighbour. The reasons for this are probably now understood (see Coulson 1970), though they need not concern us further here.

Apart from revealing many examples of hydrogen bonding in this way, crystal-structure analysis shows that the bonding is almost 'obligatory'. It is rare to find an hydroxyl group in a crystal that is not engaged in hydrogen bonding. The manner in which hydroxylic molecules pack themselves together in a crystal is such as to favour the formation of the maximum number of O—H $\cdots$ O bonds.

## 9.3. The theory of hydrogen bonding

Before proceeding further it is convenient to sketch the current theory of hydrogen bonding. A precise theory to explain hydrogen bonding is difficult—

## HYDROGEN BONDING

and for a simple reason: we are dealing with a relatively weak type of bonding. Compared with the total electronic energy of a molecular system, the change in energy when two such systems link themselves together is slight: 5–40 kJ mole$^{-1}$, perhaps, as against $10^3$ or more. In any calculation, therefore, the hydrogen-bonding energy tends to be swamped in the necessary approximations.

However, a simple qualitative view can usefully be suggested. Two mechanisms may be distinguished, even though the distinction is perhaps artificial. The first involves a direct electrostatic attraction. As the atoms $A$ and $B$ are necessarily electronegative, $A$–H must be polar in the sense pictured in formula (9.7),† whilst $B$, which is generally part of another molecule, will carry a complete, or partial, negative charge. Provided that the two

$$\overset{\delta-}{A}-\overset{\delta+}{H} \quad \overset{-}{B} \tag{9.7}$$

molecules are in the relative orientations shown, there will be an over-all attractive force between $A$H and $B$. It is now accepted that this effect is a source of bonding energy in all hydrogen bonds, and is a preponderating source in most.

The second mechanism involves the wave-mechanical effect of electron-delocalization. It can be formalized in molecular-orbital or in charge-transfer terms. For our qualitative purpose, the simplest description is in valence-bond terminology, as a 'resonance' phenomenon. We will illustrate it by reference to an O–H $\cdots$ O bond. The first approximation to an electronic formula for this system is shown in formula (9.8): there is a covalent bond between the hydrogen atom and one oxygen, whilst the second oxygen carries a lone-pair of electrons.

$$\text{O–H} \quad :\text{O} \tag{9.8}$$

$$\text{O:} \quad \text{H——O} \tag{9.9}$$

There is an alternative bond-diagram (9.9), however, and a hybrid of (9.8) and (9.9) is a better approximation. In the situation shown, the contribution of (9.9) is very small, because (9.9), with the grossly stretched bond on the right and the two non-bonded atoms on the left crushed uncomfortably together, has a much higher energy than (9.8).

Though this resonance explanation of hydrogen bonding is normally unimportant, it may become more significant in special cases, and for the following reason. In a long, and weak, hydrogen bond the hydroxyl group is little disturbed by the approach of the second oxygen; but, as this atom comes closer, the covalent bonding will be weakened, and the O–H distance will increase. (This may seem intuitively reasonable *a priori*; and there is now plenty of evidence for this trend, as we shall see later.) It follows that a situation may ultimately develop in which the two O $\cdots$ H distances are equal, with the hydrogen in the middle. This is formulated in (9.10).

† The symbol δ indicates a partial charge.

$$\left.\begin{array}{l}\text{O}-\text{H} \; :\text{O}\\ \text{O}: \; \text{H}-\text{O}\end{array}\right\} \quad (9.10)$$

Now the two contributing forms are equivalent; they will have the same energy. 'Resonance' stabilization may then become considerable, or even dominant. It is this consideration that makes 'very short' hydrogen bonds so particularly interesting.

Moving to a different system, we may cite the bifluoride anion, $FHF^-$. Most authorities now agree that the hydrogen bond here, with $F \cdots F = 2 \cdot 26$ Å, may be genuinely symmetrical. Are there any correspondingly symmetrical $O \cdots H \cdots O$ bonds? We shall examine this question in §9.8.

## 9.4. The application of neutron diffraction to crystal-structure analysis

In the present context, neutron diffraction is important because it enables us to locate hydrogen atoms much more accurately than with X-rays.

The diffraction of X-rays by crystals was discovered in 1912. It quickly led to the determination of the atomic arrangements in a few simple crystals, such as those of diamond and rock-salt. The difficulties of the method rise rapidly as more complex crystals come to be studied. But over the next half-century these difficulties have been largely overcome, and the method can be successfully applied to fairly large molecules, and even to crystalline proteins in favourable cases. We shall assume the reader to have some general acquaintance with the elementary principles of X-ray crystallography.

X-rays are scattered by the electrons. Analysis of the diffraction pattern from a crystal reveals only the electron-density distribution, and particularly the regions where this density is more highly condensed. These are the inner cores of the atoms. The higher the atomic number of the element, the greater its central electron density, and the more easily it is located. The outer parts of an atom are harder to study, and this includes the electrons principally concerned in chemical bonding. The same difficulty applied to hydrogen atoms: the minimal nuclear charge of the proton produces a relatively small electron-density condensation, compared with (say) an oxygen or a carbon atom. We have seen that hydrogen atoms can be detected given high-quality X-ray data; but, working near the limits of the method, our findings are liable to appreciable error.

We are not restricted to electromagnetic radiation. Any particle of mass $m$, moving with a velocity $v$, is associated with a wavelength $\lambda$. This is obtainable from the de Broglie equation,

$$\lambda = h/mv,$$

where $h$ is Planck's constant. Hence a beam of particles can, in principle, be diffracted by a crystal provided the wavelength is not too different from the repeat-distances in the crystal. Neutrons give suitable wavelengths when their kinetic energies are rather low, in the 'thermal' range, that is to say, when they are in thermal equilibrium with their surroundings near room temperature.

An elementary calculation of the wavelength runs as follows. The mean translational energy of gaseous molecules is equal to $(3/2)RT$. At around 25°C, the mean energy per molecule is

$$[(3/2) \times 8 \cdot 3 \times 298]/(6 \times 10^{23}) = 6 \cdot 2 \times 10^{-21} \text{ J molecule}^{-1}.$$

As this energy can also be written as $\frac{1}{2}m_n v_n^2$, and as the mass of the neutron is $1 \cdot 6 \times 10^{-27}$ kg, we have

$$v_n^2 = (2 \times 6 \cdot 2 \times 10^{-21})/(1 \cdot 6 \times 10^{-27}) \text{ m}^2 \text{ s}^{-2}$$

or $\quad v_n = 2 \cdot 8 \times 10^3 \text{ m s}^{-1}.$

Hence

$$\lambda = h/m_n v_n = (6 \cdot 6 \times 10^{-34})/(1 \cdot 6 \times 10^{-27} \times 2 \cdot 8 \times 10^3)$$
$$= 1 \cdot 5 \times 10^{-10} \text{ m} = 1 \cdot 5 \text{ Å}.$$

That neutrons really are diffracted by crystals was verified during the 1930s. Neutron diffraction, as a practicable method of supplementing X-ray crystallography, had to wait till the late 1940s, by which time the discovery of nuclear fission, and the development of the nuclear reactor provided sources of low-energy neutrons that were sufficiently intense.

For diffraction analysis we need a source of monochromatic radiation (i.e. all the radiation of the same, or very nearly the same, wavelength). This is much harder to realize with neutrons than with X-rays. The beam of thermal neutrons from the reactor-core is 'white'; it includes particles with a continuous, Maxwellian distribution of wavelengths. A nearly monochromatic beam can be obtained—though at the cost of wasting most of the neutrons—by allowing the beam to fall on to a large single-crystal, preset at such an angle, $\theta_M$, that the Bragg equation,

$$(n)\lambda = 2d \sin \theta_M, \text{ where } n = 1,$$

holds for the desired wavelength to be reflected from a suitably selected set of crystal-planes of spacing $d$. For instance, the 422-planes from a crystal of copper might be used to give a wavelength of $1 \cdot 04$ Å at $\theta_M = 45°$. The beam, thus filtered out, is not exactly monochromatic; it covers a small, though finite, band of wavelengths, and there is some admixture of neutrons of half the desired wavelength, because the Bragg equation is obeyed by such neutrons when $n = 2$. By various technical devices any effect of these unwanted wavelengths can be satisfactorily minimized (see Chapter 2).

After all this filtration, the monochromatic beam has a low intensity. Even from a large modern reactor, the number of monochromatic neutrons per second is less than the number of photons per second from an X-ray generator by a factor of at least $10^4$. Without special contrivances, photography is impossible. Some form of counter is used, usually one based on boron trifluoride enriched with respect to the boron isotope $^{10}B$. For all the reasons given, or implied, above, neutron diffraction is a more expensive technique than X-ray diffraction. Normally it would be applied only to crystals whose structures had already been determined with X-rays, and which had been shown to have special features of interest to justify the follow-up with neutrons. The more precise location of hydrogen atoms is the application we are concerned with here.

A further restriction is that much larger crystals are needed for neutron work. A larger crystal placed in a wider beam helps by increasing the counting-rate. (A large crystal would be a serious disadvantage with X-rays, since the loss of radiation by absorption within the crystal would be serious. The true absorption of neutrons is very much less.) Whereas the X-ray crystallographer prefers to use a crystal with linear dimensions $\sim 0{\cdot}02$ cm each way, he must try to grow one with dimensions nearer to $0{\cdot}2$ cm if he is to collect equally good neutron data. This is nearly always difficult; for many substances, impossible.

Neutrons are scattered by the nuclei of the various atoms in the crystal. As the size of the nucleus is $\sim 10^4$ times smaller than the atom, the scattering-region— the atom 'as seen by the neutrons'—would be virtually a mathematical point, were it is not that the atom is 'smeared out' by vibrational movements. Despite these movements, the peaks obtained by applying Fourier synthesis to neutron data are sharper than the electron-density peaks from X-ray analysis.

There are more important differences. X-ray diffraction is governed by the electron density. The *scattering amplitude* of any atom depends on the element's atomic number. Expressed in the units most convenient to crystallographers, the amplitude is equivalent to the atomic number when the Bragg angle, $\theta_B$, is zero, though it diminishes as $\theta_B$ increases. The scattering amplitude for neutrons is a property of the nuclide, rather than of the chemical element. It increases, in a general way, with atomic number; but the increase is proportional only to a fractional power of atomic number and, in any case, the increase is often overrridden by special effects. The apparently arbitrary nature of these can be seen in Table 9.2 which compares X-ray and neutron scattering amplitudes for some important elements. We may offer a note on the different units in Table 9.2. As the X-ray crystallographer is interested in electron density, it is logical for him to base his scattering amplitudes on electron units. The scattering of neutrons is characterized by the effective area (the cross-section) of the nucleus from which scattering occurs. A natural unit for this cross-section is $10^{-28}$ m$^2$. A corresponding length is conveniently expressed in terms of the unit $10^{-15}$ m. This is now usually known as a *fermi*. (Theory shows that these quite disparate units can be brought into near-congruence by use of the factor $e^2/mc^2$, where $e$ is the electronic charge, $m$ the electronic mass, and $c$ the velocity of light.)

The most important point for our purpose is that hydrogen is much more favourably placed for neutron than for X-ray analysis. The neutron-scattering amplitude for ordinary hydrogen (or the proton) is about half that for carbon or oxygen. In practice, the benefit of using neutrons is even better than this. A casual inspection of Table 9.2 might suggest that the ratio in favour of using neutrons, as compared with X-rays, for distinguishing hydrogen from oxygen is $3·7/5·8 : 1/8 = 5·3$. X-ray scattering factors diminish with Bragg angle; at a moderate value of $\theta_B$ — representing a rough average over the range covered — the amplitude for hydrogen will have fallen to 0·25, that for oxygen only to 4·1. Hence the favourable ratio, when we change from X-rays to neutrons, is better represented as $3·7/5·8 : 0·25/4·1 = 10·5$.

TABLE 9.2

*Comparison of nuclear neutron-scattering amplitudes (fermis) and atomic X-ray-scattering factors (electron units)*

| Element | $f_N$ (does not change with $\theta_B$) | $f_X$ (falls as $\theta_B$ rises) |
|---------|------------------------------------------|-----------------------------------|
| H       | −3·74                                    | 1                                 |
| D       | 6·67                                     | 1                                 |
| C       | 6·65                                     | 6                                 |
| O       | 5·80                                     | 8                                 |
| K       | 3·7                                      | 19                                |
| U       | 8·5                                      | 92                                |

The negative sign attached to the scattering length for the proton signifies that there is a phase difference of 180° between the neutron wave scattered by the proton compared with waves scattered from the other nuclides shown in Table 9.2. So, if we take the scattering length of most nuclides as positive that for the proton (and a few other nuclides) must be negative. The practical outcome, for the chemical crystallographer, is that, when he organizes his Fourier syntheses to yield positive peaks (maxima) for the ordinary atoms, he will find hollows (minima) for the hydrogens. An example will be given in Fig. 9.5 which follows the convention of representing negative contours by broken lines.

The neutron-scattering length of deuterium is positive and its amplitude is as large as that of carbon and larger than that of oxygen. When we are searching for hydrogen atoms, it is often an advantage to use deuteriated crystals. The obvious reason is the higher amplitude of deuterium; but there is another, at least as important. Because of the phenomenon of incoherent scattering from the protons, ordinary materials containing hydrogen give a large background of neutrons underlying the peaks in which we are interested. This effect is greatly reduced from deuterium.

An example illustrating the difference between X-ray and neutron results is shown in Fig. 9.2. The first diagram (a) shows the electron density in the mean molecular plant of naphthalene, $C_{10}H_8$, as calculated by Robertson and his co-workers from their X-ray measurements (Abrahams, Robertson, and White 1949).

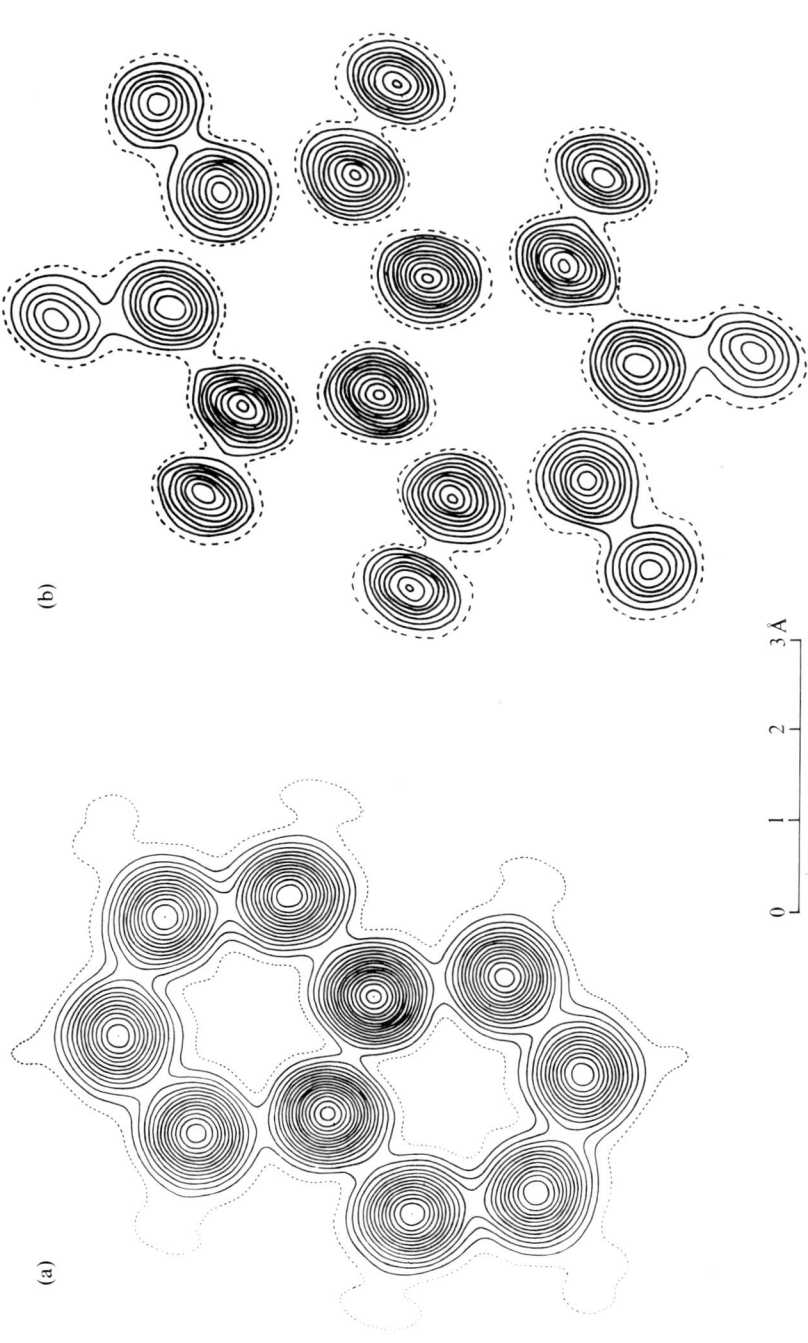

FIG. 9.2. Mean plane of the naphthalene molecule. (a) Electron density based on X-ray measurements on $C_{10}H_8$. First half-electron level shown dotted. (After Abrahams *et al.* 1949.) (b) Neutron-scattering density for $C_{10}D_8$. (Map computed by Hodder from neutron data of Pawley and Yeats 1969.) Zero level indicated by broken line.

## HYDROGEN BONDING

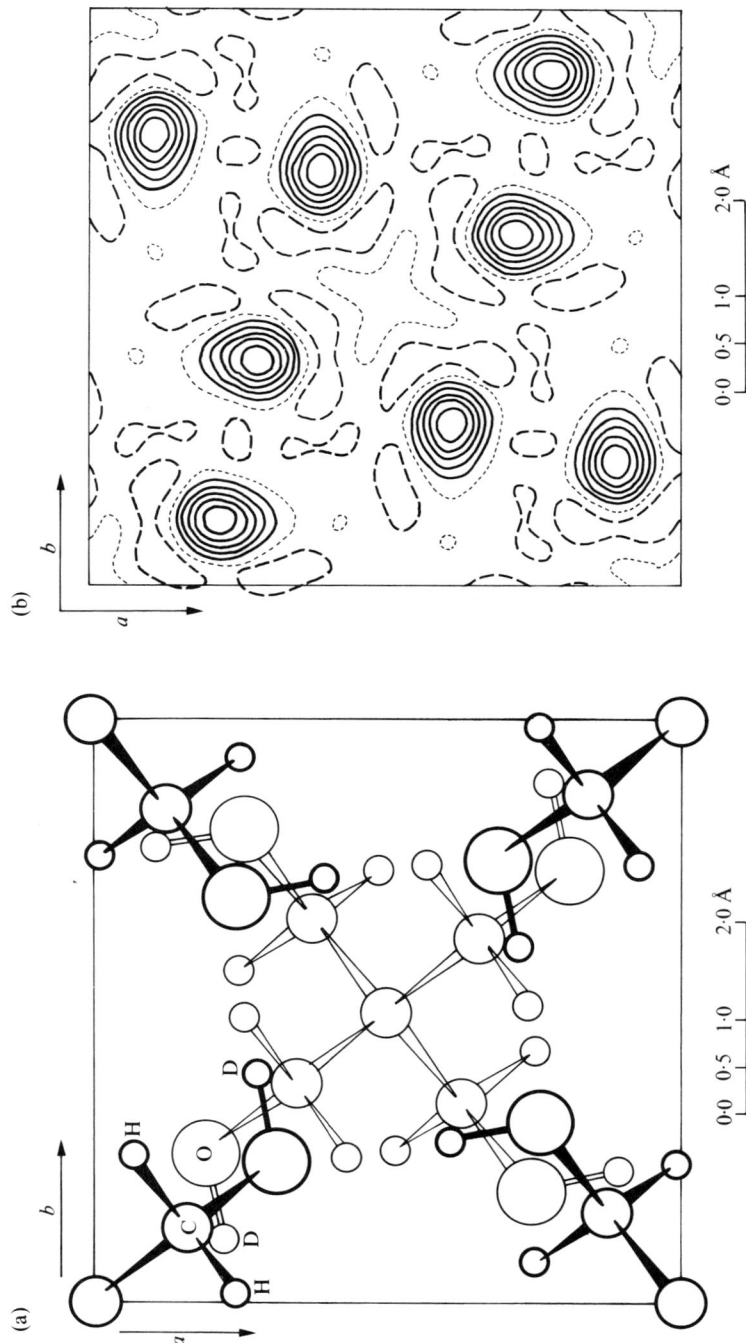

FIG. 9.3. The crystal structure of deuteriated pentaerythritol, C(CH$_2$OD)$_4$, seen along the $c$-axis. (a) Atomic arrangement showing molecules at $z = 0$ (heavy lines) and at $z = \frac{1}{2}$ (weak lines). (b) Fourier difference projection showing deuterium atoms only. (Zero contour dotted; negative contours shown as broken lines.) (After Hvoslef 1958)

The carbon atoms show up well, but the hydrogen atoms attached to eight of the carbons appear in (a) only as outward-facing buttresses. (They can be made more prominent by computing an electron-density 'difference' map, like that in Fig. 9.1(b).)

The second diagram, Fig. 9.2(a), was calculated by Hodder from data based on neutron-diffraction measurements, by Pawley and Yeats (1969), on a single crystal of perdeutero-naphthalene, $C_{10}D_8$. In this diagram the (heavy) hydrogen atoms are represented by peaks nearly as high as those for carbon. Since the neutron-scattering lengths of D and C are virtually the same, we might expect the peaks to be the same also. This would be so if the molecule was not vibrating. In fact, such a molecule, at room temperature, is executing quite lively vibrations. Part of the vibrational movement is torsional: the molecule is oscillating about its centre of mass. Movement of this sort 'smears' out the atomic peaks, and does so more the further the atom is from the centre. A careful inspection of the contour lines will confirm this; and, in particular, it makes the peaks for deuterium distinctly lower than those for carbon.

The opposite signs of the neutron scattering lengths of H and D make possible the tactic of the 'null matrix'. As the scattering length for deuterium is roughly double that for hydrogen, a crystal which is about 35 per cent deuteriated should 'lose' its hydrogen atoms so far as neutron diffraction is concerned. Provided the partial deuteriation is random, then each hydrogen site is, statistically, occupied by 0·35 of a deuteron and 0·65 of a proton, and their diffracting effects would exactly cancel one another. (In fact, for chemical reasons, incomplete deuteriation does not always occur randomly. Some sites may be preferentially substituted: an example of this is given in the next paragraph.) This intriguing possibility does not seem to have been attempted with hydrogen, though it has been applied to other isotopic species.

However, an ingenious experiment by Hvoslef (1958) made use of the opposite effects of H and D in a study of penta-erythritol $C(CH_2OH)_4$. When this compound is recrystallized from heavy water $C(CH_2OD)_4$ results, as ready deuteriation occurs only at the hydroxylic sites. Collection of full three-dimensional neutron data was impracticable at that time, so Hvoslef based his study on restricted data which would lead to a projected view of the structure along the 8·0 Å $c$-axis (see Fig. 9.3(a)); but he had data for both of the isotopic variants formulated above. He calculated a special kind of 'difference' synthesis, in which the coefficients were the differences between the two observed structure factors, $(F_o^D - F_o^H)$. The contributions from the carbon, oxygen and the eight unchanged hydrogen atoms to $F_o^D$ and $F_o^H$ are exactly the same, since the two crystals are isomorphous; but the differences between the contributions of the deuterium to the one and of hydrogen to the other, these being of opposite sign, are substantial. In the 'difference' map (Fig. 9.3(b)), the replaceable hydrogen sites are marked by high peaks, whilst everything else is lost. This provided firm locations for the hydroxylic hydrogen atoms, which had not been found in the earlier X-ray studies.

HYDROGEN BONDING

## 9.5. A solution of the phase problem by neutron diffraction

A fundamental difficulty obstructs the crystallographer who wishes to determine the positions of the atoms in a crystal by diffraction methods. This is the phase problem. The electron-density in the unit cell can be generated by Fourier series; simplified for the one-dimensional case, this may be written

$$\rho(x) = \text{constant} \times \sum_{h=-\infty}^{h=+\infty} |F(h)| \cos[2\pi(hx/a) + \alpha(h)].$$

This summation, needed to find $\rho$ at a particular point distance $x$ from the origin, uses as coefficients the structure amplitudes, $|F(h)|$, measured from a number of Bragg reflexions of successive order, $h$. However, to fit the Fourier terms together so that they add up to the correct electron-density values along $x$, we must place them together in the correct phase-relationship, which is measured by the respective phase-angles, $\alpha(h)$. Unhappily, we do not yet have any method for measuring $\alpha(h)$ directly,† and until we can determine the $\alpha(h)$'s with at least some degree of truth, the above formula is useless. The same is true in the more realistic cases of two- or three-dimensional series. It is also true for the neutron-scattering density which we might hope to determine from a set of neutron-diffraction intensities.

In a centro-symmetrical crystal the phase problem remains, though it is reduced in difficulty. Given a suitable choice of origin, the phase-angles are restricted to values of 0 and $\pi$; or, in other words, we have to decide whether to attribute a positive or negative sign to each $|F(h)|$.

There are several ways of solving the phase problem—some more difficult than others. One of the simplest can be applied when the molecule contains one atom of relatively high atomic number, a 'heavy' atom. A classical example was the solving of the structure of platinum phthalocyanine, $PtC_{32}H_{16}N_8$, by Robertson and Woodward (1940). The platinum atom (atomic number = 78) lies at a centre of symmetry in the middle of the molecule; the $b$-axis of this crystal is very short, which allows us to get an unobstructed view of the planar molecule in projection. X-ray intensities had been estimated for about 300 relevant reflections, yielding the same number of structure amplitudes. A double Fourier series with positive signs attributed to these terms gave an electron-density map, from which the molecular structure in projection could clearly be recognized. This was an *absolute* structure determination in the sense that nothing had been assumed about this structure beforehand. It was derived simply from measured amplitudes.

Macdonald has produced a similar solution of the phase problem with neutron data. He had collected full neutron-diffraction data for two isomorphous salts, potassium hydrogen di-trifluoroacetate $(KH(TFA)_2$, where $HTFA = CF_3.CO_2H)$

---

† The 'direct method' of measuring phases (Woolfson 1961) is, in fact, *indirect* in that it requires a measurement of the intensities first, which are then analysed to yield some of the phases.

and its deuteriate, KD(*TFA*)$_2$. The structures of these crystals are virtually identical, except that the hydrogen-atom site is occupied by a proton in one case, and by a deuteron in the other. If we imagine the H-structure subtracted from the D-structure, all that should remain—neutron-wise—is a 'difference' atom, (D–H), at the centre. The neutron-scattering lengths of D and H being +6·5 and −3·8 fermis, that of (D–H) will be +10·3, as explained in our account of Hvoslef's experiment in the previous section. It is a 'heavy' atom to neutrons, and at a centre of symmetry. Macdonald employed a rather different procedure from that used by Robertson and Woodward, though the basic principle is the same. He computed a three-dimensional *Patterson function*, which uses $F^2$ rather than ±$F$ as coefficients, and produces peaks corresponding to the vectors between

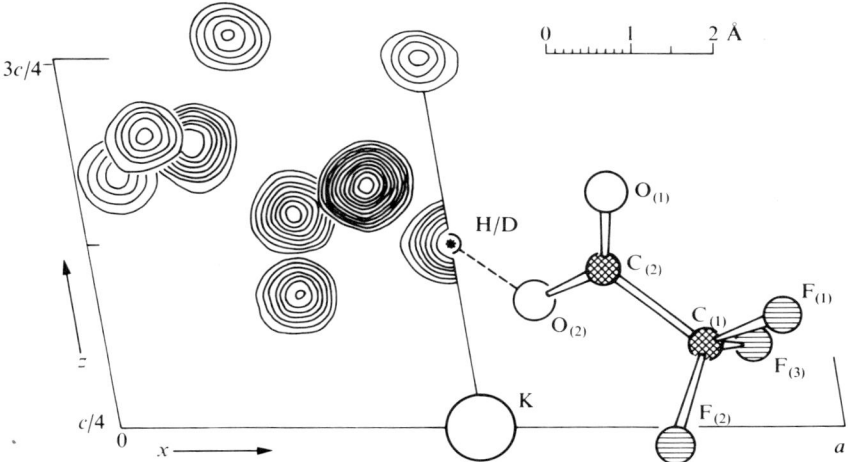

FIG. 9.4. Potassium hydrogen (or deuterium) di-trifluoroacetate: left-hand side, sections from the three-dimensional 'difference' Patterson function; right-hand side, the structure as determined independently by X-ray analysis. (The centre of symmetry, at which the hydrogen atom is located, is here symbolized by an asterisk.)

every possible pair of atoms in the structure. The 'difference' Patterson function, with $(F_D^2 - F_H^2)$ as coefficients, contains vector-peaks only between (D–H) at the centre and each other atomic nucleus; all other vectors are eliminated. In fact this function gave the view of the structure shown in the left-hand side of Fig. 9.4. On the right we represent, by conventional drawing, the structure as it had been found by ordinary X-ray methods. We see that the 'difference' Patterson map shows all the atoms of the trifluoroacetate group and the potassium atom in their true positions in the unit cell. Like that of platinum phthalocyanine, this is an absolute structure determination. In this case the phase problem had in fact already been solved. Though academic, this application of neutron diffraction is an elegant one (Macdonald, Robertson, and Speakman 1971).

## 9.6. The location of the hydrogen atom in hydrogen bonds

This is the most inportant application of neutron diffraction so far as the chemical crystallographer is concerned. It is no longer necessary to follow Hvoslef's tactic. When X-ray analysis finds two oxygen atoms in a crystal (say) 2·7 Å apart, the presence of a hydrogen bond is *inferred*. It is confirmed when an appropriately situated proton is located by neutrons. A classical example may be discussed with the help of Fig. 9.5. This is concerned with α-resorcinol, m-dihydroxybenzene, the crystal structure of which was discovered by Robertson (1936) by X-ray methods. In diagram (a) is shown the full structure, including the hydrogen atoms (which were not, of course, found in this early analysis), as viewed along the c-axis of the crystal. The electron-density map is shown in (b); the two oxygen atoms and two carbons of each molecule are resolved, and two pairs of carbons are not. But, with the aid of another projection, the general structure was firmly established. Intermolecular O · · · O distance, estimated at 2·66 and 2·75 Å, indicated hydrogen bonds. In 1956 Bacon and Curry published a c-axial projection of this structure based on 125 neutron-diffraction data. This is shown in Fig. 9.5(c). The data being more numerous than those available for the electron-density projection, the carbon atoms are now all resolved, and all six hydrogen atoms are duly represented by negative peaks. In particular, there is such a negative peak along each of the inferred hydrogen bonds, nearer to one oxygen than to the other. The situation of the hydrogen atoms is more clearly seen in Fig. 9.5(d), which is a 'difference' map, with Fourier coefficients $(F_o - F'_c)$, where $F'_c$ is the neutron structure factor calculated on the basis of the carbon and and oxygen atoms only.

Careful measurements, based on this map (supplemented by some information from the X-ray study, since the map gives only a projection†), show that the O · · · O distances in the hydrogen bonds average 2·72 Å, that the O—H distance is 1·02 Å, and the non-bonded H · · · O about 1·71 Å, and that the proton is slightly off the direct O · · · O line, as can be seen in Fig. 9.5(c) and (d).† A great deal of information of this kind is now available. It is summarized in the monographs by Pimental and McClellan (1960), and by Hamilton and Ibers (1968). Certain trends are clear, and we shall discuss them briefly here.

In an isolated hydroxyl group, such as in the gaseous water molecule, the O—H distance is 0·95–0·97 Å. It seems intuitively plausible that, when a second oxygen atom approaches in the sense O—H · · · O, the original O—H distance may increase. Since hydrogen bonding is to occur, there must be some attraction, as suggested by the formulation H · · · O. This should weaken the O—H bond and lengthen it. Alternatively, we may regard it as a repulsion between the two oxygen atoms, which are now closer together than the sum of their van der Waals radii.

It would be nice if we could observe an isolated hydrogen bond and actually

---

† G. E. Bacon and R. J. Jude have recently completed a three-dimensional neutron-diffraction study of α-resorcinol.

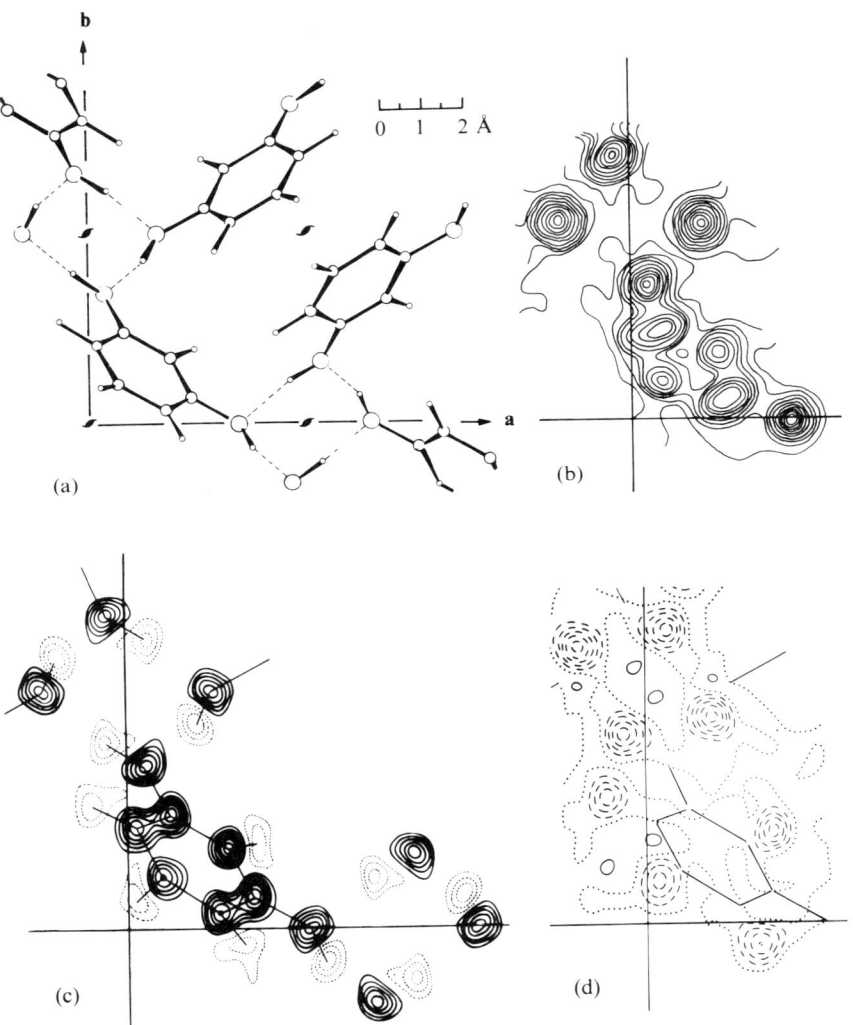

FIG. 9.5. The crystal structure of α-resorcinol, and the hydrogen bonding, seen in projection along the c-axis. (a) Schematic view of the structure and bonding. (b) Projection of the electron density from X-ray diffraction. (c) Projection of the neutron-scattering density. (d) 'Difference' map based on the neutron-diffraction study. (Negative contours are shown by broken lines; zero contour dotted.) ((b) after Robertson 1936; (c) and (d) after Bacon and Curry 1956).

measure how the various interatomic distances vary as the second oxygen atom is brought closer. At present this is quite impossible. The next best thing we can do is to collect as many examples as we can, from seemingly comparable structures, and to plot the O—H distances, determined from neutron diffraction, against O · · · O. One of the earlier attempts at this was by Nakamoto, Margoshes, and

Rundle (1955). As we are dealing with different crystalline materials in each case, and as the results are liable to experimental errors anyway, we must not be surprised to find a scatter of points on our graph. Despite this, an unmistakable trend can be discerned. This trend is represented in Fig. 9.6, the smooth curve being the one obtained by Hamilton and Ibers from a consideration of the data available in 1968. For bonds with O $\cdots$ O greater than, perhaps, 2·55 Å, the curve is firmly based within narrow limits. For bonds shorter than this,† the relationship is less clear because there are fewer reliable determinations in this range, and because the location of the proton is inherently more difficult—for reasons we shall discuss shortly.

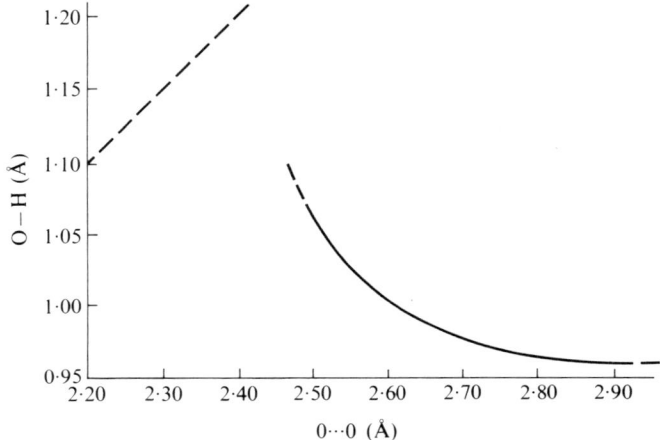

FIG. 9.6. Nakamoto–Margoshes–Rundle curve, obtained by plotting values of the O–H distances against O $\cdots$ O for a series of different hydrogen bonds.

The broken diagonal line drawn at the left-hand side of Fig. 9.6 represents hypothetical symmetrical bonds, with the proton at the mid-point, and therefore with O–H = $\frac{1}{2}$(O $\cdots$ O). Whether there really are hydrogen bonds of this sort; whether the curve to the right ever makes contact with the diagonal line; in what manner and at what O $\cdots$ O distance it may do so; and, indeed, whether the proton can be meaningfully located in this region, are questions as yet incompletely answered. This is the problem of the quest for the genuinely symmetrical O $\cdot\cdot$ H $\cdot\cdot$ O bond. We shall return to it in §9.8.

## 9.7. Ice

Early X-ray studies revealed the positions of the oxygen atoms in ice. Each oxygen is surrounded immediately by four others. These are arranged in directions corresponding nearly to the corners of a regular tetrahedron relative to its centre.

† Such bonds are often referred to as 'very short'.

The O ⋯ O ⋯ O angles are therefore near to 109°. There are two crystallographically distinguishable O ⋯ O distances: three, from any one oxygen atom, of 2·770 and one of 2·760 Å, at 0°C. These distances imply that each oxygen is involved in four rather weak hydrogen bonds.

The location of the hydrogen atoms presented difficulties. In the gaseous state the structure of the water molecule is accurately known from spectroscopic measurements. The (equilibrium) O–H distance is 0·957 Å, and the H–O–H angle is 104·5°. Various properties of ice, including its latent heat of sublimation, are such as to make it unlikely that the water molecule is greatly changed in passing from vapour to solid. The overall crystal symmetry could be satisfied by placing a proton at the mid-point of each O ⋯ O bond, but this is clearly unrealistic. To stretch the O–H bond from 0·96 to 1·38 Å would be too severe a distortion of the molecule. Rather, the water molecule must retain more of its identity: each oxygen must be covalently bonded to only two hydrogen atoms, and in any O–H ⋯ O bond the proton must be nearer to one oxygen than to the other, as is required by the Nakamoto-Margoshes-Rundle curve. But when we keep to these rules, it proves impossible to compose a structure for the unit cell consistent with the required symmetry.

The solution of this problem was found by Pauling in 1936. The absolute entropy of water vapour at 25°C can be estimated in two ways. First, we may calculate it from spectroscopic data by the methods of statistical thermodynamics. Secondly, we may use the Third Law of Thermodynamics in conjunction with measurements of specific and latent heats. In principle, we suppose the entropy of ice to be zero at 0 K, and then use our thermal data to calculate the increase of entropy as the ice is gradually warmed up to yield water vapour at 298·2 K. When this is done, the thermodynamic value is lower by a small but highly significant amount: 0·82 cal/mole/deg C (in SI units, 3·4 J mol$^{-1}$ K$^{-1}$). The difference can be attributed to there being a 'residual entropy' in ice at 0 K, and this can be interpreted by supposing it has an element of disorder in its structure, which persists at very low temperature.

Pauling was able to calculate this residual entropy on the basis of a simple model for the disorder, as follows. Every hydrogen atom lies between two oxygens; there is only one hydrogen along each O ⋯ O direction; every oxygen is covalently bonded to two hydrogens, and acts as the proton acceptor (O ⋯ H) from two other oxygens. The ordered arrangement required by these rules persists only over minute regions of the crystal: as a whole, the crystal is disordered, and the unit cell found by diffraction methods is an average of all the actual (ordered) cells of the crystal. The 'average' situation along any O ⋯ O contact can thus be represented, with 'half-protons', in this way:

$$\text{O} \text{———} \tfrac{1}{2}\text{H} \text{-----} \tfrac{1}{2}\text{H} \text{———} \text{O}$$
$$\leftarrow 1\cdot01 \rightarrow \leftarrow 0\cdot74 \rightarrow \leftarrow 1\cdot01 \text{ Å} \rightarrow$$
$$\leftarrow \text{———} 2\cdot76 \text{ Å} \text{———} \rightarrow \qquad (9.11)$$

HYDROGEN BONDING

A fairly simple statistical calculation based on this model shows that the residual entropy of ice should be

$$R \ln(\tfrac{3}{2}) = 0\cdot 81 \text{ cal/mole/deg C},$$

which is in striking agreement with experiment.

This explanation won general acceptance. It received its confirmation in 1957, when Peterson and Levy published an account of a neutron-diffraction study of heavy ice ($D_2O$), this being chosen in preference to ordinary ice for reasons already suggested. Along each O · · · O direction, in their neutron-scattering density map (see Fig. 9.7), they found two peaks of density appropriate to 'half-deuterons'.

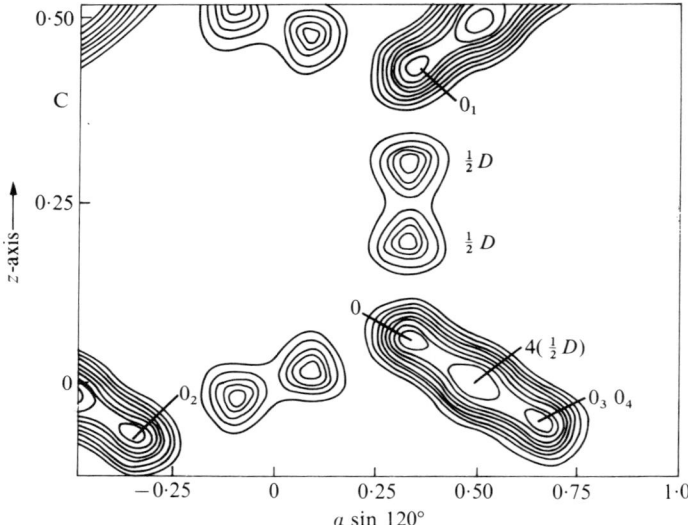

FIG. 9.7. Fourier projection of the neutron-scattering density in heavy ice at $-50°C$, on a plane at right-angles to the $b$-axis. The oxygen atom O is tetrahedrally surrounded by oxygen atoms $O_1$, $O_2$, $O_3$, $O_4$. There is a disordered distribution of deuterium atoms between pairs of possible positions (each marked as $\tfrac{1}{2}D$) between the oxygen atoms. (After Peterson and Levy 1957)

The actual dimensions found by Peterson and Levy are shown in formula (9.11). The positions of the 'half-protons' in ordinary ice will not differ appreciably. Neutron-diffraction analysis has now been applied to the other, polymorphic forms of ice which exist at high pressures (see Hamilton and Ibers 1968; Kamb et al. 1971).

## 9.8. Symmetrical hydrogen bonds and their problems

To the crystallographer, the essential feature of a crystal is a self-consistent set of symmetry elements. These form a *group* in the mathematical sense. In principle, an ideal crystal is generated by the operation of these elements upon a minimum amount of the material of which the crystal is composed. This amount is the *asymmetric unit*. The atoms within an asymmetric unit may be related to one another in a chemically significant way, but not by any symmetry requirement. On the other hand, corresponding atoms of different asymmetric units are symmetry-related. Fig. 9.8 gives a simplified example, with only a single symmetry element—a centre of inversion marked by the asterisk. The asymmetric unit may be taken as the three different objects labelled *a*, *b* and *c*. The centre of inversion generates the three equivalent objects labelled *a'*, *b'* and *c'*. (These labels are not part of the pattern.) Objects *a* and *a'* are symmetry related; *a* and *c'* are not. The only way in which an object can avoid duplication by the centre is for it to coincide with the asterisk. It would then be said to be

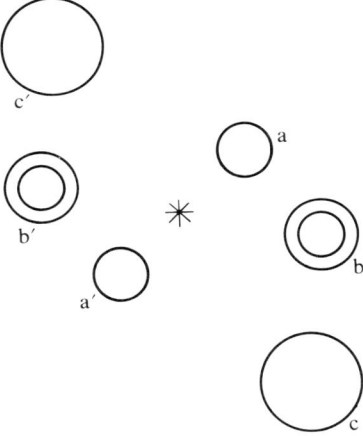

FIG. 9.8. Operation of a centre of symmetry on a set of three atoms.

in a *special position*, like the asterisk. The objects *a*, *b* and *c* are each in a *general position*.

In the great majority of OHO bonds that have been detected in crystals by diffraction methods, the two oxygen atoms are in a relationship like *a* and *b'*; they are not symmetry-related. This applied, for example, to the bonds illustrated in Figs. 9.1 and 9.3. It is rare to find two oxygen atoms symmetry-related, as are *a* and *a'*, and close enough together to imply that they are hydrogen-bonded. Cases were not discovered till 1948, but a number of examples are now known. They occur particularly in certain acid salts (including some acid salts of mono-basic acids, whose existence would not be predicted by elementary chemical rules). Such are sodium hydrogen di-acetate, $NaH(CH_3CO_2)_2$, potassium hydrogen malonate, $KH(O_2C.CH_2.CO_2)$.

In all of these, we find that the two oxygen atoms, related by a centre of symmetry (or other symmetry element), are only about 2·45 Å apart. These

## HYDROGEN BONDING

hydrogen bonds are always 'very short'. At least a score of such structures are known, and, taking those crystals which have been studied at Glasgow, we find that the average of the over-all O · · · O distances, weighted each according to its standard deviation, is 2·4468 ± 0·0014 Å.† Now the formal consequence of this situation is that the hydrogen atom must be exactly at the mid-point, for, if it were elsewhere, the symmetry element would generate a second hydrogen, and we know that there is only the one, for chemical reasons. Even though this argument is not unshakeable, these cases are at least candidates for consideration as genuinely symmetrical OHO bonds.

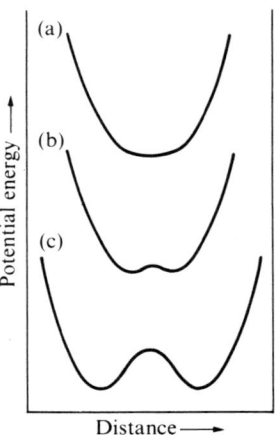

FIG. 9.9. Potential-energy curves for different positions of the proton in O–H · · · O bonds. (The right and left extremities of the diagram correspond to the situations of the two oxygen atoms, but the diagram is not to scale.)

Very relevant is the analogous case involving fluorine, which we have already mentioned: the $HF_2^-$ ion in the acid potassium and sodium salts of hydrofluoric acid. The compound $NaHF_2$ has been studied with especial care by McGaw and Ibers (1963), who find that the F · · H · · F bond lies across a crystallographic centre of symmetry, with the F · · · F distance 2·264 ± 0·003 Å. From their neutron-diffraction study of this compound, in both ordinary and deuterated forms, McGaw and Ibers came to the following conclusions:

(1) The F · · H · · F bond is without doubt very nearly symmetrical.
(2) However, from the diffraction measurements alone, it is impossible to distinguish between two situations:
    (a) The proton vibrates about a single potential-energy minimum situated at the mid-point between the two fluorines, as is represented by the curve in Fig. 9.9(a).

† This merely represents formal precision, not true accuracy. But the low value does indicate a well defined class of 'very short' OHO bonds.

(b) The curve takes the form sketched in diagram (b); the proton randomly occupies alternative sites slightly to either side of the mid-point, and its vibration along the bond-direction is of rather smaller amplitude than it would be in (a).

Either model, with appropriate adjustment of the vibrational parameter, will equally well account for the experimental intensity measurements from neutron diffraction.

We must emphasize that situation (b) is essentially different from that in ice, where the two potential-energy minima are well separated (as suggested in diagram (c)), so that the two 'half-hydrogen' peaks, about 0·7 Å apart, are easily resolved. If the FHF bond is indeed statistically disordered, the alternative sites cannot possibly be more than 0·1 Å on either side of the centre, and may well be much less. The neutron-diffraction peak will, of course, be 'smeared' out by vibration of the proton, along the bond and transversely to it; it is impossible to tell how far the smearing along the bond is due to vibration and how far (if at all) it is due to disorder. However, from an argument based on spectroscopic considerations, McGaw and Ibers concluded that situation (a) obtains, i.e. that this FHF bond is genuinely symmetrical.

None of the 'very short' OHO bonds has been so carefully considered. Inevitably, nearly all OHO bonds occur in more complex crystal structures.† Perhaps some of the bonds in the acid salts we have mentioned may also be genuinely symmetrical. The distribution between the potential-energy curves (a) and (b) of Fig. 9.9 is confused because of the phenomenon of zero-point energy. The molecule can never, in fact, exist in the state where the proton is at rest in the trough of its energy well. The level of minimum vibrational energy lies some distance above the bottom of the trough. If this level is near to, or slightly above, the top of the energy barrier between the two minima (shown in (b)), the quantum-mechanical picture becomes complicated by splitting of the energy levels, and the proton cannot be regarded as occupying either of its supposedly alternative sites.

Hamilton's view is that, in these circumstances, the potential-energy curve becomes flattened near its minimum. As we have tried to indicate in Fig. 9.9(a), the curve is no longer parabolic, and the vibration of the proton becomes anharmonic. When the environment of the bond is wholly symmetrical (as it is in the sodium bifluoride and in the crystalline acid salts of carboxylic acids), the curve, though shallow, will have its minimum at the mid-point. The bond is then really symmetrical. But if (as in some other known crystal structures) the environment is not symmetrical, the flattened potential-energy curve is easily distorted to force its minimum to one side. Hamilton and his co-workers

† The simplest OHO bonds that are short and apparently symmetrical occur in $HCrO_2$ and $HCoO_2$, which have been studied by Ibers and Hamilton. Their findings are rather complex (see Hamilton and Ibers, *Hydrogen bonding in solids* 1968). These bonds, which are rather longer than those in the acid salts, may possibly constitute a special case, transitional between short and 'very short' bonds.

(Schlemper, Hamilton, and LaPlaca 1971) have recently made a very precise neutron-diffraction study of a crystal involving a situation of this sort. They find the dispositions of the atoms directly involved in this 'very short' (chelated) OHO bond to be those given in Formula (9.12) below.

$$\begin{array}{l} a = 1{\cdot}187 \text{ Å} \\ b = 1{\cdot}242 \text{ Å} \\ c = 2{\cdot}420 \text{ Å} \\ \text{O–H–O angle} = 170° \end{array} \right\} \quad (9.12)$$

This bond deviates from symmetry by amounts that are highly significant.

## 9.9. References

ABRAHAMS, S. C., ROBERTSON, J. M. and WHITE, J. G. (1949). *Acta Crystallogr.* **2**, 238.
BACON, G. E. and CURRY, N. A. (1956). *Proc. R. Soc.* **A235**, 552.
BROADLEY, J. S., CRUICKSHANK, D. W. J., MORRISON, J. D., ROBERTSON, J. M. and SHEARER, H. M. M. (1959). *Proc. R. Soc.* **A251**, 441.
COULSON, C. A. (1970). Chapter 5 in *Thermal neutron diffraction*. (Ed. B. T. M. Willis). Oxford University Press.
HVOSLEF, J. (1958). *Acta Crystallogr.* **11**, 383.
KAMB, B., HAMILTON, W. C., LA PLACA, S. J. and PRAKASH, A. (1971). *J. chem. Phys.* **55**, 1934.
MACDONALD, A. L., ROBERTSON, J. M. and SPEAKMAN, J. C. (1971). *Acta Crystallogr.* **B. 27**, 2289.
McGAW, B. L. and IBERS, J. A. (1963). *J. chem. Phys.* **39**(10), 2677.
NAKAMOTO, K., MARGOSHES, M. and RUNDLE, R. E. (1955). *J. Am. chem. Soc.* **77**, 6480.
PAWLEY, G. S. and YEATS, E. A. (1969). *Acta Crystallogr.* **B25**, 2009.
PETERSON, S. W. and LEVY, H. A. (1957). *Acta Crystallogr.* **10**, 70.
ROBERTSON, J. M. (1936). *Proc. R. Soc.* **A157**, 79.
ROBERTSON, J. M. and WOODWARD, I. (1940). *J. chem. Soc.* **36**.
SCHLEMPER, E. O. HAMILTON, W. C. and LAPLACA, S. J. (1971). *J. chem. Phys.* **54**, 3990.
WOOLFSON, M. M. (1961). *Direct methods in crystallography.* Oxford University Press.

*General*
On hydrogen bonding in general:
HAMILTON, W. C. and IBERS, J. A. (1968). *Hydrogen bonding in solids.* Benjamin.
PIMENTEL, G. C. and McCLELLAN, A. L. (1960). *The hydrogen bond.* Freeman.

On neutron diffraction in more detail:
BACON, G. E. (1967). *Neutron diffraction.* Clarendon Press, Oxford.

For a very elementary outline of crystal-diffraction methods:
BRAND, J. C. D. and SPEAKMAN, J. C. (1964). *Molecular structure.* Arnold, London.

# 10 Structural Studies on Non-Stoichiometric Compounds by the Bragg Scattering of Neutrons

*By* A. K. CHEETHAM

*Inorganic Chemistry Laboratory, Oxford University*

## 10.1. Introduction

Non-stoichiometric compounds are characterized by their existence as a single phase over a range of composition. Ferrous oxide, for example, is not found as stoichiometric FeO, but as a non-stoichiometric compound with a phase field from $Fe_{0.94}O$ to $Fe_{0.89}O$ at $800°C$. Stability over a range of composition in binary compounds naturally implies a continuous variation of valency in one of the constituent elements. It is not surprising to find, therefore, that binary non-stoichiometric compounds are particularly common in the chemistry of the transition-series elements. Thus, the monoxides of Ti, V, Mn, Fe, Co, and Ni all display non-stoichiometry to varying degrees, and the elements Ti, Zr, Hf, V, Nb, and Ta form non-stoichiometric hydrides, carbides, and nitrides. Single phases with ranges of composition are also found in systems which form solid solutions. Particularly interesting are the so-called 'anomalous solid solutions' in which the dopant cation has a different charge from that of the cation in the parent compound (e.g. $YF_3$ in $CaF_2$). Striking similarities are often observed between such solid solutions and binary non-stoichiometric compounds with the same structure (see §10.4 below).

The existence of single phases with variable composition can be rationalized by postulating the presence of defects in the crystal structure. In their statistical thermodynamic treatment of ionic crystals, Schottky and Wagner (1930) showed that lattice defects emerged as a natural consequence of the thermodynamic equilibrium. This concept rapidly led to a greater understanding of many solid-state phenomena such as transport processes and chemical reactions, and in particular to the recognition of non-stoichiometry as a natural feature of all chemical compounds. For practical purposes, only those compounds with a readily detectable range of composition are considered to be non-stoichiometric. In such compounds. one or more of several different types of defect may be present. The simplest imperfections are the lattice vacancy (a missing atom or ion) as illustrated in Fig. 10.1(a) and the interstitial (an atom occupying a hole in the structure) shown in Fig. 10.1(b). In solid solutions and alloys, substitutional defects are also found in which atoms are replaced by atoms of another element (Fig. 10.1(c)). This and the following chapter are concerned with the identification of point defects in different types of materials, and the way in which they are distributed throughout the crystal.

In the Schottky-Wagner treatment of crystals, the defects are assumed to be randomly distributed and non-interacting, and it has been customary to interpret the structures of non-stoichiometric compounds in these terms. This approach works satisfactorily for compounds containing very small concentrations (< 0·1 atomic per cent) of defects, but it does not account for the properties

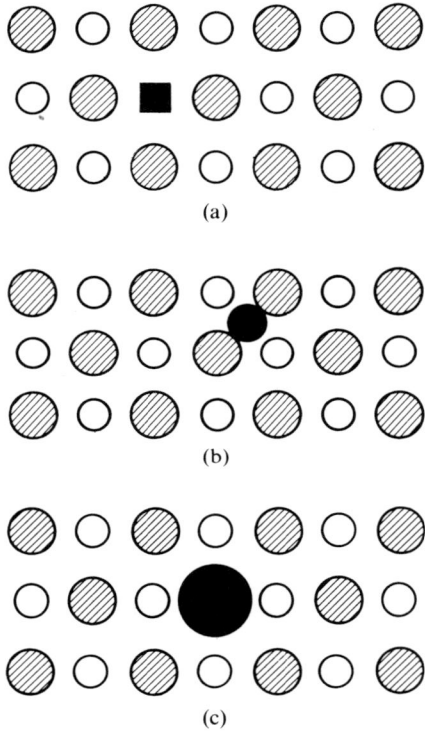

FIG. 10.1. (a) A cation vacancy ■ in an ionic crystal. (b) A cation interstitial ● in an ionic crystal. (c) A substitutional defect ● in an ionic crystal. Open circles represent cations, and cross-hatched circles anions.

of the many systems which can incorporate high concentrations of point defects (several atomic per cent). At large defect concentrations, the configurational entropy term alone is not sufficient to stabilize the solid unless the overall enthalpy of defect formation is low. In ionic compounds, this is only possible when attractive defect interactions occur, leading to the clustering of point defects. There is, indeed, a growing body of information from both structural and thermodynamic measurements to show that in many systems strong defect interactions are present. For instance, in many materials orginally thought to contain random point defects, detailed crystallographic studies have revealed the existence of closely related intermediate phases with small or indetectable ranges of composition. In these phases, long-range ordering or elimination of the defects has taken place, thus demonstrating the presence of defect interactions in the crystal. Some examples of this type of behaviour are discussed in the following paragraphs.

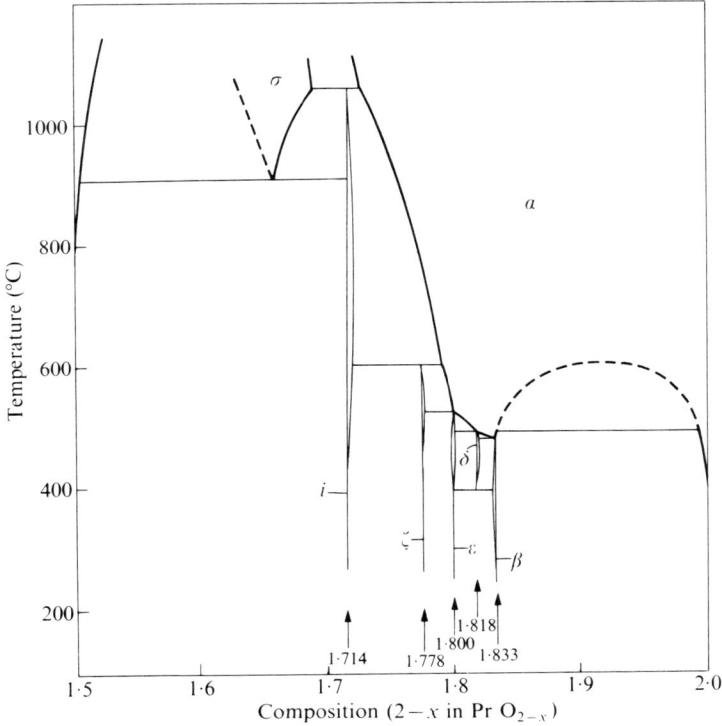

FIG. 10.2. The phase diagram of $PrO_{2-x}$. $\alpha$ and $\sigma$ denote non-stoichiometric phases which lie above the ordered phases $\beta, \delta, \epsilon, \zeta, \iota$. The compositions of these ordered phases, indicated in the diagram, correspond to the homologous series $Pr_nO_{2n-2}$, where $n$ = 7, 9, 10, 11, 12. (After Hyde *et al.* 1966).

*The long range ordering of point defects* has been found in the $Cr_{1-x}S$ system which is based upon the nickel arsenide structure. Early investigations of this system (Haraldsen 1937) indicated that the range from CrS to $Cr_{0.67}S$ contained several grossly non-stoichiometric sulphide phases in which the defects were incorporated as chromium vacancies. However, later measurements (Jellinek 1957) revealed that under equilibrium conditions there exists an homologous series of well-defined phases $Cr_nS_{n+1}$ in which the chromium vacancies are ordered. Similar series of ordered phases have been discovered in the oxygen-deficient fluorite systems $CeO_{2-x}$ (Bevan 1955; Bevan and Kordis 1964), $PrO_{2-x}$ (Hyde *et al.* 1966) and $TbO_{2-x}$ (Baenziger *et al.* 1961). The phase diagram for $PrO_{2-x}$ is shown in Fig. 10.2. An X-ray diffraction study of the ordered phase $Pr_7O_{12}$ (Eyring and Baenziger 1962) showed that the oxygen vacancies are ordered along a $\langle 111 \rangle$ direction in the parent fluorite cell (Fig. 10.3). The similarity between this ordering pattern and that found in the C-type rare-earth oxide structure (also based upon the fluorite cell) suggests that the pattern is a particularly stable configuration for oxygen-deficient fluorite compounds. At high temperatures, long-range order in the $MO_{2-x}$ systems is lost and the non-stoichiometric $\alpha$ and $\sigma$ phases are formed (Fig. 10.2).

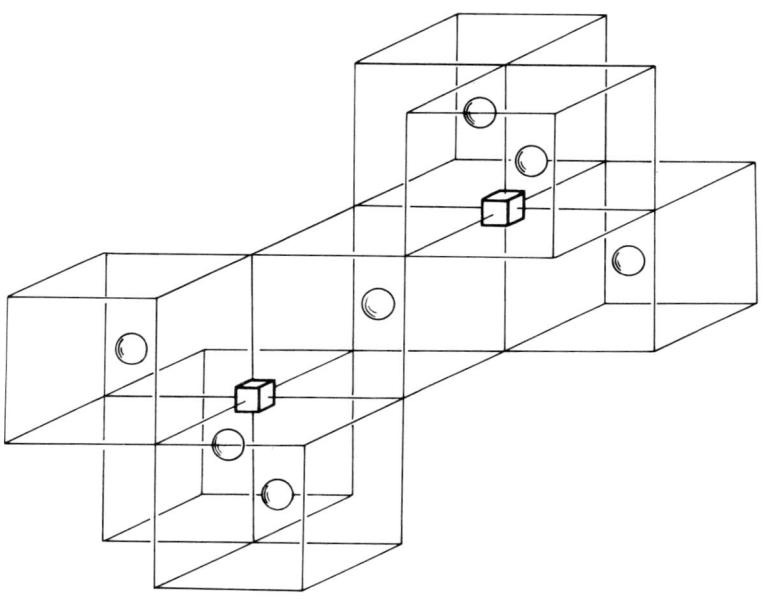

FIG. 10.3. A section of the ordered $Pr_7O_{12}$ structure. Oxygen atoms are at the corners of the cubes. ▢ represent oxygen vacancies along the $\langle 111 \rangle$ direction. The cube edge corresponds to $\frac{1}{2}a_0$ in the fluorite structure: see Fig. 10.5. ◯ are the 6- and 7- co-ordinated cations.

*The elimination of point defects* by the formation of 'shear structures' (see Anderson 1969) also leads to homologous series of intermediate compounds. This is commonly found in transition-metal oxides adopting the $ReO_3$ and $TiO_2$ structures, in which $MO_6$ octahedra share corners and edges respectively. For example, in the $MoO_{3-x}$ system (based upon an oxygen-deficient $ReO_3$ structure), there is a series of intermediate compounds which have the general formula $Mo_nO_{3n-m}$ (Kihlborg 1963). The compounds contain blocks of the perfect $ReO_3$ structure, separated by shear planes of $MoO_6$ octahedra sharing edges

(Fig. 10.4). In this way, the oxygen vacancies are eliminated, and the repetition of the shear planes at regular intervals gives rise to the ordered intermediate compounds. The intervals are often large and may be observed directly in the electron microscope (see, e.g. Hyde and Bursill 1970).

The ordered intermediates discussed above are not genuine non-stoichiometric compounds, and their structures, in principle, can be determined by conventional crystallographic methods. In truly non-stoichiometric materials, long-range ordering of the defects is absent, and such systems present an unusual crystallographic problem. Whereas in a perfectly ordered crystal, the observed

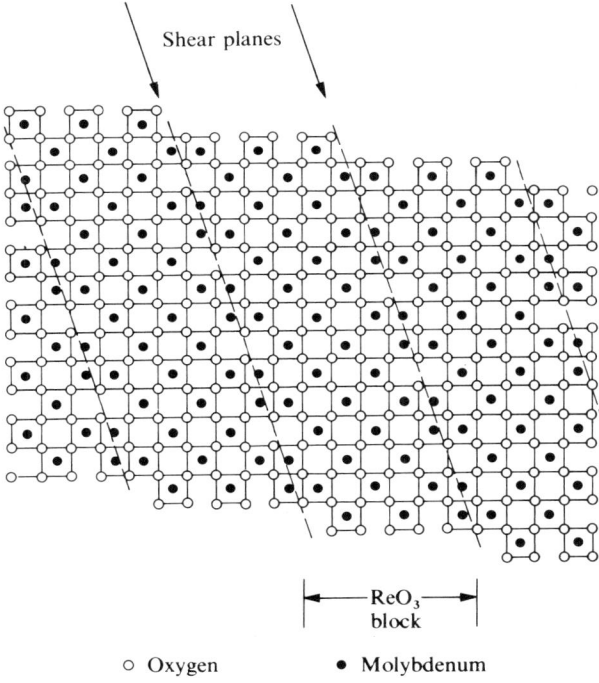

FIG. 10.4. The shear structure principle in $Mo_8O_{23}$. Blocks of unit cells with the perfect $ReO_3$ structure are separated by shear planes, across which the $MoO_6$ octahedra share edges. (After Anderson 1969).

structure factors define the contents of a unique unit cell, in a non-stoichiometric compound, the unit cell contents vary throughout the crystal in a way which depends upon the distribution of the defects. The structure factors obtained from a Bragg scattering experiment define only the contents of the 'average unit cell', so that the positions in the average cell occupied by the atoms, and their occupancy numbers, are known, but the distribution of the defects throughout the crystal is not revealed directly. However, we shall see later that some information about this distribution can be obtained by using the results of the Bragg experiment in combination with plausible values of the ionic radii. (More direct knowledge of the defect distribution is derived from diffuse scattering studies described in the next chapter). The particular interest in the defect distribution stems from the understanding which it gives of the interactions between the defects in crystals.

For materials containing less than 2–3 atomic per cent of defects, the diffraction method is not sufficiently sensitive to detect this small defect concentration. Our limited structural knowledge of such systems is usually obtained indirectly from conductivity, spectroscopic, and resonance experiments. For non-stoichiometric compounds containing more than 2–3 atomic per cent of defects, direct structural measurements are possible, and it is with these materials that the present chapter is concerned. They include, for example, the $UO_{2+x}$ phase which has a composition range from $UO_{2.00}$ to $UO_{2.25}$ at $1000°C$, and the cerium hydride phase which apparently has a range from $CeH_{1.95}$ to $CeH_{3.00}$, even at room temperature. We shall consider first, however, the theory and practice of the Bragg scattering method, with particular reference to its use in examining the structures of non-stoichiometric compounds.

## 10.2. Theory

The coherent elastic differential scattering cross-section, $d\sigma/d\Omega$, for an assembly of $M$ atoms is given by (see eqn (1.11) of Chapter 1)

$$\frac{d\sigma}{d\Omega} = \sum_{m,m'=1}^{M}\sum b_m b_{m'} \exp[i\mathbf{Q} \cdot (\mathbf{r}_m - \mathbf{r}_{m'})] \tag{10.1}$$

where $b_m$ is the scattering length of the $m$th atom, and $\mathbf{r}_m$ is its vector distance from an arbitrary origin. The scattering vector $\mathbf{Q}$ is defined as

$$\mathbf{Q} = \mathbf{k} - \mathbf{k}', \tag{10.2}$$

in which $\mathbf{k}$ and $\mathbf{k}'$ are the wavevectors of the incident and scattered beams respectively.

If we now consider the scattering from a crystal in which the atoms are represented by an assembly of $N$ identical cells on a basis lattice, eqn (10.1) becomes

$$\frac{d\sigma}{d\Omega} = \sum_{n,n'=1}^{N}\sum F_n F_{n'} \exp[i\mathbf{Q} \cdot (\mathbf{r}_n - \mathbf{r}_{n'})], \tag{10.3}$$

where the vector $\mathbf{r}_n$ defines the $n$th point on the basis lattice, and $F_n$ is the structure factor of the $n$th cell:

$$F_n = \sum_{\kappa} b_{\kappa} \exp(i\mathbf{Q} \cdot \mathbf{r}_{\kappa}). \tag{10.4}$$

The structure factor summation is over all the atoms in the cell, each labelled by a symbol $\kappa$; $\mathbf{r}_\kappa$ is the vector to the $\kappa$th atom in the cell from the origin of the cell, so that

$$\mathbf{r}_\kappa = \mathbf{r}_m - \mathbf{r}_n \equiv \mathbf{r}_{m'} - \mathbf{r}_{n'}.$$

For a non-stoichiometric compound the unit cells are not all identical. If $\langle F \rangle$ represents the structure factor averaged over all cells, eqn (10.3) can be re-written as

$$\frac{d\sigma}{d\Omega} = \underbrace{\sum_n \sum_{n'} \langle F \rangle^2 \exp[i\mathbf{Q} \cdot (\mathbf{r}_n - \mathbf{r}_{n'})]}_{\text{BRAGG INTENSITY}}$$

$$+ \underbrace{\sum_n \sum_{n'} (F_n F_{n'} - \langle F \rangle^2) \exp[i\mathbf{Q} \cdot (\mathbf{r}_n - \mathbf{r}_{n'})]}_{\text{DIFFUSE INTENSITY}}. \quad (10.5)$$

The first term in eqn (10.5) gives rise to sharp Bragg peaks, and it is this scattering which is the subject of the present chapter. The second term gives diffuse intensity, and is discussed further in the following chapter.

The differential scattering cross-section for Bragg scattering is therefore

$$\left(\frac{d\sigma}{d\Omega}\right)_{\text{Bragg}} = \langle F \rangle^2 |\sum_n \exp(i\mathbf{Q} \cdot \mathbf{r}_n)|^2$$

$$= N_0 \frac{(2\pi)^3}{v} \sum_\tau \delta(\mathbf{Q} - \tau) \times \langle F \rangle^2 \quad (10.6)$$

from eqn (1.15), where $N_0$ is the total number of unit cells, each of volume $v$. The $\delta$-function ensures that Bragg scattering occurs only when $\mathbf{Q}$ coincides with a reciprocal lattice vector $\tau_{hkl}$. For all other values of $\mathbf{Q}$, the differential scattering cross-section is zero. Thus, for those values of $\mathbf{Q}$ which satisfy the Bragg condition, peaks of scattered radiation are observed whose intensities are related to the average structure factors $\langle F(\mathbf{Q}) \rangle$ defining the contents of the average cell.

By analogy with eqn (10.4), $\langle F(\mathbf{Q}) \rangle$ may be expressed in the form

$$\langle F(\mathbf{Q}) \rangle = \sum_\kappa m_\kappa b_\kappa \exp(i\mathbf{Q} \cdot \mathbf{r}_\kappa) \exp(-W_\kappa(\mathbf{Q})). \quad (10.7)$$

$\mathbf{r}_\kappa$ is the vector to the $\kappa$th atom in the average cell, and $m_\kappa$ is the occupancy number of the $\kappa$th site in the average cell: in a perfect crystal $m_\kappa$ would be unity for an occupied site, but it is included here as a variable because in a non-stoichiometric compound some of the sites have only partial occupancy. The term $\exp(-W_\kappa(\mathbf{Q}))$ is the Debye-Waller temperature factor term which modifies the structure-factor expression to allow for the effect of thermal motion. In the harmonic approximation, the exponent $W_\kappa$ is given by

$$W_\kappa(Q) = \tfrac{1}{2}\overline{(\mathbf{Q}\cdot\mathbf{u}_\kappa)^2}, \tag{10.8}$$

where $\overline{u_\kappa^2}$ is the mean-square thermal displacement of the $\kappa$th atom from its equilibrium position. For Bragg scattering (see eqn (1.6))

$$Q = 4\pi \sin\theta_B/\lambda$$

so that

$$W_\kappa(\mathbf{Q}) = 8\pi^2 \overline{u_\kappa^2}\,\frac{\sin^2\theta}{\lambda^2} \tag{10.9}$$

$$= B_\kappa \sin^2\theta/\lambda^2. \tag{10.10}$$

$B_\kappa (= 8\pi^2 \overline{u_\kappa^2})$ is the 'temperature factor' which is normally quoted in the analysis of (X-ray or neutron) diffraction data.

In some non-stoichiometric compounds an increase in the observed temperature factors with increasing defect concentration can occur (e.g. Cheetham *et al.* 1971a). Part of this increase arises from the static relaxation of atoms around the defects, and in this respect it is similar to the size effect which is apparent in alloys (Huang 1947). In addition, the thermal contribution itself will be modified by changes in both the interatomic spacing and in the lattice energy upon introducing defects. To account for the thermal and static contributions to $W$, eqn (10.8) should be modified to

$$W_\kappa(\mathbf{Q}) = \tfrac{1}{2}\overline{[\mathbf{Q}\cdot(\mathbf{u}_\kappa + \mathbf{u}_\kappa^s)]^2} \tag{10.11}$$

in which $\mathbf{u}_\kappa^s$ is the average static displacement of the $\kappa$th atom from its normal position. By estimating the thermal component of $W_\kappa(\mathbf{Q})$ it should, therefore, be possible to assess the extent of any relaxations. This could be done, for example, by comparing the observed temperature factors with those of the parent (non-defective) compound at low temperatures where differences between the thermal contributions are minimized.

The structural parameters $m_\kappa$, $\mathbf{r}_\kappa$ and $B_\kappa$ may be obtained from the experimental data by either least-squares or Fourier methods. In the least-squares method, the magnitudes of the observed structure factors $|F_{hkl}^{obs}|$ are compared with those calculated from eqn (10.7) and the structural parameters are adjusted to minimize a function of the type

$$\sum_{hkl} w_{hkl}(|F_{hkl}^{obs}| - |F_{hkl}^{calc}|)^2$$

($w_{hkl}$ is the weight given to the $hkl$ reflexion. Note that $F_{hkl}^{calc}$ is identical with $\langle F \rangle$ in eqns (10.6) and (10.7).) For a satisfactory refinement it is essential that the approximate positions of all the atoms are known. The Fourier method involves the inversion of the structure-factor data to give the scattering density $\rho(xyz)$ at the point in the average cell with fractional co-ordinates $xyz$,

$$\rho(xyz) = \frac{1}{v} \sum_{h,k,l=-\infty}^{\infty} F_{hkl}^{obs} \exp[-2\pi i(hx + ky + lz)], \qquad (10.12)$$

where $v$ is the volume of the unit cell. The quantities $F_{hkl}^{obs}$ are given by

$$F_{hkl}^{obs} = |F_{hkl}^{obs}| \exp(i\alpha) \qquad (10.13)$$

where $|F_{hkl}^{obs}|$ is derived from the intensity of the $hkl$ reflexion and $\alpha$ is the phase of the reflexion. The problem of determining $\alpha$ in the general case is discussed in §9.5, but for a non-stoichiometric compound a good approximation to the phases can be obtained simply by using the known phases of the parent structure. At the end of the Fourier analysis, a least-squares refinement must be carried out since it is difficult to estimate occupancy numbers (and their standard deviations) from scattering density maps.

For a conventional crystal-structure analysis, the positions $xyz$ of each atom $\kappa$ in the unit cell can usually be found from structure factor measurements of only moderate precision (errors 5 per cent). The determination of the fractional occupancy numbers $m_\kappa$ of non-stoichiometric compounds requires data of exceptionally high accuracy, and so we shall consider next a few points concerning the achievement of the required accuracy.

## 10.3. Achievement of required accuracy in measuring Bragg intensities

In principle, most crystallographic studies may be carried out using either X-rays or neutrons. In both cases it is possible, if sufficient care is taken, to reduce the uncertainty in measuring the intensity to as little as 1 per cent. Raccah and Henrich (1969), for example, have discussed how this can be done by X-ray diffraction with powders, and Cooper et al. (1968) have achieved this accuracy in their single-crystal neutron study of anharmonicity in barium fluoride. It must always be borne in mind, of course, that X-ray experiments are quicker, because of the higher beam flux, and that neutron diffraction is a very expensive technique. Nevertheless, neutrons have several properties which are invaluable for the type of problem under discussion:

(a) Because the variation of scattering factor with atomic number is much less with neutrons than with X-rays, only with neutrons is one able to locate easily the positions of light atoms in the presence of heavy atoms. Many non-stoichiometric compounds contain heavy metal atoms and light non-metal atoms—for example metal hydrides, carbides, oxides, and fluorides—and it is the positions and occupation numbers of these light atoms that we may wish to determine.

(b) A large number of non-stoichiometric compounds are only stable at temperatures above room temperature, for example $Fe_{1-x}O$ and $UO_{2+x}$, and if these are to be studied under equilibrium conditions, high-temperature diffraction methods are necessary. Most materials absorb neutrons to a much less extent than X-rays, and this property makes accurate high-temperature measurements (where the incident beam must pass through the structural components of a furnace) easier with neutrons.

(c) The absence of a form-factor dependence of the scattering of neutrons leads to a less rapid decrease of intensity with scattering angle than for X-rays. This feature is particularly important at high temperatures, where the Debye–Waller temperature factors are large, since it enables a larger number of observations to be made at high values of $Q$ with neutrons.

TABLE 10.1

$(Ca/Y)F_{2.25}$: *Structural Parameters in Average Cell*
*(From powder study of Cheetham et al. 1971a)*

| Atom | Co-ordinates in average cell | | | Contribution to $2 + x$ in $(Ca/Y)F_{2+x}$ |
|---|---|---|---|---|
| | $x$ | $y$ | $z$ | |
| Normal $F$ | 0.25 | 0.25 | 0.25 | 1.62 (0.02) |
| Interstitial $F'$ | 0.5 | $v$ | $v$ | 0.47 (0.02) |
| Interstitial $F''$ | $w$ | $w$ | $w$ | 0.16 (0.03) |

$v = 0.368\ (0.005)$     $w = 0.410\ (0.004)$

$UO_{2.25}(U_4O_9)$: *Structural Parameters in Average Cell*
*(From single-crystal study of Willis 1964)*

| Atom | Co-ordinates in average cell | | | Contribution to $2 + x$ in $UO_{2+x}$ |
|---|---|---|---|---|
| | $x$ | $y$ | $z$ | |
| Normal $O$ | 0.25 | 0.25 | 0.25 | 1.77 (0.02) |
| Interstitial $O'$ | 0.5 | $v$ | $v$ | 0.29 (0.05) |
| Interstitial $O''$ | $w$ | $w$ | $w$ | 0.19 (0.04) |

$v = 0.372\ (0.005)$     $w = 0.378\ (0.005)$
*Estimated standard deviations in parentheses.*

A further question concerns the relative merits of measurements on single crystal and polycrystalline samples. When studying systems of low symmetry, single crystals are clearly necessary if severe overlapping of the Debye–Scherrer lines is to be avoided (see, for instance, the powder diffraction profile of monoclinic naphthalene given in Fig. 1.6 (Chapter 1)). The single-crystal technique also has the advantage that it enables a larger number of independent observations to be made, and so, in principle, should lead to a greater precision in the final structural parameters. However, Bragg intensities recorded from crystals are subject to several systematic effects, such as extinction, multiple Bragg scattering, and thermal diffuse scattering, for which allowance must be made if reliable values of the structural parameters are to be obtained (Cooper and Rouse 1970). Although in a few instances powders may be prone to preferred orientation effects, particularly under the conditions of an X-ray experiment, extinction is small or negligible and the thermal diffuse scattering contribution to the Bragg peaks is less than with single crystals (Suortti 1967). Consequently, it is possible to measure structure factors from powders with at least the same accuracy as from single crystals and to obtain useful structural information, even from a limited number of powder observations. Furthermore, polycrystalline materials are usually more readily available, diffraction measurements are less time-consuming, and high- or low-temperature experiments can be performed with relative ease. The structural parameters obtained from a polycrystalline sample of $(Ca/Y)F_{2.25}$ and a single crystal of $U_4O_9$ (i.e. $UO_{2.25}$) are compared in Table 10.1. A similar level of accuracy was achieved in each experiment.

## 10.4. Applications

### 10.4.1. Anion-deficient compounds with the fluorite structure

The structural chemistry of ionic non-stoichiometric compounds based on the fluorite structure (Fig. 10.5) has been reviewed by Roberts (1971). The predominant

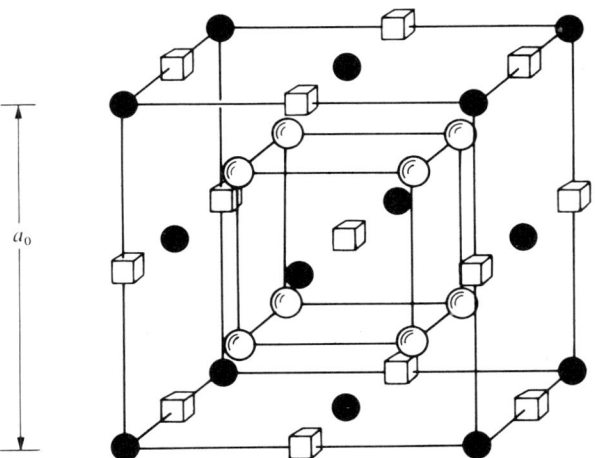

FIG. 10.5. The fluorite structure. ● represent cations, ◯ are anions and ▭ represent interstitial positions.

defects in these materials are always confined to the anion sublattice, and so it is convenient to distinguish those systems containing an excess of anions and those which are anion deficient. In the latter group are the oxides of rare-earth elements which may adopt the +4 valence state—cerium, praseodymium, and terbium.

## STUDIES ON NON-STOICHIOMETRIC COMPOUNDS

Such oxides exhibit wide ranges of stoichiometry at high temperatures although, as described in § 10.1, at low temperatures they are characterized by a series of ordered phases (Fig. 10.2). Similar behaviour is exhibited by the fluorite solid solutions formed by the dioxides of the elements Zr, Hf, Ce, Th and U with the lanthanide sesquioxides ($Ln_2O_3$) and alkaline earth oxides (MO). However, in the ternary systems, ordered structures at low temperatures are less prevalent (although see Bevan *et al.* 1964); no doubt the formation of ordered phases depends to a large extent upon the mobility of the altervalent cations which in ternary systems is less than in binary compounds.

Although several investigations of the ordered phases have been carried out, our structural knowledge of the disordered systems is very limited. We have seen that for a non-stoichiometric compound the results of a Bragg experiment define only the contents of the 'average unit cell', and as a consequence of this, Bragg diffraction experiments are often uninformative about systems in which the defects arise simply as clustered or isolated vacancies. For example, in 1972 Steele carried out a Bragg neutron diffraction study on a solid solution of $Y_2O_3$ in $ThO_2$. No superlattice reflexions were detected and the observed structure factors were in excellent agreement with a model in which the Th and Y atoms are distributed at random over all the cation sites and vacancies are present in the oxygen sublattice. No atomic displacements or new sites were observed. Subbarao *et al.* (1965) performed accurate lattice parameter and density measurements on the same compound and proposed an identical model. The Bragg experiment provides no additional information, since the Bragg scattering from the partially-filled oxygen sub-lattice is independent of the actual distribution of vacancies unless long-range ordering of vacancies gives rise to superlattice reflexions. It is known that in ordered structures the vacancies exhibit a strong tendency to form chains along the $\langle 111 \rangle$ directions in the fluorite cell (see § 10.1), and one might expect, therefore, that small regions of such order exist in the disordered phase. However, to make any definitive conclusions about the defect distribution an examination of the diffuse scattering is necessary.

Bragg studies on systems containing vacancies are not always so uninformative. For example, a solid solution of CaO in $ZrO_2$ adopts the fluorite structure, and lattice parameter/density measurements (Tien and Subbarao 1963) again indicate an oxygen-vacancy model. Bragg neutron diffraction measurements, by Carter and Roth (1968), revealed an unexpected displacement of the oxygen atoms from their normal positions and in samples annealed at low temperatures these displacements assumed a long-range ordered arrangement, giving rise to superlattice reflexions in the diffraction pattern. More detailed powder neutron diffraction studies by Steele (1971) suggested that only those oxygen atoms which are adjacent to anion vacancies are displaced. Such displacements are only apparent in solid solutions formed with $ZrO_2$ (and presumably $HfO_2$), and are probably a consequence of the small size of the $Zr^{4+}$ ion.

### 10.4.2. Anion-excess compounds with the fluorite structure

A large number of compounds are known in which the fluorite structure accommodates an excess of anions. Examples are the $UO_{2+x}$ phase (see § 10.1), and the non-stoichiometric fluorides of samarium and europium. The alkaline-earth fluorides $CaF_2$, $SrF_2$, and $BaF_2$ also form non-stoichiometric fluorite phases ($MF_{2+x}$) with all the rare-earth trifluorides (Ippolitov et al. 1967). The Bragg scattering of neutrons has proved a particularly valuable technique for studying the defect structures of these systems.

Until quite recently it was thought that the excess anions in such systems occupied the large holes at the $(\frac{1}{2}, \frac{1}{2}, \frac{1}{2})$ position in the fluorite structure (Fig. 10.5). Electron spin resonance experiments on a single crystal containing 1 mole per cent $Nd^{3+}$ in $CaF_2$ certainly gave strong evidence for this model (Bleaney et al. 1956). However, an alternative model was proposed by Willis (1963) for the disordered $UO_{2+x}$ system. He examined a single crystal of composition $UO_{2.12}$ at 800°C by neutron diffraction, and the analysis of the observed structure factors revealed two types of interstitial site, and a number of vacancies in the normal oxygen sub-lattice. One of the new interstitial sites (O') was located along a $\langle 110 \rangle$ direction from the $(\frac{1}{2}, \frac{1}{2}, \frac{1}{2})$ position, whilst the other (O") was located along a $\langle 111 \rangle$ direction from this site. No interstitials at the $(\frac{1}{2}, \frac{1}{2}, \frac{1}{2})$ position were detected. The observed occupation numbers and positional parameters are given in Table 10.2. By a simple consideration of ionic radii, it can be shown that the defects must be arranged in clusters, and various defect complexes have been proposed which are compatible with the observed structural parameters. The distances from the O' and O" interstitials to the nearest normal oxygen atoms are about 1·5 Å. This is prohibitively low since a normal O-O interionic distance is 2·6-2·7 Å (Shannon and Prewitt 1969). This anomaly may be resolved if it is assumed that the interstitial atoms are always associated with vacancies in the normal anion sublattice. The $\langle 110 \rangle$ displacement of the O' interstitial is accounted for by association with two anion vacancies, whilst the $\langle 111 \rangle$ displacement of the O" interstitials arises from association with just one vacancy (Fig. 10.6). Using these defect units, Willis proposed a cluster model—the so-called 2:2:2 cluster (Fig. 10.7)—for which the expected occupation numbers for the different oxygen atoms ($UO_{1.88}O'_{0.12}O''_{0.12}$) are close to the experimental values. Other cluster models, equally compatible with the observed occupation numbers, have also been proposed for the $UO_{2+x}$ system, and extremely accurate intensity measurements are clearly necessary if one is to distinguish between these different models.

Neutron diffraction (Willis 1964) and X-ray studies (Belbeoch et al. 1961) have also been carried out on the $U_4O_9$ phase. At low temperatures the range of non-stoichiometry is small and the diffraction patterns reveal superlattice reflexions which index on a $4 \times a_0$ cubic cell in the space group $I\bar{4}3d$. The intensities of the fundamental reflexions have been analysed (see Table 10.1) and it is particularly interesting to find that the average cell contains the same type of interstitial sites as those observed in $UO_{2+x}$. Such a close relationship

STUDIES ON NON-STOICHIOMETRIC COMPOUNDS

TABLE 10.2

$(Ca/Y)F_{2\cdot 10}$: *Structural parameters in average cell*
*(After Cheetham et al. 1970)*

| Atom | Co-ordinates in average cell | | | Contribution to $2+x$ in $(Ca/Y)F_{2+x}$ |
|---|---|---|---|---|
| | $x$ | $y$ | $z$ | |
| Normal $F$ | 0·25 | 0·25 | 0·25 | 1·88 (0·04) |
| Interstitial $F'$ | 0·5 | $v$ | $v$ | 0·14 (0·03) |
| Interstitial $F''$ | $w$ | $w$ | $w$ | 0·08 (0·03) |

$v = 0\cdot36\ (0\cdot01)$, $\quad w = 0\cdot42\ (0\cdot01)$, $\quad R_{\text{index}} = 0\cdot85$ per cent.
*Estimated standard deviations in parentheses.*

$UO_{2\cdot 12}$: *Structural parameters in average cell*
*(After Willis 1963)*

| Atom | Co-ordinates in average cell | | | Contribution to $2+x$ in $UO_{2+x}$ |
|---|---|---|---|---|
| | $x$ | $y$ | $z$ | |
| Normal O | 0·25 | 0·25 | 0·25 | 1·87 (0·03) |
| Interstitial O′ | 0·5 | $v$ | $v$ | 0·08 (0·04) |
| Interstitial O″ | $w$ | $w$ | $w$ | 0·16 (0·06) |

$v = 0\cdot38\ (0\cdot01)$, $\quad w = 0\cdot41\ (0\cdot01)$, $\quad R_{\text{index}} = 3\cdot5$ per cent.
*Estimated standard deviations in parentheses.*

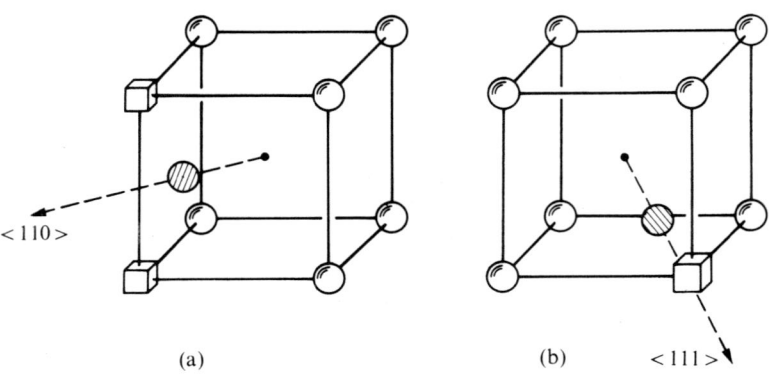

FIG. 10.6. (a) Association of an O′-type interstitial with two normal anion vacancies. (b) Association of an O″-type interstitial with one normal anion vacancy.

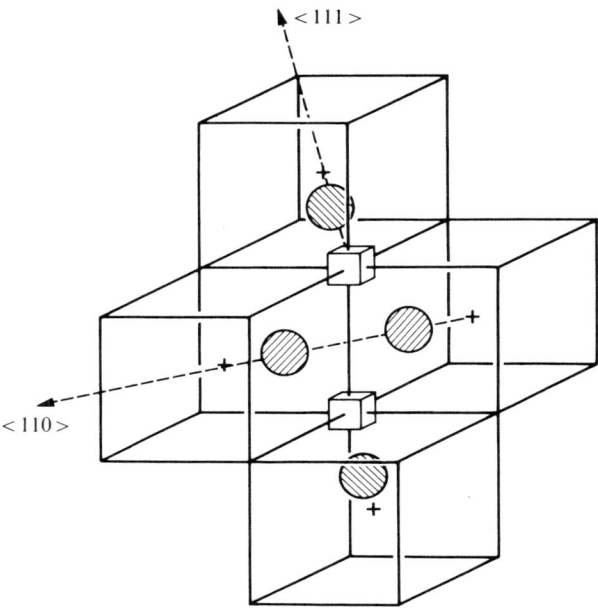

FIG. 10.7. A 2:2:2 cluster. Normal anions are at corners of cubes. ☐ are normal anion vacancies, ⊘ are anion interstitials along the ⟨110⟩ directions from the $\frac{1}{2}, \frac{1}{2}, \frac{1}{2}$ sites represented by +, and ⊘ are anion interstitials along ⟨111⟩.

between the $UO_{2+x}$ and ordered $U_4O_9$ structures is supported by entropy measurements made by Roberts (1963). The structure of fully ordered $U_4O_9$ has, however, not yet been solved. Recent proton channelling experiments (Matzke et al. 1971) on $U_4O_9$ are compatible with the average cell determined by Willis, and also reveal small displacements of the uranium atoms from their normal positions. This no doubt accounts for the strong superlattice reflexions observed with X-rays.

In spite of this structural knowledge of the $UO_{2+x}$ system, it was still assumed by later workers that the interstitials in alkaline-earth fluoride solid solutions were located at the $(\frac{1}{2}, \frac{1}{2}, \frac{1}{2})$ position. Indeed, a large number of electron spin resonance and electron nuclear double resonance experiments (e.g. Baker et al. 1968) on solid solutions containing less than 1 mole per cent $LnF_3$ have shown the presence of interstitials at only this position. On the other hand, a recent powder neutron diffraction study by Cheetham et al. (1970) on a solid solution containing 10 mole per cent $YF_3$ in $CaF_2$ revealed a defect structure showing a remarkable similarity to that of $UO_{2+x}$. Again, two types of interstitial were observed, and vacancies were detected in the normal fluorine sub-lattice. The structural parameters are tabulated in Table 10.2 for comparison with those of $UO_{2 \cdot 12}$. It was suggested that this type of defect structure is a general feature of fluorite-type compounds containing excess anions. These systems always have solubility

## STUDIES ON NON-STOICHIOMETRIC COMPOUNDS

limits between about $MX_{2.20}$ and $MX_{2.50}$, and such limits would be a natural consequence of this structure. Measurements at a series of compositions in the $CaF_2/YF_3$ system are compatible with small clusters of the 2:2:2 type (Fig. 10.7) or slightly larger 3:4:2 type (Fig. 10.8) at defect concentrations of less than 10 mole per cent $YF_3$, but at higher defect concentrations there is evidence that even larger clusters are present (Cheetham *et al.* 1971*a*). The occupation numbers of a solid solution containing 25 mole per cent $YF_3$, $(Ca/Y)F_{2.25}$, are noticeably

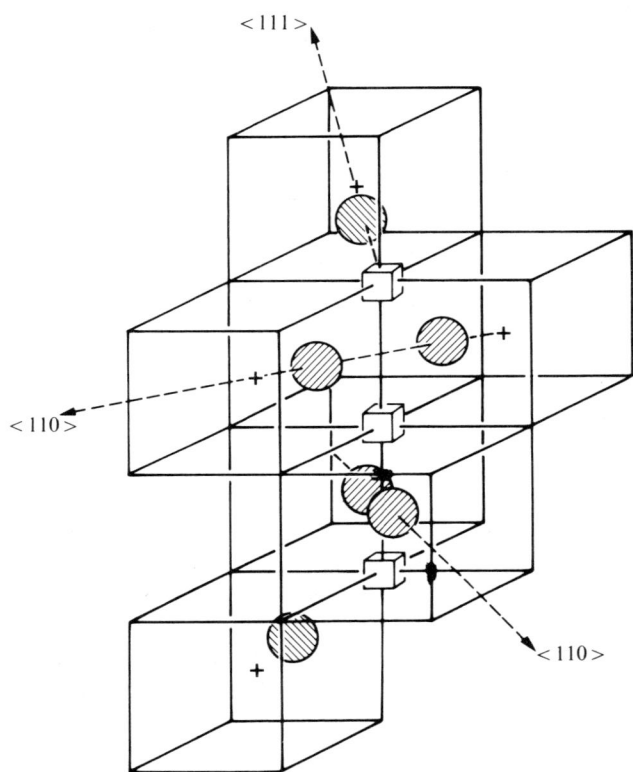

FIG. 10.8. A 3:4:2 cluster. Same symbols as for Fig. 10.7.

different from those of $U_4O_9$ (Table 10.1). In the ternary system no long-range order is observed (Fig. 10.9) and we have noted already (§ 10.4.1) that long-range order is less prevalent in ternary defect compounds.

The difference between the defect structure at $<$ 1 mole per cent $LnF_3$ revealed by resonance measurements and that observed at higher defect concentrations by diffraction techniques is undoubtedly genuine. The structural change presumably stems from the change in configurational entropy associated with the distribution of $F^-$ ions on $(\frac{1}{2}, \frac{1}{2}, \frac{1}{2})$ sites. As the extra fluoride content increases, this term will diminish in importance and, in consequence, attractive defect interactions begin to dominate, leading to the clustering of interstitial ions and the occupation of new types of interstitial site. The role played by the trivalent cations in the defect structure is revealed not by the Bragg scattering, but by diffuse scattering measurements, and is discussed in Chapter 11.

Structural knowledge is often essential for the interpretation of data from

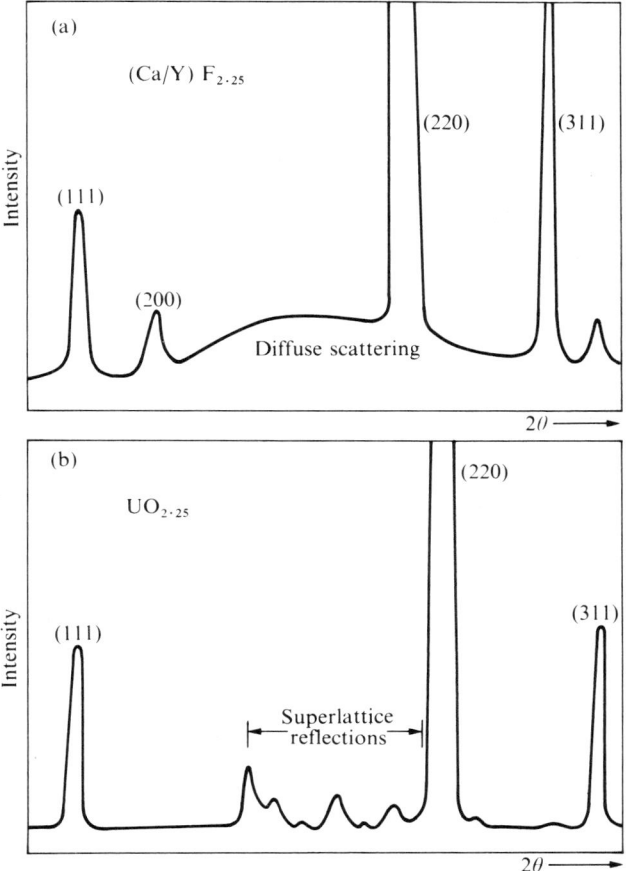

FIG. 10.9. Neutron powder patterns of (a) (Ca/Y)$F_{2.25}$ showing the diffuse scattering from the defects. (After Cheetham *et al.* 1971a). (b) $UO_{2.25}$ (i.e. $U_4O_9$) showing the superlattice reflexions arising from long-range ordering of the defects.

other experiments. In the present case, the neutron diffraction results have led to a greater understanding of electron spin resonance measurements on a single crystal of 10 mole per cent $CaF_2/YF_3$ irradiated with X-rays (Hall *et al.* 1970). A strong signal from an $F_2^-$ molecular ion with a $\langle 110 \rangle$ orientation was observed, in contrast to studies on crystals containing low defect concentrations ($< 1$ mole

## STUDIES ON NON-STOICHIOMETRIC COMPOUNDS

per cent $LnF_3$) in which $F_2^-$ units with a variety of orientations were detected. The preference of the $F_2^-$ ion for a ⟨110⟩ orientation has been attributed to the formation of a molecular ion from an adjacent pair of $F'$ interstitials in the 2:2:2 or 3:4:2 clusters (Figs. 10.7 and 10.8). This experiment also provides indirect support for the authenticity of the defect structure deduced from Bragg experiments. A recent single-crystal X-ray study of cerium-doped $CaF_2$ (Aleksandrov and Garashina 1970) revealed only one type of interstitial, similar to the $F''$ site in the models described above, although using X-rays it would be hard to detect all the fluorine-atom positions in the presence of the heavier cerium and calcium atoms.

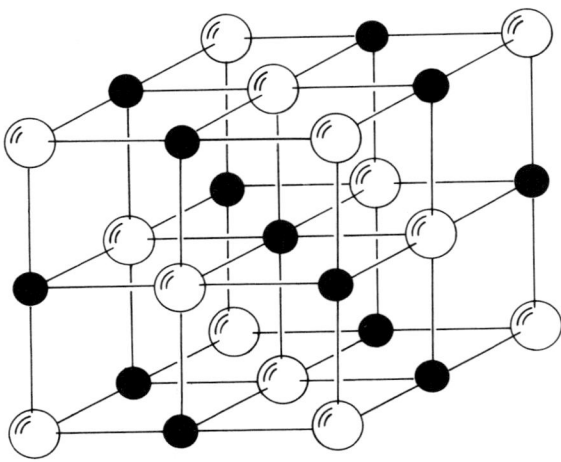

FIG. 10.10. The sodium chloride structure. ○ represents an anion, and ● a cation.

### 10.4.3. Defective sodium-chloride structures

Non-stoichiometry is frequently exhibited by materials which adopt the sodium chloride or rocksalt structure (Fig. 10.10). Most non-stoichiometric materials with this structure are metallic, as exemplified by the carbides, nitrides and monoxides of Ti, Zr, Hf, V, Nb and Ta. Ionic examples are comparatively rare although the $Mn_{1-x}O$ and $Fe_{1-x}O$ phases are well-known. In contrast to the fluorite systems, solid solutions with wide ranges of composition are not very common although many 'stoichiometric' solid solutions of the type MgO/MnO are found. The present discussion is concentrated mainly on the $Fe_{1-x}O$ system which has been extensively studied by diffraction, thermodynamic and transport methods, but reference is also made to other materials.

The phase diagram for $Fe_{1-x}O$, or wustite, (Fig. 10.11) shows that it is only stable under equilibrium conditions above 570°C. Early lattice parameter measurements at a series of compositions (Jette and Foote 1933) suggested that the structure was an iron-deficient sodium-chloride type, a result later confirmed by Willis and Rooksby (1953). Iron self-diffusion measurements (Himmel *et al.* 1953) also support the iron-vacancy model. A neutron diffraction study by Roth (1960) on quenched powder samples revealed a new and unexpected feature in the structure. Roth's intensity measurements showed that the number of iron

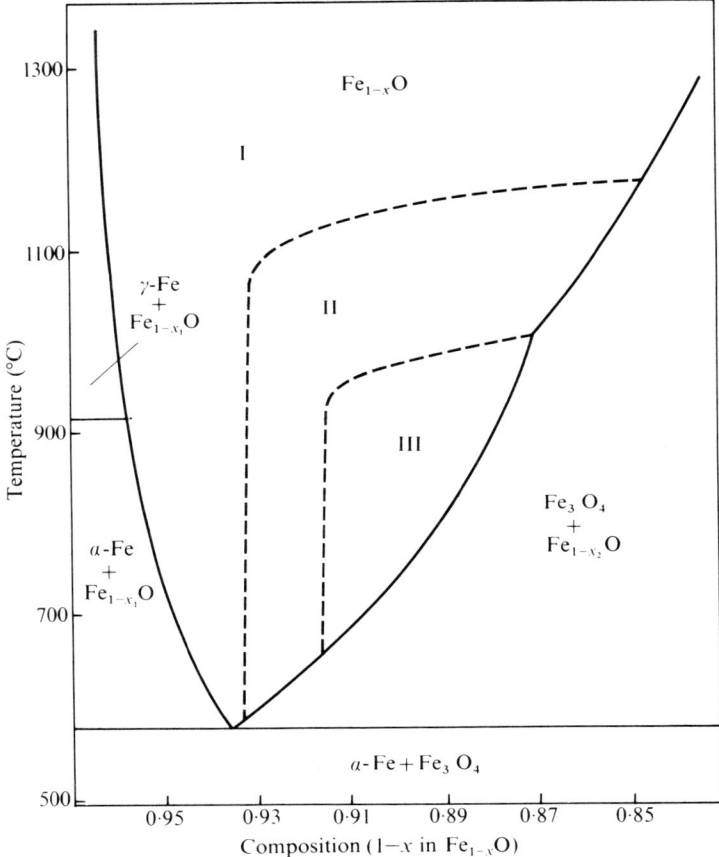

FIG. 10.11. The $Fe_{1-x}O$ phase diagram. The broken lines separate the three regions I, II, and III observed by Fender and Riley (1968).

vacancies in the metal sublattice was greater than expected from a knowledge of the composition alone, and that the atoms from these extra vacancies were located in tetrahedral interstitial sites. The ratio of the number of iron vacancies to the number of tetrahedral interstitials was found to be approximately 2:1, and on this basis a simple defect complex containing two vacancies and one tetrahedral interstitial was proposed. More recently, Koch and Cohen (1969) have studied quenched powders and single crystals of $Fe_{1-x}O$ by X-ray diffraction. With a

## STUDIES ON NON-STOICHIOMETRIC COMPOUNDS

quenched crystal of composition $Fe_{0.902}O$ they observed a vacancy: interstitial ratio rather larger than that given by Roth. Superlattice reflexions were also detected and an analysis of their intensities led them to propose a new defect complex containing thirteen vacancies and four interstitials, i.e. a ratio of 3·25 (Fig. 10.12). In contrast to the $Fe_3O_4$ structure, this complex contains iron vacancies at a separation of $a_0$ along the $\langle 100 \rangle$ directions (a feature revealed by the X-ray Patterson analysis), thus conflicting with the popular idea that $Fe_{1-x}O$ is composed of small regions (microdomains) of the $Fe_3O_4$ structure distributed in a perfect FeO matrix (Anderson 1964). In a high-temperature neutron diffraction study, Cheetham et al. (1971b) measured the vacancy: tetrahedral ratio at a series of temperatures and compositions (Fig. 10.13). The results suggested the persistence of Koch-Cohen clusters in the

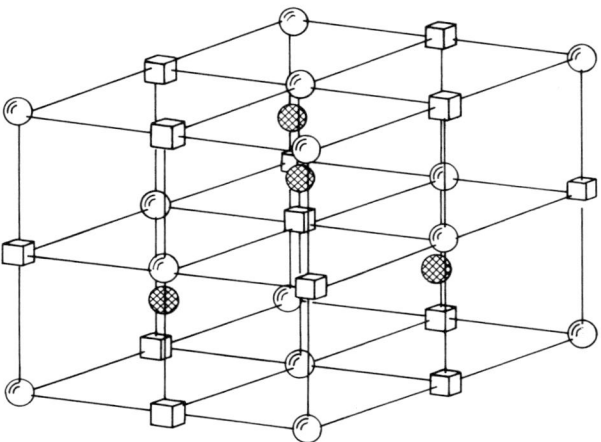

FIG. 10.12. The Koch–Cohen cluster in $Fe_{1-x}O$. ◯ are oxygen atoms, ◉ are tetrahedral-iron interstitials, and ▢ are octahedral-iron vacancies.

equilibrium region, but no support for the Roth model was found. However, there is some evidence to suggest that at the highest temperatures and lowest defect concentration the ratio is rather closer to 4:1, perhaps indicating that the Koch-Cohen clusters are dissociating into smaller units containing four vacancies and one tetrahedral interstitial (Fig. 10.14). Such a cluster preserves, of course, the local environment of a tetrahedral ion common to both $Fe_3O_4$ and the Koch–Cohen clusters. Mössbauer studies (Greenwood and Howe 1972) are also consistent with a small tetrahedral cluster at high temperatures. The dissociation of the larger clusters may also account for the complex thermodynamic behaviour observed by Fender and Riley (1968) in which the $Fe_{1-x}O$ phase was divided into three regions separated by order-disorder transitions (Fig. 10.11).

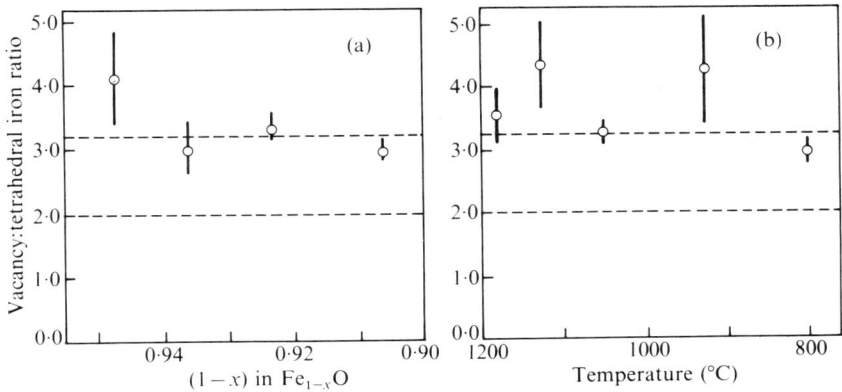

FIG. 10.13. The ratio of vacancies to tetrahedral-iron atoms in $Fe_{1-x}O$ as a function of (a) composition, (b) temperature (for the fixed composition $Fe_{0.904}O$). The vertical lines through the experimental points indicate estimated standard deviations. The lower horizontal broken line at 2·0 gives the predicted ratio for the defect model of Roth (1960), and the upper horizontal line at 3·25 the predicted ratio for the defect model of Koch and Cohen (1969).

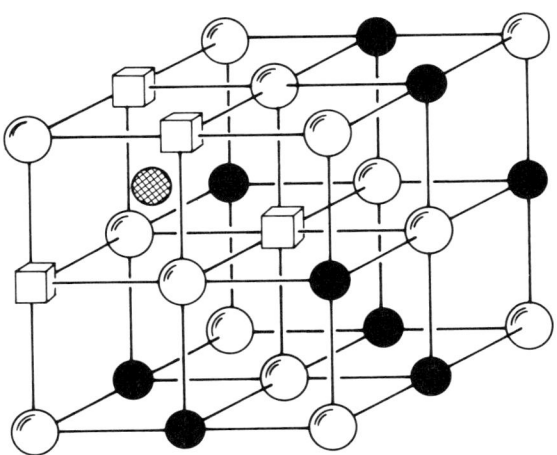

FIG. 10.14. Possible 4:1 cluster in $Fe_{1-x}O$. ◯ represent oxygen atoms, ◉ tetrahedral-iron interstitials, ☐ octahedral-iron vacancies, and ● undisplaced iron atoms.

Neutron diffraction has also been used to study the V–C system. The $VC_x$ phase, which adopts the NaCl structure, is characterized by a complete vanadium sub-lattice and vacancies in the carbon sub-lattice, and, like many other metallic systems, it forms ordered phases at low temperatures. X-ray diffraction studies on $V_8C_7$ (de Novion et al. 1966) suggested that the carbon vacancies are ordered in a spiral arrangement, a result recently confirmed by neutron diffraction

(Henfrey and Fender 1970), whilst electron diffraction measurements on $V_6C_5$ (Venables et al. 1968) have shown that in this phase, too, the vacancies assume a spiral configuration. Neutron diffraction work on the disordered region has not been very informative (Henfrey 1970), since again one is faced with a system in which the defects are present simply as vacancies in one sub-lattice. The presence of a broad, diffuse peak in the position of the forbidden 100 reflexion was thought to be indicative of short-range order. This postulate is strongly supported by the results of nuclear magnetic resonance studies on $VC_x$ using the $^{51}V$ nucleus (Froidevaux and Rossier 1967), in which evidence of short range ordering of the carbon vacancies is found over the whole of the $VC_x$ composition range. Diffuse scattering studies on this and the Nb–C system are described in the following chapter.

*10.4.4. Non-stoichiometric hydrides*

The power of neutron diffraction in the detection of light atoms is most apparent in the study of metal hydrides. Non-stoichiometry is exhibited in many of these systems, for example, by the hydrides of the group IVA and VA elements,

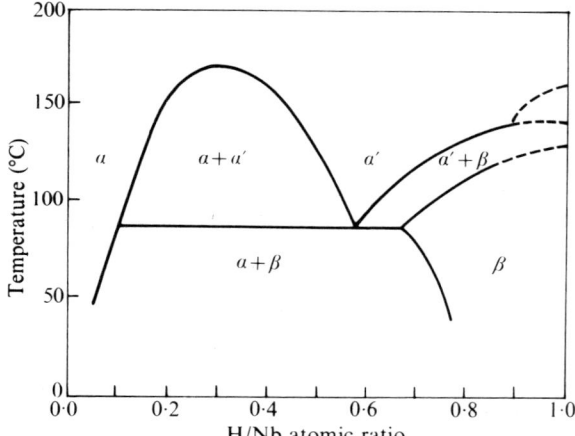

FIG. 10.15. The niobium–hydrogen phase diagram. The phases $\alpha$, $\alpha'$ and $\beta$ have wide ranges of composition (After Walters and Chandler 1965).

and also by those of the rare-earth and actinide metals (Mackay 1966). These compounds are characterized by metallic behaviour, and it is interesting to examine the influence of this property on their defect structure.

The niobium–hydrogen system has been extensively studied and there are three non-stoichiometric phases ($\alpha$, $\alpha'$ and $\beta$), all based upon the body-centred cubic niobium structure (Fig. 10.15). Neutron diffraction studies on the $\beta$-phase (Somenkov et al. 1969) have revealed that the hydrogen atoms occupy tetrahedral sites in the body-centred cubic structure. At room temperature in a sample of composition $NbD_{0.95}$ (deuterides were preferred to hydrides because of the superior neutron scattering properties of deuterium) the hydrogen atoms were ordered giving rise to tetragonal superlattice lines, but on warming to 200°C these lines disappeared indicating the loss of long range order. On cooling a sample from the disordered $\alpha$-phase, new ordered structures have been detected at 270 K and 170 K (Somenkov *et al.* 1970). This system exhibits unusual behaviour on cooling in that successive ordering processes take place in preference to disproportionation into separate phases. Thermodynamic measurements on the $\alpha$-phase (McQuillan 1970) showed marked departures from ideal behaviour at very dilute concentrations of hydrogen and these deviations were attributed to the clustering of defects. Diffuse scattering measurements on the disordered $\alpha$-phase are described in Chapter 11.

The increased scope of neutron diffraction which is possible with higher beam fluxes is illustrated by experiments on non-stoichiometric cerium hydride, $CeH_{2+x}$. Early neutron diffraction experiments by (Holley *et al.* 1955) on a deuteride of composition $CeD_{2.48}$ confirmed that $CeH_{2+x}$ adopts the fluorite structure, and it was proposed that the excess hydrogen atoms are incorporated as $(\frac{1}{2}, \frac{1}{2}, \frac{1}{2})$ interstitials in this structure. More recent work (using a higher neutron flux) by Cheetham and Fender (1972) on a sample of composition $CeD_{2.46}$ has shown that the interstitial deuterium atoms occupy sites slightly displaced from this position, and that at room temperature they are ordered giving rise to weak superlattice reflexions which index on a $2a_0 \times a_0 \times a_0$ tetragonal cell. This system also exhibits successive ordering phenomena similar to those in niobium hydride. Ordering of defects in hydride systems in quite common, and indeed is not unexpected since the hydrogen mobility is high and the conduction-band electrons provide a mechanism by which long-range interactions may be readily transmitted. Clearly, much is yet to be learnt about these interesting systems and further neutron diffraction studies will certainly be rewarding.

## 10.5. References

ALEKSANDROV, V. B. *and* GARASHINA, L. S. (1970). *Soviet Phys. Dokl.* **14**, 1040.
ANDERSON, J. S. (1964). *Proc. chem. Soc.* 166.
ANDERSON, J. S. (1969). *Bull. Soc. chim. Fr.* 2203. Masson et Cie, Paris.
BAENZIGER, N. C., EICK, H. A., SCHULDT, M. S. *and* EYRING, L. (1961). *J. Am. chem. Soc.* **83**, 2219.
BAKER, J. M., DAVIES, E. R. *and* HURRELL, J. P. (1968). *Proc. R. Soc.* **A308**, 403.
BELBEOCH, B., PIEKARSKI, C. *and* PERIO, P. (1961). *Acta Crystallogr.* **14**, 837.
BEVAN, D. J. M. (1955). *J. inorg. nucl. Chem.* **1**, 49.
BEVAN, D. J. M. *and* KORDIS, J. (1964). *J. inorg. nucl. Chem.* **26**, 1509.
BEVAN, D. J. M., BARKER, W. W., MARTIN, R. L. *and* PARKS, T. C. (1964). *Proc. 4th conference rare-earth research.* 441.

BLEANEY, B., LLEWELLYN, P. M. and JONES, D. A. (1956). *Proc. phys. Soc.* **B69**, 858.
CARTER, R. E. and ROTH, W. L. (1968). In *E.M.F. measurements at high temperatures (Ed. by C. B. Alcock)*.
CHEETHAM, A. K., FENDER, B. E. F., STEELE, D., TAYLOR, R. I. and WILLIS, B. T. M. (1970). *Solid St. Commun.* **8**, 171.
CHEETHAM, A. K., FENDER, B. E. F. and COOPER, M. J. (1971a). *J. Phys. C.* **4**, 3107.
CHEETHAM, A. K., FENDER, B. E. F. and TAYLOR, R. I. (1971b). *J. Phys. C.* **4**, 2160.
CHEETHAM, A. K. and FENDER, B. E. F. (1972). *J. Phys. C.* **5**, L35.
COOPER, M. J., ROUSE, K. D. and WILLIS, B. T. M. (1968). *Acta Crystallogr.* **A24**, 484.
COOPER, M. J. and ROUSE, K. D. (1970). In *Thermal neutron diffraction. (Ed. by B. T. M. Willis)*. Oxford University Press.
EYRING, L. and BAENZIGER, N. C. (1962). *J. appl. Phys. Supp.* **33**, 428.
FENDER, B. E. F. and RILEY, F. D. (1969). *Physics Chem. Solids.* **30**, 793.
FROIDEVAUX, C. and ROSSIER, D. (1967). *Physics Chem. Solids.* **28**, 1197.
GREENWOOD, N. N. and HOWE, A. T. (1972). *J. Chem. Soc.(Dalton)* 110.
HALL, T. P. P., LEGGEAT, A. and TWIDELL, J. W. (1970). *J. Phys. C.* **3**, 2352.
HARALDSEN, H. (1937). *Z. anorgallg. Chem.* **234**, 372.
HENFREY, A. W. (1970). *D. Phil. Thesis.* University of Oxford.
HENFREY, A. W. and FENDER, B. E. F. (1970). *Acta Crystallogr.* **B26**, 1882.
HIMMEL, L., MEHL, R. F. and BIRCHENALL, C. E. (1953). *Trans. Am. Inst. Min. Engrs.* 827.
HOLLEY, C. E., MULFORD, R. N. R., ELLINGER, F. H., KOEHLER, W. C. and ZACHARIASEN, W. H. (1955). *J. Phys. Chem.* **59**, 1226.
HUANG, K. (1947). *Proc. R. Soc.* **A190**, 102.
HYDE, B. G., BEVAN, D. J. M. and EYRING, L. (1966). *Phil. Trans. R. Soc.* **A259**, 583.
HYDE, B. G. and BURSILL, L. A. (1970). *The chemistry of extended defects in non-metallic solids.* North-Holland Publishing Co., Amsterdam.
IPPOLITOV, E. G., GARASHINA, L. S. and MAKLASKLOV, A. G. (1967). *Inorganic Materials.* **3**, 59.
JELLINEK, F. (1957). *Acta Crystallogr.* **10**, 620.
JETTE, E. R. and FOOTE, F. (1933). *J. chem. Phys.* **1**, 29.
KIHLBORG, L. (1963). *Ark. Kemi.* **21**, 443.
KOCH, F. and COHEN, J. B. (1969). *Acta Crystallogr.* **B25**, 275.
MACKAY, K. M. (1966). *Hydrogen compounds of the metallic elements.* Spon, London.
MATZKE, H., DAVIES, J. A. and JOHANSSON, N. G. E. (1971). *Can. J. Phys.* **49**, 2215.
MCQUILLAN, A. D. (1970). *J. chem. Phys.* **53**, 156.
DE NOVION, C. H., LORENZELLI, R. and COSTA, P. (1966). *C. r.hebd. Séanc. Acad. Sci., Paris.* **263B**, 775.
RACCAH, P. M. and HENRICH, V. E. (1969). *Phys. Rev.* **184**, 607.
ROBERTS, L. E. J. (1963). *Adv. chem. ser.* **39**, 66.

ROBERTS, L. E. J. (1971). *Essays in structural chemistry* (*Ed. by A. J. Downs, D. A. Long and L. A. K. Staveley*.) Macmillan, London.
ROTH, W. L. (1960). *Acta Crystallogr.* **13**, 140.
SCHOTTKY, W. *and* WAGNER, C. (1930). *Z. phys. Chem.* **B11**, 163.
SHANNON, R. D. *and* PREWITT, C. T. (1969). *Acta Crystallogr.* **B25**, 925.
SOMENKOV, V. A., ZEMLYANOV, M. G., KOST, M. E., CHERNOPLEKOV, M. A. *and* CHERTKOV, A. A. (1969). *Soviet Phys. Dokl.* **13**, 669.
SOMENKOV, V. A., PETRUNIN, V. F., SHIL'SHTEIN, S. Sb. *and* CHERTKOV, A. A. (1970). *Soviet Phys. Crystallogr.* **14**, 522.
STEELE, D. (1971). *D. Phil. Thesis.* University of Oxford.
SUBBARAO, E. C., SUTTER, P. H. *and* HRIZO, J. (1965). *J. Am. Ceram. Soc.* **48**, 443.
SUORTTI, P. (1967). *Ann. Acad. Scient. Fenn.* A VI, 240.
TIEN, T. V. *and* SUBBARAO, E. C. (1963). *J. chem. Phys.* **39**, 1041.
VENABLES, J. D., KAHN, D. *and* LYE, R. G. (1968). *Phil. Mag.* **18**, 177.
WALTERS, R. J. *and* CHANDLER, W. T. (1965). *Trans Am. Inst. Min. Engrs.* **233**, 762.
WARREN, B. E. (1969). *X-ray diffraction.* Addison-Wesley, New York.
WILLIS, B. T. M. (1963). *Nature, Lond.* **197**, 755.
WILLIS, B. T. M. (1964). *J. de Physique.* **25**, 431.
WILLIS, B. T. M. *and* ROOKSBY, H. P. (1953). *Acta Crystallogr.* **6**, 827.

*General*

ANDERSON, J. S. (1969). The constitution of non-stoichiometric compounds. *Bull. Soc. Chim. Fr.* 2203.
GREENWOOD, N. N. (1968). *Ionic crystals, lattice defects, and non-stoichiometry.* Butterworth, London.
WADSLEY, A. D. (1964). Chapter 3. Inorganic non-stoichiometric compounds, in *Non-stoichiometric compounds* (*Ed. L. Mandelcorn.*) Academic Press, London.

# 11 Diffuse Scattering and the Study of Defect Solids

*By* B. E. F. FENDER

*Inorganic Chemistry Laboratory, Oxford University*

## 11.1. Introduction

If grossly defect solids, including non-stoichiometric compounds (e.g. $Fe_{1-x}O$, $VC_{1-x}$), solid solutions (e.g. $Ca(Y)F_{2+x}$, $Ti(Nb)O_{2+x}$), and radiation-damaged solids, contained only random point defects (i.e. non-interacting vacant sites, interstitial or substitutional atoms), then the average unit cell would be a typical representation of the crystal as a whole and analysis of Bragg reflexions alone might be sufficient to characterize the structure of the solid (see Chapter 10). In fact, in most solids with an appreciable equilibrium concentration of defects (greater than about 1 per cent), the creation of random point defects will be energetically unfavourable. The free energy change associated with the configurational entropy of randomly distributed defects ($\approx Tn \ln n$ where $T$ is the absolute temperature and $n$ the fractional defect concentration) will usually be insufficient, even at high temperatures, to overcome the enthalpy of defect formation, unless this is minimized by the clustering or short-range ordering of point defects. For convenience, the association of point defects on near-neighbour sites is referred to as a 'cluster' and the spacings at greater distance arising primarily from repulsion between point defects as short-range order. The probability of strong interactions between defects makes it clear why the observation of clustering and other short-range ordering effects is central to our understanding of the stability and existence ranges of a very large number of inorganic compounds and solid solutions. In irradiated crystals, particularly when the radiation source is of high energy, clustering of defects again appears to be common, though the reasons are different.

Local atomic ordering has no effect on the intensity of the Bragg peaks[†] and it influences the elastic scattering merely by modulating the background between Bragg peaks. This *diffuse scattering*[‡] is of low intensity and is invariably contaminated with that from the other processes which contribute to the total background scattering. Evaluation of defect scattering thus poses theoretical

---

[†] This implies that the occupancy of the sites in the average cell is unchanged by the ordering process (see Chapter 10).

[‡] Throughout this book we use the word *incoherent* to describe scattering which is associated with random distributions of isotopes or with random orientations of nuclear spin. Scattering due to other random effects, such as those discussed in this chapter, is called *diffuse*.

and experimental problems which are much more difficult than those involved in the analysis of Bragg peaks. We are now, in effect, enquiring about the atomic arrangements as viewed from a particular site in the interior of the crystal, and the increased sophistication of this question is reflected in the enhanced complexity of the measurements and their interpretation. Nevertheless, X-ray diffuse scattering has provided otherwise inaccessible information on short-range ordering in alloys, as exemplified by the classical experiments on the Cu–Au system (Cowley 1950a, Moss 1964). For some diffuse scattering experiments, neutrons have certain advantages over X-rays and these have been exploited in the study of irradiated solids (graphite: Antal et al. 1955, Martin and Henson 1964; silica: Mitchell and Wedepohl 1958; beryllium oxide: Sabine et al. 1963; magnesium oxide: Martin 1968; germanium and silicon: Clark, Mitchell, and Stewart 1971). Neutron-scattering techniques, developed for the study of radiation-damaged solids, are also applicable to solids in which the defect content is controlled by chemical manipulation and it is the scope of these studies which form the major part of this chapter. §11.2 deals with the main theoretical considerations, §11.3 with experimental details and §11.4 with illustrative results and the prospects for future study.

## 11.2. Theory

The basic equations appropriate to diffuse scattering are applicable to both neutrons and X-rays, though the symbols we shall use are those commonly adopted in dealing with neutron scattering. The elastic scattering from an assembly of $N$ identical atoms gives rise to a scattering amplitude of

$$\psi = \sum_{n=1}^{N} b_n \exp(i\mathbf{Q}\cdot\mathbf{R}_n), \tag{11.1}$$

where $\mathbf{R}_n$ is the position vector of the $n$th nucleus, $\mathbf{Q}$ the scattering vector, and $b$ is the atomic scattering length. For neutrons, where scattering is by the nucleus (i.e. excluding magnetic scattering), $b$ is independent of $\mathbf{Q}$ but for X-rays the scattering length decreases with $\mathbf{Q}$ and the $Q$-dependence (form factor) must be known for accurate evaluation of intensities. The differential scattering cross-section $d\sigma/d\Omega$, the intensity of scattering per unit solid angle for unit incident intensity, is

$$\frac{d\sigma}{d\Omega} = \psi\psi^* = \sum_{m,n=1}^{N}\sum b_m b_n \exp[i\mathbf{Q}\cdot(\mathbf{R}_m - \mathbf{R}_n)], \tag{11.2}$$

where $\mathbf{R}_m$ and $\mathbf{R}_n$ are the vector distances referred to the same arbitrary origin. All possible pairs of atoms contribute to the total scattering but only for a

# DIFFUSE SCATTERING: STUDY OF DEFECT SOLIDS

perfect crystal is this confined to the Bragg peaks. This is seen by rewriting eqn (11.2), generalized to include several different atomic species ($\nu$) including vacancies, as

$$\frac{d\sigma}{d\Omega} = \sum_{i,}^{\nu} \sum_{j,}^{\nu} \sum_{m,\,n=1}^{N_\nu\,N_\nu} \langle b_m^i \rangle \langle b_n^j \rangle \exp[i\mathbf{Q}\cdot(\mathbf{R}_m - \mathbf{R}_n)]$$

$$+ \sum_{i,}^{\nu} \sum_{j,}^{\nu} \sum_{m,\,n=1}^{N_\nu\,N_\nu} (b_m^i b_n^j - \langle b_m^i \rangle \langle b_n^j \rangle) \exp[i\mathbf{Q}\cdot(\mathbf{R}_m - \mathbf{R}_n)], \quad (11.3)$$

where $\langle\,\rangle$ indicates an average over all species at sites equivalent to $\mathbf{R}_m$ (or $\mathbf{R}_n$) in the unit cells comprising the total assembly. $N_\nu$ is the total number of atoms of type $\nu$. The first term contains contributions from all atoms and is the Bragg intensity. The second term represents the diffuse scattering. This scattering is seen to arise when, for a particular value of $\mathbf{R}_m - \mathbf{R}_n$, the sum of the scattering from individual atom pairs differs from that expected for the average occupancy of the lattice, i.e. $\langle b_m^i b_n^j \rangle \neq \langle b_m^i \rangle \langle b_n^j \rangle$. There is no reason, in principle, why differences should not occur up to quite large values of $\mathbf{R}_m - \mathbf{R}_n$, making the distinction beteen Bragg and diffuse scattering a formal one, but, in practice, the mutual influence of point defects will be confined to small values of $\mathbf{R}_m - \mathbf{R}_n$ and the separation is easy to visualize.

The normally low concentration of defects, and the relatively small number of $\mathbf{R}_m - \mathbf{R}_n$ terms which contribute to the scattering, account for the small cross-sections observed for diffuse scattering. Evaluation of the diffuse term in terms of the atomic distribution involves a knowledge of positions and occupation numbers for all atoms (obtained from the Bragg scattering experiment), but the major obstacle to the calculation of local order is the formidable complexity of the description when the short-range ordering involves several defects, possibly distributed on two or more sub-lattices (see §11.2.3). For a random distribution of defects, of course, the calculations are simple and include the cases of random solid solutions (§11.2.1) and random vacancies and interstitials (§11.2.2).

## 11.2.1. Random solid solutions

For the random distribution of atoms A and B on a lattice, the differential cross-section per atom is readily derived as

$$\left(\frac{d\sigma}{d\Omega}\right)_{\text{diffuse}} = x_A x_B (b_B - b_A)^2 \quad (11.4)$$

which is the Laue expression (Laue 1918) for monotonic scattering. $x_A$ and $x_B$ are the mole fractions of A and B respectively. For a number of different atomic species

$$\left(\frac{d\sigma}{d\Omega}\right)_{\text{diffuse}} = \langle b^2 \rangle - (\langle b \rangle)^2 \quad (11.5)$$

which is also the familiar expression for neutron elastic incoherent scattering, arising from a random distribution of isotopes or spin states (eqn 1.10c).

### 11.2.2. Random vacancies and interstitials

It is equally valid to consider one of the species, say B, to be a vacant site, in which case eqn (11.4) becomes

$$\left(\frac{d\sigma}{d\Omega}\right)_{\text{diffuse}} = c(1-c)b_A^2 \qquad (11.6)$$

with $c$ the fractional vacancy concentration. If interstitial atoms (A) are introduced, we can view the complete set of symmetry-related interstitial sites (whether occupied or not) as constituting a new sub-lattice, so that eqn (11.4) is again valid except that now

$$\left(\frac{d\sigma}{d\Omega}\right)_{\text{diffuse}} = c(1-c/q)b_A^2 \qquad (11.7)$$

where $c$ represents the fraction of interstitial *atoms* per atom of the host lattice and $q$ is the number of interstitial *sites* per host atom. When the interstitial and host atoms are identical, eqns (11.6) and (11.7) differ only in the $c^2$ term (when $q \neq 1$) and therefore the scattering from isolated vacancies and interstitials cannot normally be distinguished. Eqns (11.4), (11.6), and (11.7) demonstrate that it is *deviations* from perfect order which give rise to diffuse scattering (see Cochran 1956).

### 11.2.3. Short-range ordering

As soon as correlations are introduced between defects, the diffuse scattering rapidly becomes more difficult to calculate explicitly. Short-range ordering in binary (AB) solid solutions is, however, still quite readily dealt with. In this case, the local atomic environment can be described in terms of the probability $P_A(r_i)$ of finding an A atom at a distance $r_i$ (usually referred to as the $i$th shell) from B, or the probability $P_B(r_i)$ of a B atom being at $r_i$ from A. Eqn (11.4) is now modified to

$$\left(\frac{d\sigma}{d\Omega}\right)_{\text{diffuse}} = x_A x_B (b_B - b_A)^2$$
$$+ \frac{1}{N} \sum_{m,n=1}^{N} \sum x_A x_B (b_B - b_A)^2 \alpha_i^A \exp[i\mathbf{Q}.(\mathbf{R}_m - \mathbf{R}_n)], \qquad (11.8)$$

where $\alpha_i^A = 1 - P_A(r_i)/x_A$ is a short-range order parameter (Cowley 1950b). The second term in eqn (11.8) modulates the monotonic scattering in a way which depends on the different values of $\alpha_i$. This expression is inadequate if the atoms are displaced from their special positions either because of a difference in atomic size, or because of relaxation around a vacancy or interstitial site. A number of authors (see Warren 1969), have considered the appropriate modification for near-neighbour effects, which can be expressed in an additional term given by the expression:

$$\frac{d\sigma}{d\Omega}\text{(size effect)} = \frac{1}{N} \sum_{m,n=1}^{N} \sum [(x_A - x_A P_B) b_A^2 \epsilon_i^{AA} + 2x_B P_A b_A b_B \epsilon_i^{AB}$$

$$+ x_B(1 - P_A) b_B^2 \epsilon_i^{BB}] \, iQ \cdot (R_m - R_n)$$

$$\times \exp[iQ \cdot (R_m - R_n)]. \qquad (11.9)$$

$\epsilon_i^{AA}$ is the ratio $(r_i^{AA} - r_i)/r_i$ where $r_i^{AA}$ is the distance between a given A atom and another A atom; $\epsilon_i^{AB}$ and $\epsilon_i^{BB}$ are defined in a similar way. Eqns (11.8) and (11.9) form the basis of the successful interpretation of diffuse scattering arising from atomic disorder in substitutional alloys like $Cu_3Au$. Because we are dealing with short-range order and thus only small values of $R_m - R_n$, eqn (11.8) for such simple structures may be reduced for a single crystal to

$$\frac{d\sigma}{d\Omega}\text{(single crystal)} = x_A x_B (b_B - b_A)^2 \sum_{n=1}^{N} \alpha_n \cos(Q \cdot r_n), \qquad (11.10)$$

(where $\alpha_n$ is the short-range order parameter expressed in terms of the vector $r_n$) or for a powder, where the cosine term is averaged over all possible orientations, to

$$\frac{d\sigma}{d\Omega}\text{(polycrystal)} = x_A x_B (b_B - b_A)^2 \sum_{n=0}^{N} s_i \alpha_i \frac{\sin(Qr_i)}{Qr_i}, \qquad (11.11)$$

where $Q = 4\pi \sin\theta/\lambda$, and $s_i$ is the number of atoms at a distance $r_i$ from the origin. Generally we expect $d\sigma/d\Omega$ to converge rapidly over a small number of shells.

Can we extend these equations to deal with more complicated situations? In principle this is possible, but we should soon lose any simple physical picture in the face of the confusing array of parameters which would be necessary to describe the atomic correlations. A more direct, though approximate, approach is that developed by Martin (1967) to deal with defect clustering in irradiated solids. Martin makes the assumption, which is one we can also attempt to apply to non-stoichiometric compounds, that the defect or defect cluster together with the disturbed region of crystal around it can be treated in isolation. If all possible orientations in space are allowed, and the defects do not interact, then the diffuse scattering from such a cluster would be exactly that from a cluster or 'molecule' made up of atoms and vacant sites not found in the perfect crystal. The scattering amplitude for such a defect cluster is given by eqn (11.1) except that the summation now extends over $n'$, the number of point defects in the cluster. This leads to the differential cross-sections per cluster,

$$\left(\frac{d\sigma}{d\Omega}\right)_{\text{diffuse}} = \sum_{i,j=1}^{n'\ n'} b_i b_j \cos(\mathbf{Q}\cdot(\mathbf{r}_i - \mathbf{r}_j)) \text{ (single crystal)} \quad (11.12)$$

and

$$\left(\frac{d\sigma}{d\Omega}\right)_{\text{diffuse}} = \sum_{i,j=1}^{n'\ n'} b_i b_j \frac{\sin(Qr_{ij})}{Qr_{ij}} \text{ (polycrystal)}, \quad (11.13)$$

and to the total cross-sections per cluster,

$$\sigma_{\text{total}}^{\text{diffuse}} = 4\pi \sum_{i,j=1}^{n'\ n'} b_i b_j \frac{\sin(2\pi r_{ij}/\lambda)}{2\pi r_{ij}/\lambda} \overline{\cos(2\pi(\mathbf{r}_i - \mathbf{r}_j)\cos\phi/\lambda)}$$

(single crystal) (11.14)

and

$$\sigma_{\text{total}}^{\text{diffuse}} = 4\pi \sum_{i,j=1}^{n'\ n'} b_i b_j \frac{\sin^2(2\pi r_{ij}/\lambda)}{(2\pi r_{ij}/\lambda)^2} \text{ (polycrystal)}. \quad (11.15)$$

The bar in eqn (11.14) indicates that the cross-section must be averaged over all equivalent orientations of the cluster; $\phi$ is the angle between the line joining two defects and the incident beam direction.

Obviously this treatment is valuable only when there is a strong tendency for defects to associate, but in such cases we see the advantage in retaining a simple physical model. A further advantage implicit in this approach is the ease with which atomic relaxation is incorporated. Each displaced atom creates a vacancy–interstitial pair in the defect cluster.

Not all positions or orientations of the cluster are possible so that, to describe the scattering per atom of the perfect host lattice, eqns (11.12) to (11.15) must be multiplied by $c(1 - c/\nu)$ where $c$ is the fractional concentration of defects per atom of host and $\nu$ is the number of available sites for the cluster (again per host atom). The quantity $\nu$ has to be assessed, but the term $1 - c/\nu$ is in any case usually close to unity. Serious consideration must be given, however, to the possibility of interactions between clusters. In general, unless the defect concentration is very small, mutual influence on the spacing and orientation of defect clusters is to be expected, either from medium-range forces (e.g. coulombic) or from the effect of successively relaxed atoms. Fortunately, though the exact form of the diffuse scattering may be affected by these interactions, in many cases its dominant features will still reflect the geometry of the cluster. It turns out that the larger the cluster the more the diffuse scattering is dominated by the characteristics of the individual cluster. On the other hand, small defect clusters with strong short-range ordering interactions give rise to scattering which is heavily influenced not merely by the properties of the individual cluster but also by the spacing of nearest-neighbour clusters. This association of defect clusters can be envisaged as a new enlarged 'multicluster'.

Ideally we should like to be able to convert the experimentally observed diffuse scattering directly into atomic or point-defect positions and, in principle, this is possible by a Fourier inversion of the cross-section data. For example, eqn (11.11) can be transformed to

$$g(r) = \frac{2r}{\pi} \int_0^\infty Q \left( \left[ \frac{d\sigma(Q)}{d\Omega} \bigg/ x_A x_B (b_B - b_A)^2 \right] - 1 \right) \sin(rQ) \, dQ \quad (11.16)$$

where $g(r)$, the radial distribution function, is finite only when $r$ is close to $r_i$ and the areas of the corresponding peaks in $Q$-space give the various values of $s_i \alpha_i$. Mitchell and Stewart (1967) have discussed one-dimensional Fourier inversion of total cross-sections, but the draw-back to applying Fourier methods is that the range of experimental data over $\sin\theta/\lambda$ (differential cross-sections) or $\lambda$ (total cross-sections) is usually too restricted to allow accurate evaluation of the radial distribution function. As a result, we are forced to fall back on a comparison of the observed scattering with that calculated from a variety of models. A unique solution cannot now be proved and the trial of a wide range of alternative descriptions of the local order will usually be necessary. Nevertheless, particularly in non-stoichiometric compounds, the choice of models will often be narrowed by the results of Bragg experiments and by indirect structural or chemical evidence, so that we need not be too inhibited by this limitation.

Before comparisons can be carried out, it is necessary to separate the diffuse scattering from other contributions to the background. These form a formidable list which, depending on the radiation used, may include: isotopic- and spin-incoherent scattering, elastic and inelastic paramagnetic scattering (neutrons), Compton modified scattering and fluorescence (X-rays), thermal diffuse scattering, multiple Bragg scattering, close-neighbour size effect, and Huang diffuse scattering (see Warren 1969), as well as air and stray background scattering (X-rays and neutrons). A number of these processes can be eliminated or assessed experimentally, but contributions from thermal diffuse scattering (see Warren 1969), scattering caused by static atomic displacements and multiple Bragg effects are likely to be troublesome to estimate. Thermal diffuse scattering, which results from thermal vibration of the atoms, increases towards larger values of $\sin\theta/\lambda$ (with peaks in the TDS under the Bragg reflexions). An increase in scattering with $Q$ is also observed with static displacements, arising from short-range atomic relaxations (see neutron powder diffraction patterns illustrated Fig. 11.1) as well as from the variations in remote-neighbour distances (Huang 1947; Borie 1957, 1959, 1961). Minimization of these effects points to the desirability of measurements at low $Q$ which can be achieved either by the use of long wavelengths or

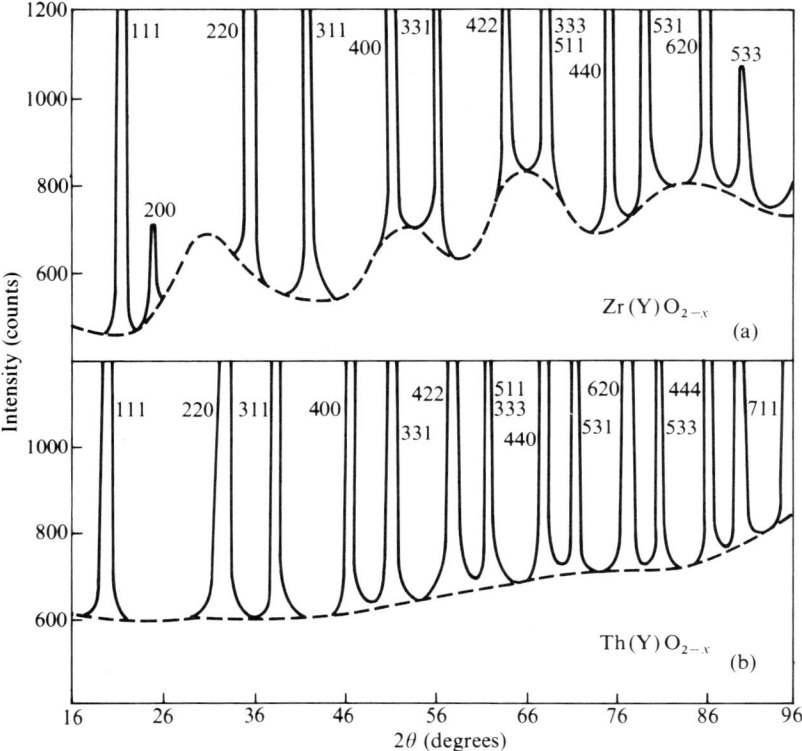

FIG. 11.1. (a) Neutron diffraction pattern of $Zr(Y)O_{2-x}$. Atom displacements cause modulated diffuse scattering which increases with scattering angle. (b) Neutron diffraction pattern of $Th(Y)O_{2-x}$. There are no atomic displacements and the background is nearly smooth. (After Steele and Fender 1971).

with conventional wavelengths of about 1 Å by observations at low $\theta$. The latter alternative is usually less satisfactory because this is experimentally the most difficult region in which to reduce the effects of air scattering and other stray scattering from the beam. Employment of long wavelengths leads, however, exclusively to the use of neutrons because the absorption of X-rays is prohibitively high above 3 Å. In addition, the use of long wavelength neutrons may allow the Bragg cut-off to be exceeded (i.e. $\lambda/2d > 1$ for the largest inter-lattice spacing), thus removing both the need to subtract Bragg peaks from the diffuse scattering and multiple Bragg effects.

## 11.3. Experimental

Two alternative experimental approaches are possible; either we can observe the differential scattering from a sample at constant wavelength, or follow changes in transmitted flux, and therefore total scattering, as a function of wavelength. Both methods are employed. A schematic lay-out of an apparatus suitable for transmission experiments is shown in Fig. 11.2. Near the centre of the reactor, a liquid hydrogen–deuterium moderator increases the proportion of long-wavelength neutrons which pass down the beam tube. A polycrystalline beryllium filter scatters neutrons with a wavelength shorter than the Bragg cut-off for beryllium (3·95 Å), and the remaining beam after passing through a bismuth filter to remove fast neutrons, is monochromated by a mechanical wavelength selector. This selector, which consists of a series of discs with slits cut to simulate a helical slot parallel to the beam, may be rotated at different speeds, each rotational velocity allowing only neutrons of a particular velocity to pass through the slit

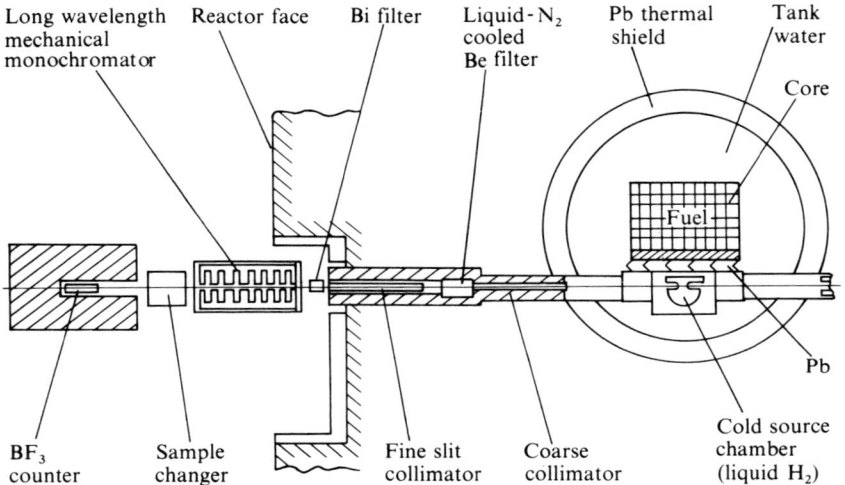

FIG. 11.2. Apparatus for the study of total scattering by defects. The sketch is based on the original apparatus installed at the HERALD reactor, Aldermaston. (After Clark, Mitchell, and Stewart 1971).

system. The transmitted neutron flux is measured with and without the sample in the beam by a well-shielded $BF_3$ counter. The success of such an experiment depends on the sample absorption cross-section being suitably small; many nuclei have acceptable absorption properties, though boron, cadmium and gadolinium are notable exceptions.

The differential scattering apparatus, Fig. 11.3, is similar except that a multi-counter bank replaces the single counter in the straight-through position. The neutron beam is also now chopped into pulses of about 300 $s^{-1}$ duration by a rotor and this provides the basis for wavelength selection. Scattered neutrons are only recorded for a certain short time after a pulse leaves the rotor, and the chosen delay time determines a particular and relatively narrow wavelength range for the detected neutrons. Variation in delay time, rotor speed, and the counting period determine $\lambda$ and $\Delta\lambda$. The total neutron flux falling on the sample is monitored by a fission counter so that the observed scattering can be related to a known incident flux. For these long wavelength studies the diffuse intensity, of course, is relatively low, but differential measurements are commonly carried out at 7 or 8 Å and transmission measurements from 4 Å to at least 14 Å.

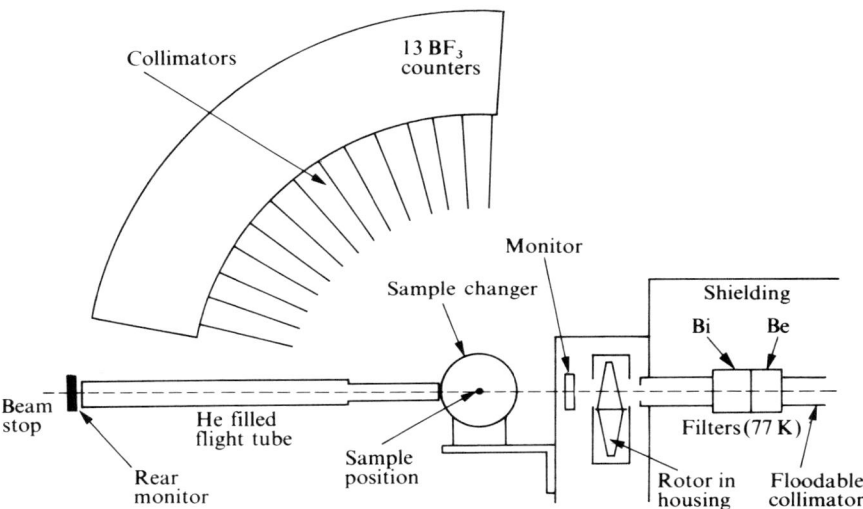

FIG. 11.3. Differential scattering apparatus. The sketch is based on the original apparatus installed at the PLUTO reactor, Harwell. (After Low and Collins 1963).

## 11.4. Results

The scope of the diffuse scattering technique is illustrated in this section by reference to results from systems of differing complexity.

### 11.4.1. Interstitial atoms

The transition elements of groups IIIA, IVA and VA all form metallic hydrides, but V, Nb and Ta, in particular, dissolve hydrogen to a considerable extent in the body-centred structure of the parent metal. It is generally assumed that the hydrogen is located in tetrahedral sites (see Fig. 11.4) but how are the atoms distributed? Some of the thermodynamic evidence appears to point to clustering of hydrogen atoms (Kofstad and Wallace, 1959; McQuillan and Wallbank, 1970), but neutron diffuse scattering demonstrates a different result, at least in the niobium–deuterium system at room temperature, as Fig. 11.5 indicates. The observed scattering of deuterium in a single crystal of niobium lies very close to the cross-section calculated by eqn (11.7) assuming a random distribution of deuterium atoms (Fender and Henfrey 1970). Fig. 11.5 shows the build up of diffuse scattering in the forward-$Q$ direction that would have been observed if the deuterium had been associated in pairs of nearest neighbour or next-nearest

259

# DIFFUSE SCATTERING: STUDY OF DEFECT SOLIDS

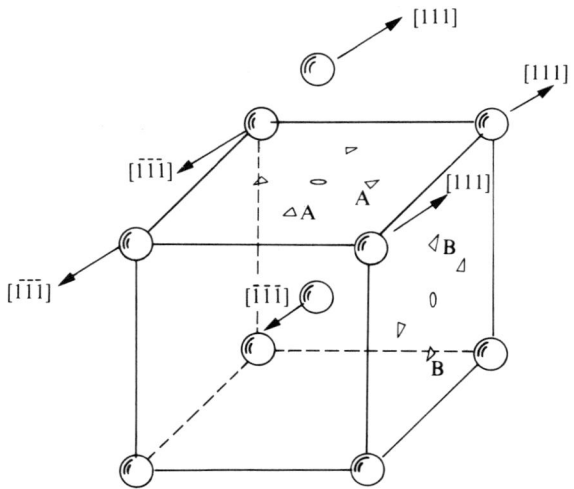

FIG. 11.4. The body-centred cubic structure. ○ octahedral sites; △ tetrahedral sites; A–A nearest-neighbour sites; B–B next-nearest-neighbour sites.

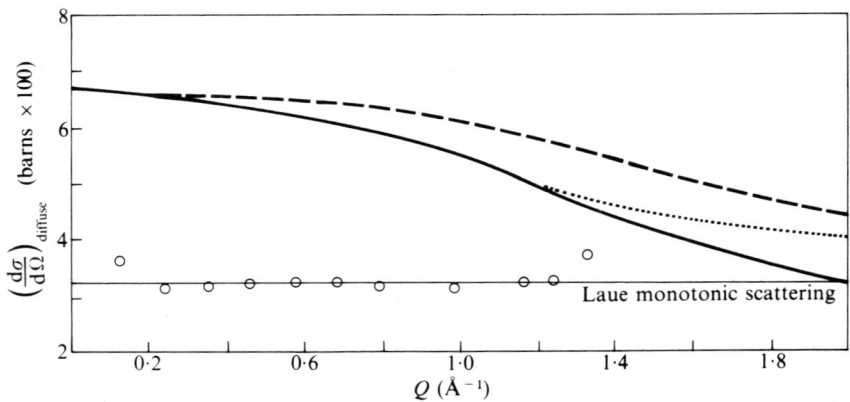

FIG. 11.5. Diffuse scattering intensity versus $Q$ ($=4\pi \sin\theta/\lambda$) for $NbD_{0.079}$. ○ Experimental. Horizontal straight line is calculated (Laue monotonic) scattering for random distribution of isolated deuterium atoms. —— Calculated for a random distribution of deuterium atom pairs at BB (see Fig. 11.4). ··· Calculated for random pair clusters at AA. – – – Calculated for random pair clusters at AA, with niobium atoms displaced by 0·1 Å along $\langle 111 \rangle$ as shown in Fig. 11.4.

neighbour clusters. The figure also shows the results of a calculation using eqn (11.13) which demonstrates the effect on the diffuse scattering of a 0·1 Å displacement of niobium atoms.

Irradiated crystals pose more difficult problems because the high-energy bombardment can lead to the trapping of atoms in highly metastable sites. It is therefore more difficult to assess the location of any interstitial atom and the displacements forced on neighbouring atoms may be considerable. In consequence, the range of models which needs to be explored is very large. An idea of the difficulty of this exercise is seen from the experiments on graphite by Martin (1968). The graphite sample was irradiated with nearly $10^{20}$ fast neutrons at 30°C: the magnitude and the approximate form (i.e. dependence on $Q$) of the observed diffuse scattering were reproduced by a calculation based on a model consisting of 2% interstitials, plus a corresponding number of random isolated vacancies, associated in clusters of about four atoms inserted between the carbon layers—but only when nearly 100 normal carbon atoms were relaxed.

*11.4.2. Vacancies*

The introduction of a relatively large ($> 1$ per cent) concentration of random vacancies may be a relatively rare phenomenon. Nevertheless, apart from a small number of ordered compounds, including $V_6C_5$, $V_8C_7$ and $Nb_{0.75}O_{0.75}$, it has been generally assumed that in the quasi-metallic compounds $MX_{1-x}$ (where M

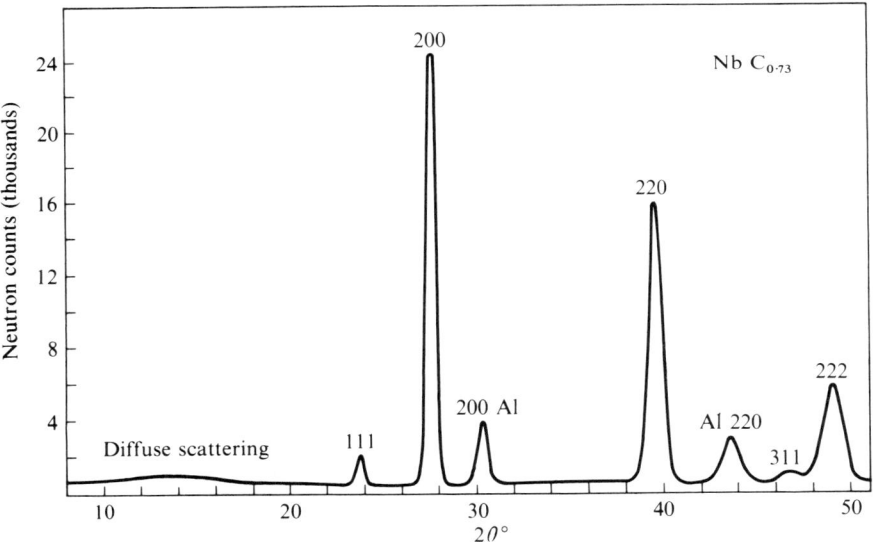

FIG. 11.6. Neutron diffraction pattern for $NbC_{0.73}$ cooled rapidly from 1850°C. The aluminium peaks are from the sample container.

is an early transition-series metal, Ti, V, Nb, and X is carbon, nitrogen or oxygen) vacancies occur randomly either on one or both sub-lattices. The true situation in these metallic systems is probably indicated by studies on niobium carbide by Henfrey and Fender (1971). This compound, which has a rock-salt structure, incorporates up to 30 per cent vacancies on the carbon sub-lattice, and careful neutron Bragg experiments give no indication of atom displacements. X-ray studies fail to reveal any evidence of short-range ordering but this is hardly

## DIFFUSE SCATTERING: STUDY OF DEFECT SOLIDS

surprising, for with more than a quarter of the carbon atoms missing the diffuse scattering around the forbidden 100 peak is barely discernible, even in the neutron diffraction pattern (Fig. 11.6). Fig. 11.7, however, shows the same region studied with long-wavelength neutrons and now, with a much lower

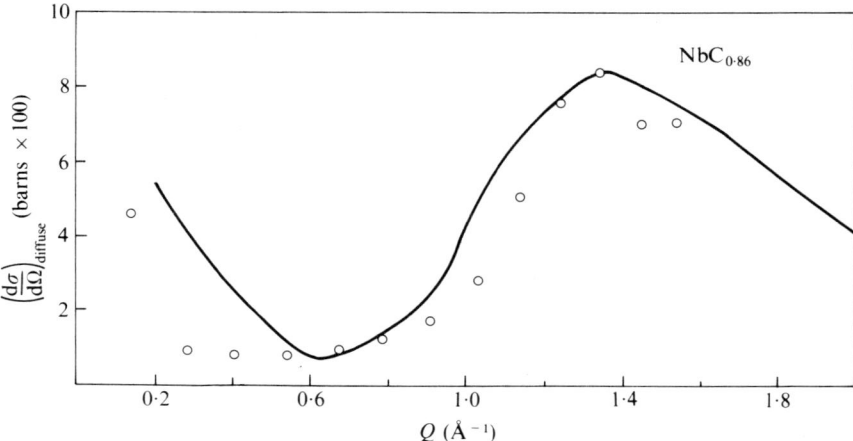

FIG. 11.7. Diffuse scattering for $NbC_{0.86}$. ○ Experimental. —— Calculated for model discussed in text.

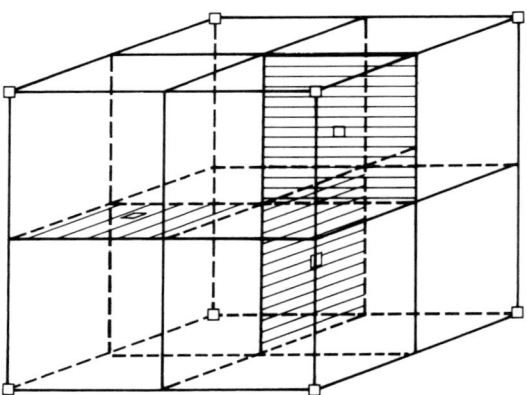

FIG. 11.8. $V_8C_7$ structure. The unit cell is formed from eight sub-cells of the rock-salt structure. There are carbon vacancies □ in a spiral array, with the vacancies inside the large cell occupying sites at the centres of the shaded planes.

vacancy concentration, the diffuse scattering is marked and a good quantitative fit is obtained (using eqn (11.8)) between the experimentally observed scattering and a calculated differential cross-section assuming ordering over three co-ordination shells. The calculated carbon occupation probabilities around a vacancy as origin are 0·96, 0·92 and 0·79 for the first, second and third shells respectively, compared with a probability of 0·86 based on a random distribution. This clear evidence for vacancy–vacancy repulsion, at least up to the third shell, demonstrates a close parallel with the ordered $V_8C_7$ phase (De Novion et al. 1966; Henfrey and Fender 1970), where carbon vacancies are placed on a spiral array in such a way that third-neighbour sites are vacant (Fig. 11.8).

A slightly more complex situation is illustrated by studies on anion-deficient oxides based on the cubic fluorite structure. Many of these compounds, which include binary systems (e.g. $CeO_{2-x}$, $ThO_{2-x}$) and solid solutions (e.g. $Zr(Ca)O_{2-x}$, $Ce(Y)O_{2-x}$), have broad existence ranges with up to 10 or 15 per cent vacant oxygen sites. Where anion vacancies are introduced by the solution of an oxide MO or $M_2O_3$ in the parent oxide, the possibility of short-range ordering of cations as well as vacancies has to be considered. Further, the local

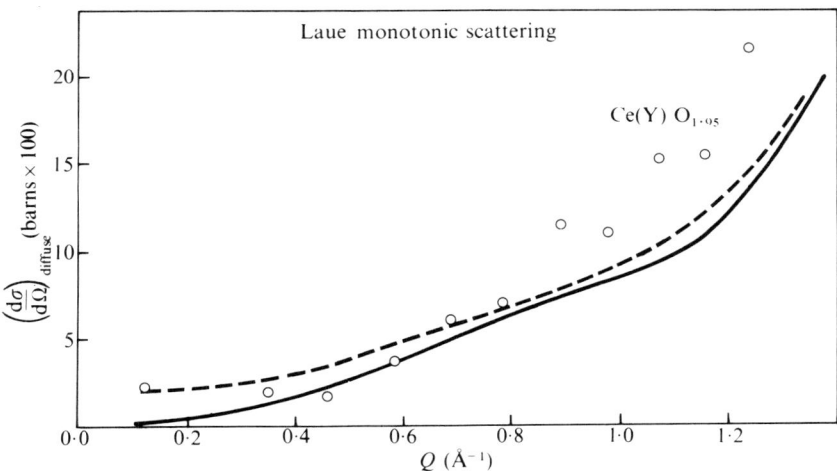

FIG. 11.9. Diffuse scattering intensity for $Ce(Y)O_{1.95}$. ○ Experimental. ——— Calculated for two oxygen vacancies and four $Y^{3+}$ ions in positions shown in Fig. 11.10. ——— Calculated for two oxygen vacancies and four $Y^{3+}$ ions in Fig. 11.10, but omitting ion marked A. The Laue monotonic scattering is calculated for a random distribution of oxygen vacancies.

charge on the defects is likely to lead to association between vacant sites and the dopant cation. The scattering obtained from a sample of $Ce(Y)O_{1.95}$ cooled rapidly from about 1700°C is shown in Fig. 11.9 (Steele and Fender 1971). The results are quite incompatible with a combination of randomly-distributed $Ce^{4+}$ and $Y^{3+}$ ions and randomly-distributed vacancies on the anion sub-lattice. Short-range ordering of vacancies alone fails to reproduce either the magnitude or the trend of the diffuse scattering but, when either three or four $Y^{3+}$ ions are brought as close as possible to a pair of vacancies in next-nearest-neighbour association (Fig. 11.10), the experimental scattering is quite well reproduced by eqn (11.13). Such a cluster, which provides a nearly octahedral environment for the central $Y^{3+}$ ion, is similar to that for one quarter of the cations in the

C-type $M_2O_3$ structure, so that the neutron results provide important evidence to corroborate the view that the wide range of non-stoichiometry in these systems depends on the ready incorporation of a structural element of a lower oxide. (The same point has been discussed in Chapter 10.)

This assumption, which is central to a microdomain theory of non-stoichiometry, is basically a way of envisaging structural conditions whereby the enthalpy of mixing associated with a large concentration of point defects is kept relatively low. In crystals where the defects are introduced by irradiation, there is no

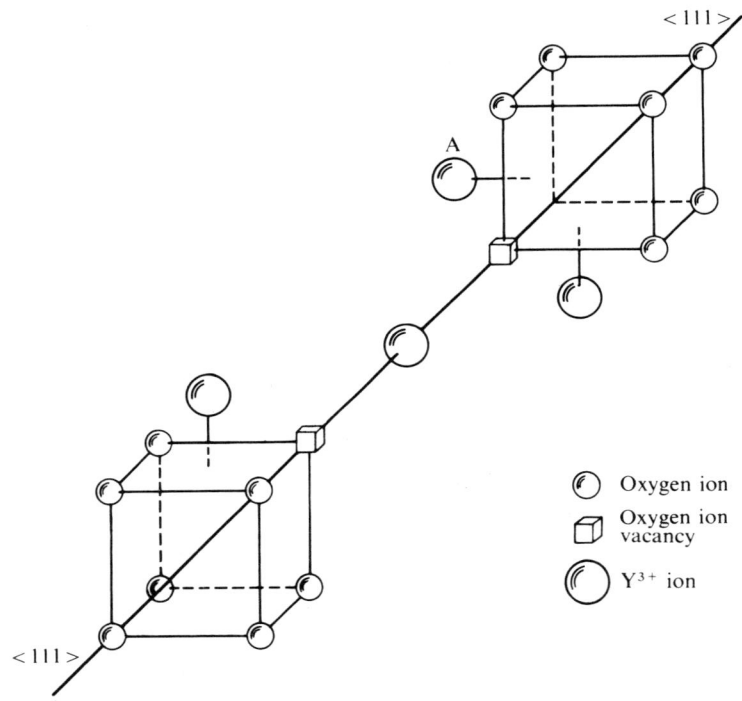

FIG. 11.10. Postulated defect structure of $Ce(Y)O_{2-x}$. Cerium ions are not shown.

equivalent guide-line and so we may expect agreement between calculated and experimental scattering to be more difficult to achieve, either because there is quite a wide distribution of defect aggregates or as a result of extensive relaxation of atoms surrounding the defects. Studies on irradiated germanium by Clark, Mitchell, and Stewart (1971) serve as an illustration of the progress made in this field. Fig. 11.11 shows the experimental transmission data for a crystal irradiated with fast neutrons. The main features of the experimentally determined cross-section are (a) rapidly falling values of $\sigma_{total}$ at 6 Å, (b) a peak at 9-10 Å and (c) a low minimum at ~13 Å. These characteristics are well reproduced by a divacancy (i.e. a nearest-neighbour pair of vacancies in the diamond structure) with an inward relaxation of atoms over five shells. However, without the benefit of complementary Bragg studies, for which the defect concentration is too low, it is in fact not possible to distinguish between a divacancy and a di-interstitial complex. Nevertheless, the neutron measurements still provide a unique indication of the structural changes consequent on radiation

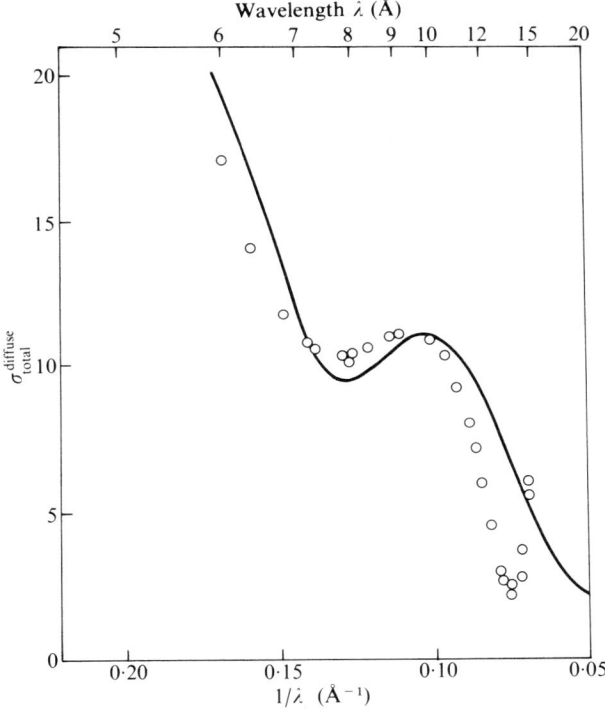

FIG. 11.11. Total cross-section versus neutron wavelength for germanium irradiated with $1 \cdot 0 \times 10^{20}$ fast neutrons per cm². ○ Experimental. ——— Calculated for a model consisting of a divacancy with 54 atoms relaxed towards the defect. (After Clark, Mitchell, and Stewart 1971).

damage. They also suggest a way in which the extent and magnitude of lattice relaxations can be explored.

### 11.4.3. Interstitials and vacancies

The non-stoichiometric compounds considered so far have been relatively simple. Can we derive useful information from diffuse scattering in more complex systems? Can we examine systems where vacancies, substitutional defects, and interstitial atoms are all present or where one of the ions is magnetic? It is certainly the case that short-range order treatments will now be particularly complicated but, if there is an association between defects, the Martin approximation (§ 11.2.3) should be helpful. Two systems, $Ca(Y)F_{2+x}$ and $Fe_{1-x}O$, illustrate the value of scattering experiments in such circumstances. In the former compound, Bragg experiments by Cheetham et al. (1970) demonstrate the presence of two closely related fluoride interstitial sites along the $\langle 110 \rangle$ and $\langle 111 \rangle$ directions from the $(\frac{1}{2}\frac{1}{2}\frac{1}{2})$ position. Vacancies (or normal anions displaced into interstitial sites) in the anion sites together with substitutional disorder also occur. As pointed out earlier, clustering of the fluoride ions and vacancies is inevitable if absurdly short F–F distances are to be

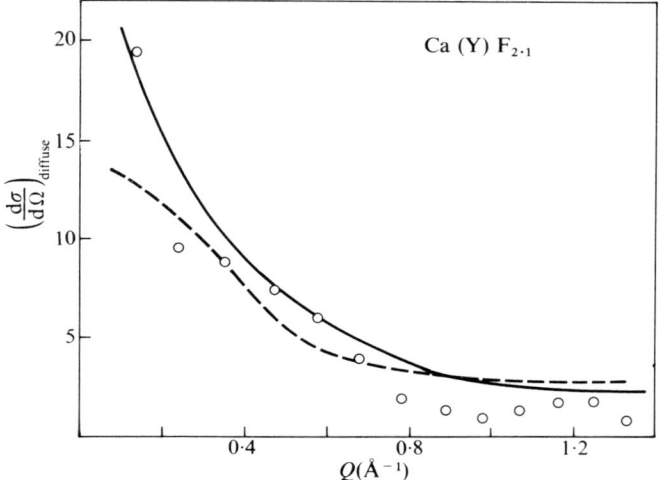

FIG. 11.12. Diffuse scattering by defects in $Ca(Y)F_{2.1}$. ○ Experimental. – – – Calculated with atoms as shown in 2:2:2 model of Fig. 10.7. ——— Calculated with atoms as shown in 3:4:2 model of Fig. 10.8. In both models, the $Y^{3+}$ ions are not in the closest cation sites to the defect cluster.

avoided, and in Chapter 10 two models compatible with the Bragg data are proposed. These were called the 2:2:2 cluster model (see Fig. 10.7) and the 3:4:2 cluster model (see Fig. 10.8). Fig. 11.12 compares the diffuse scattering to be expected from these models (again using eqn (11.13)) with the experimental data (Steele, Childs, and Fender 1971). The agreement is encouraging and, though not sufficiently sensitive to distinguish between the two suggestions, the extended cluster is slightly favoured. Each cluster creates a local region of negative charge which is neutralized overall by the pressure of $Y^{3+}$ ions, but the results also give some indication that not all these are in cation sites closest to the cluster. An important aspect of these experiments is that, in providing direct evidence for clustering (not, in this case, related to any known ordered structures) in a simple ionic compound, they pose intriguing theoretical problems which test our understanding of lattice energetics.

If some of the atoms in a solid, whether defect or not, contain unpaired electrons, magnetic scattering (see Chapter 12) will contribute to the total neutron scattering. Above the magnetic ordering temperature, paramagnetic diffuse scattering is observed which decreases with scattering angle according to the size of the orbitals occupied by unpaired electrons. For transition-metal ions such as $Mn^{2+}$ or $Fe^{2+}$, this fall in the magnetic form factor is about 20 per cent at $Q = 1.2$. When magnetic ordering occurs in a non-stoichiometric compound, the neutron scattering will usually reflect a resultant defective magnetic structure as well as the crystallographic defect structure. These effects are illustrated by the diffuse scattering observed from $Fe_{1-x}O$. Comparison of the measurements on

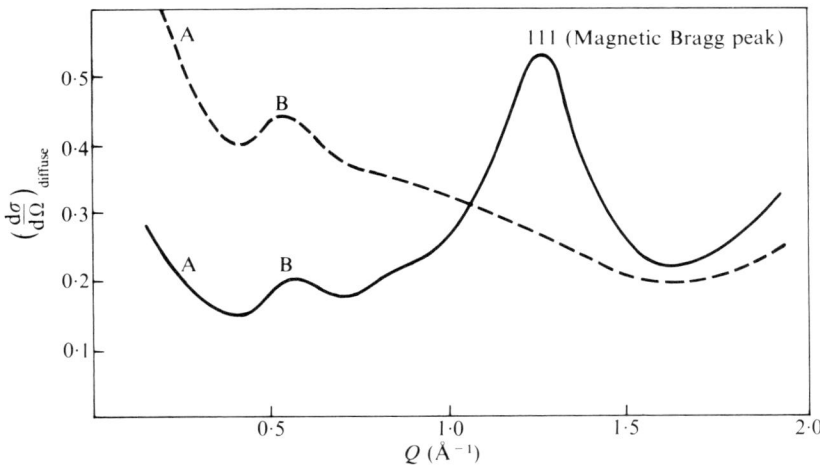

FIG. 11.13. Diffuse scattering intensity from $Fe_{1-x}O$ as a function of $Q$. —— 30 K, - - - 300 K. (After Childs 1967).

$Fe_{0.95}O$ at 30 K and 300 K (Fig. 11.13) shows two main differences—a decrease in scattering at low $Q$ for the 30 K measurements and a substantial peak at $Q = 1.25$ Å$^{-1}$. The former represents a loss of magnetic diffuse scattering as the sample is cooled below the Néel temperature and the latter a 'defect magnetic peak'. The ideal FeO structure has a magnetic structure in which the spins lie perpendicular to the (111) plane and so the 111 peak is forbidden in the perfect magnetic lattice, but appears in the defect structure. Despite the dominance of the magnetic scattering the nuclear defect structure is not totally obscured, and at 30 K most of the scattering below arises from this source. Persistent features are the small peak (B) and the rise at low $Q$(A) (Fig. 11.13). Application of eqn (11.12) demonstrates that the peak at B is not reproduced either by a random distribution of Koch–Cohen clusters (see Fig. 10.12) or by the smallest cluster likely in the $Fe_{1-x}O$ system, viz. an interstitial iron surrounded by four octahedral cation vacancies. However, the scattering is compatible with the short-range ordering of Koch–Cohen clusters on a cubic lattice at a spacing of three times the FeO unit cell—a result in line with the superlattice reflexions observed by Koch and Cohen (1969) in their single-crystal X-ray study.

The range of examples discussed above demonstrates the role diffuse neutron scattering has played in probing the structural effects of defect interactions either when defects are introduced chemically or by bombardment. Such studies can be expected to contribute quite substantially to our insight of the solid state. Combined with Bragg investigations, when possible, they provide an opportunity to observe the influence of electronic structure on local atomic arrangements and to perceive how this, in turn, may affect properties such as diffusion and reactivity. The very early stages of precipitation of a new phase or the initial annealing processes in radiation-damaged solids may also be followed on a finer scale than can be observed by electron microscopy. As higher neutron fluxes become available, single-crystal investigations and experiments at high temperatures will be facilitated, but an extension of elastic-diffuse-scattering techniques to the study of molecular species trapped in crystalline matrices may also be envisaged.

## 11.5. References

ANTAL, J. J., WEISS, R. J. and DIENES, G. J. (1955). *Phys. Rev.* **99**, 1081.
BORIE, B. (1957). *Acta Crystallogr.* **10**, 99.
BORIE, B. (1959). *Acta Crystallogr.* **12**, 280.
BORIE, B. (1961). *Acta Crystallogr.* **14**, 472.
CHEETHAM, A. K., FENDER, B. E. F., STEELE, D. and WILLIS, B. T. M. (1970). *Solid St. Comm.* **8**, 171.
CHILDS, P. E. (1967). *D. Phil. Thesis*, University of Oxford.
CLARK, C. D., MITCHELL, E. W. J. and STEWART, R. J. (1971). *Crystal lattice defects* **2**, 105.
COCHRAN, W. (1956). *Acta Crystallogr.* **9**, 259.
COWLEY, J. M. (1950a). *J. appl. Phys.* **21**, 24.
COWLEY, J. M. (1950b). *Phys. Rev.* **77**, 667.
DE NOVION, C. H., LORENZELLI, R. and COSTA, P. (1966). *C.R. Acad. Sci., Paris.* **263B**, 775.
FENDER, B. E. F. and HENFREY, A. W. (1970). *J. chem. Phys.* **52**, 3250.
HENFREY, A. W. and FENDER, B. E. F. (1970). *Acta Crystallogr.* **B26**, 1882.
HENFREY, A. W. and FENDER, B. E. F. (1971).
HUANG, K. (1947). *Proc. R. Soc.*, **A190**, 102.
KOCH, F. and COHEN. J. B. (1969). *Acta Crystallogr.* **B25**, 257.
KOFSTAD, P. and WALLACE, W. E. (1959). *J. Am. chem. Soc.* **81**, 5019.
LAUE, M. von. (1918). *Annln. Phys.* **56**, 497.
LOW, G. G. E. and COLLINS, M. F. (1963). *J. appl. Phys.* **34**, 1195.
MCQUILLAN, A. D. and WALLBANK, A. D. (1970). *J. chem. Phys.*
MARTIN, D. G. (1967). Mathematical formulation of the scattering of long wavelength neutrons by defects of solids, (*Rep. U.K. atom. Energy Auth., London*) R5479.

MARTIN, D. G. (1968). *J. Phys.* C. **1**, 333.
MITCHELL, E. W. J. *and* WEDEPOHL, P. T. (1958). *Phil. Mag.* **3**, 1280.
MITCHELL, E. W. J. *and* STEWART, R. J. (1967). *Phil. Mag.* **15**, 617.
MOSS, S. C. (1964). *J. appl. Phys.* **35**, 3547.
SABINE, T. M., PRYOR, A. W. *and* HICKMAN, B. S. (1963). *Phil. Mag.* **8**, 43.
STEELE, D. *and* FENDER, B. E. F. (1971).
STEELE, D., CHILDS, P. E. *and* FENDER, B. E. F. (1972). *J. Phys.* C. **5**, 2677.
WARREN, B. E. (1969). *X-ray diffraction.* Addison-Wesley, Reading, Massachusetts.

# 12 Neutron Diffraction and Covalency

By A. J. JACOBSON

Inorganic Chemistry Laboratory, Oxford University

## 12.1. Introduction

It is well known that to account for the structures and thermochemistry of compounds of the transition elements and B group elements, some consideration of metal–ligand interactions is necessary. The electrostatic model, in which the ions are regarded as hard spheres with characteristic charges and radii, has a limited range of validity; for example, for electrostatic interactions alone, the sodium chloride structure of CaS is predicted but not the reduction to 4:4 co-ordination in the zinc-blende structure of the closely similar CdS. For the transition metals, the simple theory has been extended to include the effect of the symmetry of the electrostatic potential due to the surrounding ligands, on the $d$-orbital energies. The crystal field theory (e.g. Figgis 1966; Ballhausen 1962) is very successful in interpreting magnetic and spectroscopic properties and also enables some thermodynamic data to be rationalized; for example, the variations in the heats of formation of the first-row transition-metal oxides. However, the failure of this extended electrostatic theory to quantitatively account for the magnitude of crystal field splittings or absorption band intensities clearly indicates the desirability of including metal–ligand interactions into the model.

Direct evidence for delocalization of the metal d electron density onto the ligand was first provided by the observation of ligand hyperfine splittings in the EPR spectrum of $IrCl_6^{2-}$ (Owen and Stevens 1953) Many subsequent observations for other transition metal complexes have been made by this method and also by NMR (for a general review of the resonance methods see Owen and Thornley 1966). In this chapter, the neutron diffraction techniques, which give the same type of information about unpaired spin distributions as the resonance methods, are examined. It should be noted with the development of the experimental approach, a number of important *ab initio* calculations of the effects of covalency have been made (e.g. Shulman and Sugano 1963; Watson and Freeman 1964; Hubbard et al. 1966; Ellis et al. 1968).

The model usually used to interpret the neutron scattering and the resonance experiments is a simple one-electron molecular-orbital description (Coulson 1961; Ballhausen and Gray 1965). The metal–ligand interactions are introduced by writing the wavefunction for the complex as a linear combination of metal and ligand wavefunctions, e.g. for a bonding orbital:

$$\psi_{MO}^b = N_b(\psi_L + a\psi_M), \qquad (12.1a)$$

$N_b$ is a normalization constant defined by $\langle \psi_{MO} | \psi_{MO} \rangle = 1$ and $\psi_M$ and $\psi_L$ are the metal and ligand wavefunctions. In this convention, the admixture or covalency parameter $a$ is a small positive quantity and $\psi_{MO}$, as a result, is mainly ligand in character. $N_b^2 a^2$ may be interpreted as the probability of transfer of an electron from the metal into the region of the ligand. The corresponding antibonding orbital is written:

$$\psi_{MO}^a = N_a(\psi_M - b\psi_L) \qquad (12.1b)$$

and the two admixture parameters $a$ and $b$ are related by the orthogonality condition $\langle \psi_{MO}^a | \psi_{MO}^b \rangle = 0$. The modification of the metal and ligand orbitals achieved in this way gives rise to lowering and raising of the energies of the bonding and antibonding partners, so that for a three electron system, with a half filled antibonding orbital, there is a net gain in energy due to the covalent interaction. A discussion of the energy changes and their theoretical relation to the admixture parameters is beyond the scope of this chapter and for further information the reader is referred to Owen and Thornley (1966).

In the next section, the molecular-orbital bonding scheme for octahedral complexes is described and related to the quantity which can be measured in a neutron experiment, namely, the magnetic form factor. Specific experimental techniques and the results so far obtained are considered in §§12.3 and 12.4.

## 12.2. Covalency

The basic theory of the effect of covalency on the magnetic scattering of neutrons was first developed by Hubbard and Marshall (1965) whose approach, though not notation, is followed here.

### 12.2.1. Molecular orbitals for octahedral complexes

Methods of constructing molecular orbitals for complexes of various symmetries using the standard techniques of group theory are discussed by a number of authors (see, e.g. Cotton (1971), Ballhausen (1962) and Figgis (1966)). Essentially, combinations of the valence orbitals for the six ligands, which transform as the irreducible representations of the molecular point group, are first found and then mixed with the appropriate metal orbitals which transform as the same irreducible representations.

For the 3d, 4s and 4p orbitals of the first row transition metals, $\sigma$ bonds can be formed using the $3d_{x^2-y^2}$ and the $3d_{z^2}$ ($e_g$ symmetry), the 4s ($a_{1g}$) and the three 4p orbitals ($t_{1u}$). The six ligand p orbitals which have lobes pointing directly towards the metal ion also give rise to combinations with $t_{1u}$, $e_g$ and $a_{1g}$ symmetry and hence are suitable for admixture. In Fig. 12.1 the overlap of the $d_{x^2-y^2}$ with the appropriate ligand combination ($\chi_e$) of $e_g$ symmetry is

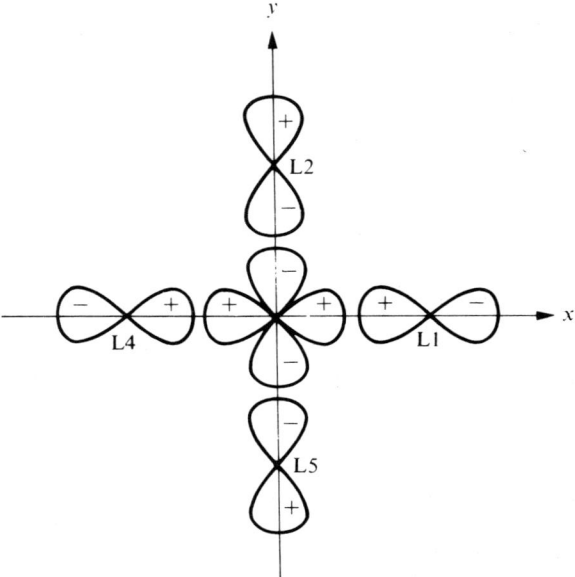

FIG. 12.1. Overlap (schematic) of the $d_{x^2-y^2}$ orbital with ligand p$\sigma$ orbitals. The numbering of the ligands is as shown, with $L_1$, $L_2$, $L_3$ along the positive $x$, $y$, $z$ axes and $L_4$, $L_5$, $L_6$ along the negative $x$, $y$, $z$.

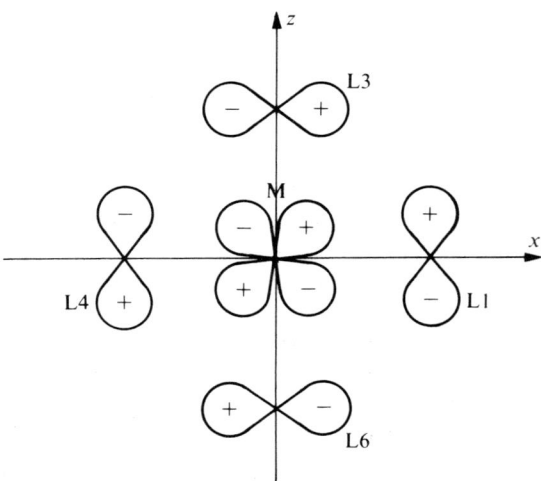

FIG. 12.2. Overlap (schematic) of the $d_{xz}$ orbital with ligand p$\pi$ orbitals. The bonding is of $\pi$ symmetry with respect to each M-L axis.

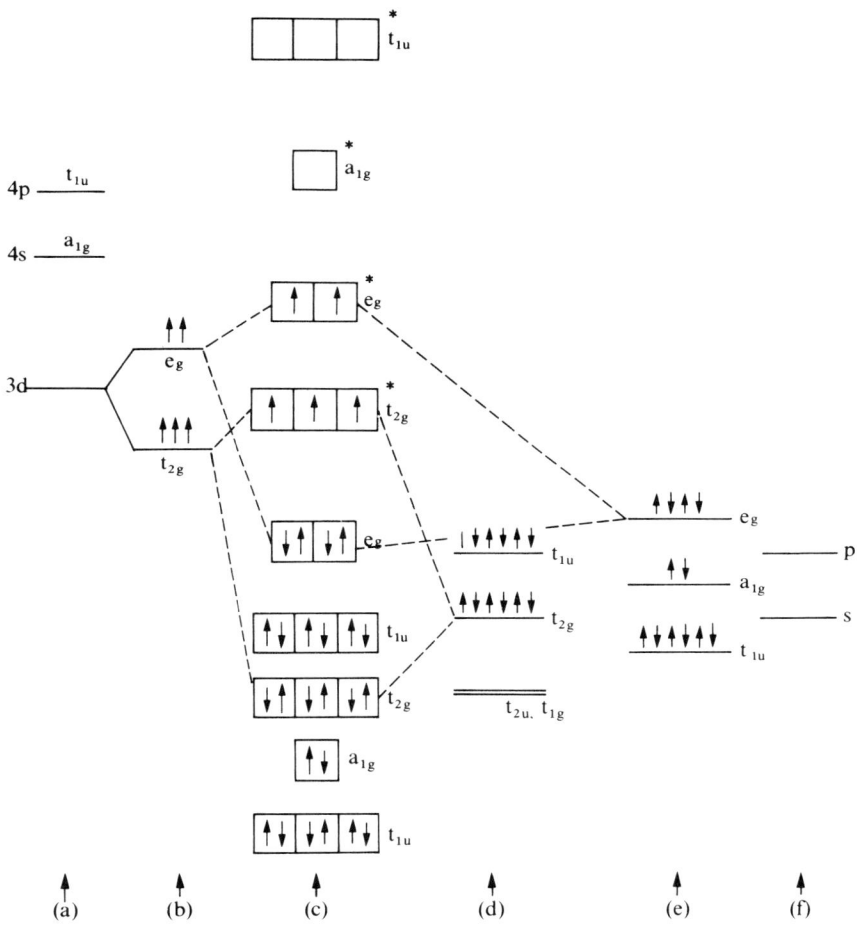

FIG. 12.3. Schematic molecular orbital energy level diagram for the octahedral complex $MnF_6^{4-}$. The occupation of the levels by electrons for the $3d^5$ configuration is shown by an arrow for each electron, the spin states being indicated by the direction of the arrow. (a) Energy levels for free $Mn^{2+}$ ion. (b) Splitting of free-ion levels by an electric field with octahedral symmetry. (c) Energy levels for complex formed by bonding of $Mn^{2+}$ with orbitals of surrounding ligands. The antibonding orbitals are marked*. (d) and (e) show, respectively, the $\pi$- and $\sigma$-bonding ligand orbitals. The $t_{1g}$ and $t_{2u}$ orbitals are non-bonding. (f) Energy levels of free ligand ions. (After Hall et al. 1963).

shown schematically. The metal wavefunctions which give $\pi$ bonding are $d_{xy}$, $d_{xz}$ and $d_{yz}$ ($t_{2g}$) and the 4p set (previously used in $\sigma$ bonding). The twelve remaining ligand p orbitals give $t_{1g}$, $t_{2g}$, $t_{1u}$ and $t_{2u}$ combinations, but there are no metal orbitals with symmetries $t_{1g}$ and $t_{2u}$, and so these ligand functions remain non-bonding. The overlap of $d_{xz}$ and one ligand $t_{2g}$ orbital is illustrated in Fig. 12.2. In Fig. 12.3 the bonding model for the complex $MnF_6^{4-}$ is summarized by means of a molecular-orbital energy level diagram. The 41 valence electrons from the six $F^-$ ions and the $Mn^{2+}$ ion have been assigned to the various energy levels using the Aufbau principle and it can be seen that the five unpaired electrons reside in the $e_g^*$ and the $t_{2g}^*$ antibonding orbitals which are primarily metal d in character.

It should be pointed out that covalency involves net charge transfer from the ligand back to the region of the metal ion *via* the electrons in the bonding orbitals which have no antibonding partners of the same spin. Magnetic scattering of neutrons, however, arises from the interaction of the neutron with unpaired spins and hence the bonding orbitals make no contribution. In considering the effect of covalency on neutron scattering, account need only be taken of the half-filled antibonding orbitals. It should be made clear, though, that when it is said that spin is transferred from metal to ligand, what is in fact being transferred is a net spin in the reverse direction, from ligand to metal, through the bonding orbitals. The explicit forms of the $e_g^*$ and $t_{2g}^*$ orbitals, which define the covalency parameters, are:

$$\psi_{z^2} = N_e[d_{z^2} - \lambda_\sigma/\sqrt{12}(-2p_{z3} + 2p_{z6} + p_{x1} - p_{x4} + p_{y2} - p_{y5})$$
$$- \lambda_s/\sqrt{12}(2s_3 + 2s_6 - s_1 - s_2 - s_4 - s_5)] \quad (12.2)$$

$$\psi_{x^2-y^2} = N_e[d_{x^2-y^2} - \lambda_\sigma/2(+p_{x4} - p_{x1} + p_{y2} - p_{y5})$$
$$- \lambda_s/2(s_1 + s_4 - s_2 - s_5)] \quad (12.3)$$

$$\psi_{xy} = N_t[d_{xy} - \lambda_\pi/2(p_{y1} - p_{y4} + p_{x2} - p_{x5})] \quad (12.4)$$

$$\psi_{xz} = N_t[d_{xz} - \lambda_\pi/2(p_{z2} - p_{z5} + p_{y3} - p_{y6})] \quad (12.5)$$

$$\psi_{yz} = N_t[d_{yz} - \lambda_\pi/2(p_{x3} - p_{x6} + p_{z1} - p_{z4})]. \quad (12.6)$$

$\lambda_\sigma$, $\lambda_s$ and $\lambda_\pi$ are the covalent admixture parameters, and $N_e$ and $N_t$ normalisation constants defined by $N^2 \langle \psi | \psi \rangle = 1$. $N_e$ and $N_t$ may be written,

$$N_e = [1 + \lambda_s^2 + \lambda_\sigma^2 - 2\lambda_s S_s - 2\lambda_\sigma S_\sigma]^{-1/2} \quad (12.7)$$

$$N_t = [1 + \lambda_\pi^2 - 2\lambda_\pi S_\pi]^{-1/2} \quad (12.8)$$

where $S_s = \langle d_{x^2-y^2} | s \rangle$, $S_\sigma = \langle d_{x^2-y^2} | p \rangle$ and $S_\pi = \langle d_{xy} | p \rangle$. The numbering of the ligands in eqn (12.2) to (12.6) is as given in Figs. 12.1 and 12.2, and for the $e_g^*$ molecular orbitals the s contribution has been included.

As there are some differences in the notations used by different authors, some brief remarks are appropriate here. The covalency parameters defined above are related to those used by Hubbard and Marshall ($A$'s) by:

$$A_\sigma^2 = \tfrac{1}{3}\lambda_\sigma^2, A_s^2 = \tfrac{1}{3}\lambda_s^2 \text{ and } A_\pi^2 = \tfrac{1}{4}\lambda_\pi^2.$$

In NMR and EPR experiments spin transfer coefficients are usually quoted and are related to our parameters by

$$f_\sigma = \tfrac{1}{3}N_e^2\lambda_\sigma^2, \ f_s = \tfrac{1}{3}N_e^2\lambda_s^2 \text{ and } f_\pi = \tfrac{1}{4}N_t^2\lambda_\pi^2.$$

## 12.2.2. Magnetic form factors

Experimentally, unpaired spin density distributions are determined by measurement of the magnetic form factor which is defined as:

$$f(\mathbf{Q}) = \int e^{i\mathbf{Q}\cdot\mathbf{r}} D(\mathbf{r})\, d\mathbf{r} \tag{12.9}$$

$\mathbf{Q}$ is the scattering vector and $D(\mathbf{r})$ is the magnetic moment density normalized so that $f(\mathbf{Q}) = 1$ at $\mathbf{Q} = 0$. The magnetic form factor is then related to the covalent spin density distribution by $D(\mathbf{r})$, which for a molecular orbital $\psi$ is $|\psi|^2$ if the orbital is half filled and zero if filled or doubly occupied.

In considering the effect of covalency on the form of $D(\mathbf{r})$, it is convenient to first write the orbitals in §12.2.1 in a simpler way:

$$e_g^*: \quad \psi_e = N_e[d_e - \lambda_s\chi_s - \lambda_\sigma\chi_{2p\sigma}] \tag{12.10}$$

$$t_{2g}^*: \quad \psi_t = N_t[d_t - \lambda_\pi\chi_{2p\pi}]. \tag{12.11}$$

$\chi_s$, $\chi_{2p\sigma}$ and $\chi_{2p\pi}$ represent the linear combinations of ligand s and p orbitals, respectively.

For $\sigma$ bonding:

$$\begin{aligned}D(\mathbf{r}) &= N_e^2[d_e - \lambda_s\chi_s - \lambda_\sigma\chi_{2p\sigma}]^2 \\ &= N_e^2 \cdot [d_e^2 + \lambda_s^2\chi_{2s}^2 + \lambda_\sigma^2\chi_{2p\sigma}^2 - 2\lambda_s d_e\chi_s \\ &\quad - 2\lambda_\sigma d_e\chi_{2p\sigma} + 2\lambda_\sigma\lambda_s\chi_{2p\sigma}\chi_{2s}]. \end{aligned} \tag{12.12}$$

An expression for $N_e^2$ may be obtained from eqn (12.7) by expanding and neglecting terms in $\lambda$ of higher order than $\lambda^2$:

$$N_e^2 = [1 - \lambda_s^2 - \lambda_\sigma^2 + 2\lambda_s S_s + 2\lambda_\sigma S_\sigma]. \tag{12.13}$$

Substituting eqn (12.13) in eqn (12.12) and again neglecting higher-order terms:

$$\begin{aligned}D(\mathbf{r}) &= d_e^2[1 - \lambda_s^2 - \lambda_\sigma^2 + 2\lambda_s S_s + 2\lambda_\sigma S_\sigma] \\ &\quad - [2\lambda_s d_e\chi_s + 2\lambda_\sigma d_e\chi_{2p\sigma}] \\ &\quad + [\lambda_s^2\chi_s^2 + \lambda_\sigma^2\chi_{2p\sigma}^2 + 2\lambda_\sigma\lambda_s\chi_{2p\sigma}\chi_s]. \end{aligned} \tag{12.14}$$

Eqn (12.14) can be rearranged to separate the metal and ligand contributions from the overlap terms:

$$\begin{aligned}D(\mathbf{r}) &= d_e^2(\mathbf{r})[1 - \lambda_s^2 - \lambda_\sigma^2] \\ &\quad + 2[\lambda_s S_s d_e(\mathbf{r})^2 + \lambda_\sigma S_\sigma d_e(\mathbf{r})^2 - \lambda_s d_e(\mathbf{r})\chi_s(\mathbf{r}') \\ &\quad - \lambda_\sigma d_e(\mathbf{r})\chi_{2p\sigma}(\mathbf{r}')] \\ &\quad + [\lambda_s^2\chi_{2s}^2(\mathbf{r}') + \lambda_\sigma^2\chi_{2p\sigma}(\mathbf{r}') + 2\lambda_\sigma\lambda_s\chi_{2s}(\mathbf{r}')\chi_{2p\sigma}(\mathbf{r}')]. \end{aligned} \tag{12.15}$$

In eqn (12.15) $\mathbf{r}$ and $\mathbf{r}'$ refer to the metal and ligand co-ordinate systems respectively.

## NEUTRON DIFFRACTION AND COVALENCY

The form factor may now be found from eqn (12.9) and eqn (12.15). For convenience, it is divided into three parts, (a), (b) and (c).

(a) metal

$$f_a(\mathbf{Q}) = \int e^{i\mathbf{Q}\cdot\mathbf{r}} d_e^2(\mathbf{r})(1 - \lambda_s^2 - \lambda_\sigma^2)\,d\mathbf{r} \qquad (12.16)$$

$$= 1 - \lambda_s^2 - \lambda_\sigma^2, \text{ at } Q = 0;$$

(b) overlap

$$f_b(\mathbf{Q}) = \int e^{i\mathbf{Q}\cdot\mathbf{r}} [2\{\lambda_s S_s d_e^2(\mathbf{r}) + \lambda_\sigma S_\sigma d_e^2(\mathbf{r})\}$$
$$- \lambda_s d_e(\mathbf{r})\chi_s(\mathbf{r}') + \lambda_\sigma d_e(\mathbf{r})\chi_{2g\sigma}(\mathbf{r}')\,]\,d\mathbf{r}; \qquad (12.17)$$

(c) ligand

$$f_c(\mathbf{Q}) = \int e^{i\mathbf{Q}\cdot\mathbf{r}} [\lambda_s^2 \chi_s^2(\mathbf{r}') + \lambda_\sigma^2 \chi_{2p\sigma}^2(\mathbf{r}')]\,d\mathbf{r} \qquad (12.18)$$

which, when spherically averaged over the ligands, becomes:

$$f_c(\mathbf{Q}) = \lambda_s^2 \sin(QR)/QR \int j_0(Qr) s^2(\mathbf{r})\,d\mathbf{r}$$
$$+ \lambda_{\sigma^2} \sin(QR)/QR \int j_0(Qr) p^2(\mathbf{r})\,d\mathbf{r}. \qquad (12.19)$$

$j_0$ is the spherical Bessel function of zero order $(\sin(Qr)/Qr)$. $R$ is the metal-ligand interatomic distance and s and p are the ligand wavefunctions. At $Q = 0$

$$f_c(\mathbf{Q}) = \lambda_\sigma^2 + \lambda_s^2.$$

The treatment for $\pi$ bonding follows from eqn (12.11), in an exactly analogous manner.

The three form factors, with their summation and the result obtained for a free-ion $d$ function are shown schematically in Fig. 12.4. The covalent form factors have three important features:

(i) the form factor associated with the metal ion is reduced by a factor $1 - \lambda_s^2 - \lambda_\sigma^2$ when compared with the free ion;

(ii) the overlap term, which is zero at $Q = 0$, rises to a maximum at finite $Q$, causing the resultant form factor to be expanded relative to the free ion;

(iii) the ligand form factor, which has a value of $\lambda_s^2 + \lambda_\sigma^2$ at $Q = 0$, falls off sharply with increasing $Q$ and is mostly concentrated between 0 and 1·3 Å$^{-1}$

(the ligand 'forward peak') for internuclear distances of 2 – 2·5 Å but has a finite value elsewhere.

For particular d-electron configurations, the metal-ion reduction factors can be deduced by inspection of the molecular orbitals in § 12.2.1. For example, for $Mn^{2+}$ $d^5$, an amount of spin, $\frac{1}{4}\lambda_\pi^2$, is transferred into each one of the twelve

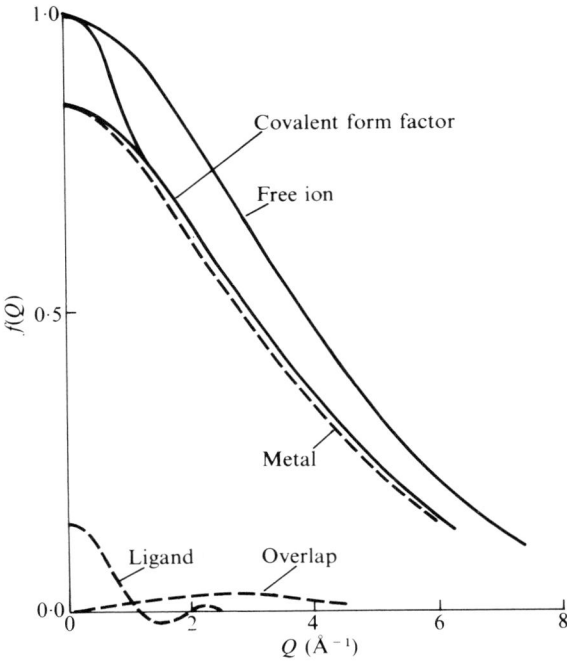

FIG. 12.4. Covalent form factor, together with its three components, shown by broken lines, consisting of the contributions from the metal ion, the ligand, and the overlap. The free-ion calculation is also shown.

ligand p-orbitals, and $\frac{1}{3}\lambda_\sigma^2$ and $\frac{1}{3}\lambda_s^2$ into the six ligand $p_\sigma$ and s orbitals. The total loss in spin density from the metal per unpaired d electron is

$$\tfrac{6}{5}[\tfrac{1}{3}\lambda_\sigma^2 + 2 \cdot \tfrac{1}{4}\lambda_\pi^2 + \tfrac{1}{3}\lambda_s^2].$$

The reduction factor in Hubbard and Marshalls' notation, then, is $1 - \tfrac{6}{5}(A_\sigma^2 + 2A_\pi^2 + A_s^2)$. Similarly, for the other two d-electron configurations often studied, $d^3$ and $d^8$, the corresponding results are $1 - 4A_\pi^2$ and $1 - 3(A_\sigma^2 + A_s^2)$.

With this information, it is now possible to examine the experimental neutron methods which have been used to investigate covalency. These fall into three groups, each of which measures a different part of $f(Q)$.

## 12.3. Experimental methods

### 12.3.1. Polarized neutron experiments

The scattered intensity† for polarized neutrons incident on a magnetically ordered solid is proportional to

$$N^2 + 2\mathbf{q} \cdot \boldsymbol{\lambda}(N'M'' + N''M') + q^2 M^2. \tag{12.20}$$

$\mathbf{q}$ is the magnetic interaction vector, defined by $\mathbf{q} = \boldsymbol{\eta} - \mathbf{Q}(\mathbf{Q} \cdot \boldsymbol{\eta})$; $\boldsymbol{\eta}$ and $\boldsymbol{\lambda}$ are unit vectors in the directions of the atomic magnetic moment and the polarization of the incident neutron beam, respectively (see Fig. 12.5); and $\mathbf{Q}$ is the scattering vector. $N = N' + iN''$ and $M = M' + iM''$ are general nuclear and magnetic structure factors each expressed as the sum of real and imaginary

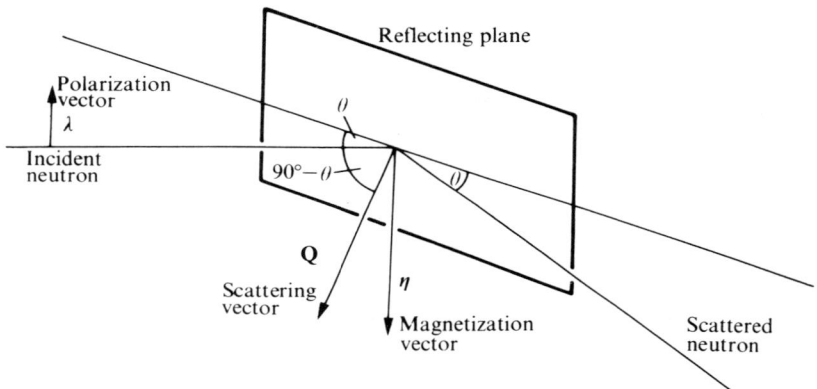

FIG. 12.5. Relationships between vectors defined in § 12.3.1. $\theta$ is the Bragg angle.

parts. The magnetic structure factor (cf. the nuclear structure factor, Chapter 1) may be written

$$F_{hkl}^{\text{mag}} = \sum \sigma_\kappa p_\kappa \, e^{2\pi i(hx_\kappa/a + ky_\kappa/b + lz_\kappa/c)} \, e^{-W_\kappa(\mathbf{Q})}. \tag{12.21}$$

$\sigma$ is $+1$ for parallel and $-1$ for antiparallel spins, and the magnetic scattering length $p = (e^2\gamma/mc^2)Sf(Q)$, where $\gamma(\approx -1\cdot 91)$ is the magnetic moment of the neutron expressed in nuclear magnetons, $S$ is the spin quantum number of the magnetic atom, and $f$ is the magnetic amplitude form factor. $e^{-W(\mathbf{Q})}$ is the Debye-Waller temperature factor, and the summation is taken over the magnetic atoms $\kappa$ in the unit cell.

† In this chapter we quote the results, without proof, for the intensity of elastic scattering from a magnetically ordered system. Their derivation follows in an analogous manner as for nuclear elastic scattering, considered in Chapter 1. For a fuller account of the subject, the reader is referred to the book by Marshall and Lovesey (see References).

For a simple centrosymmetric structure, when the direction of the magnetization ($\eta$) is perpendicular to the plane of scattering and the neutron polarization is parallel (+) or antiparallel (−) to the magnetization, eqn (12.20) becomes:

$$N^2 \pm 2NM + M^2.$$

Experimentally the ratio $R$ of the scattered intensities of a Bragg reflection is measured for the two neutron polarization states:

$$R = (N+M)^2/(N-M)^2.$$

From the polarization ratio R, the magnetic structure factor M can be obtained, and hence the value of $Sf(\mathbf{Q})$, if the nuclear structure is accurately known.

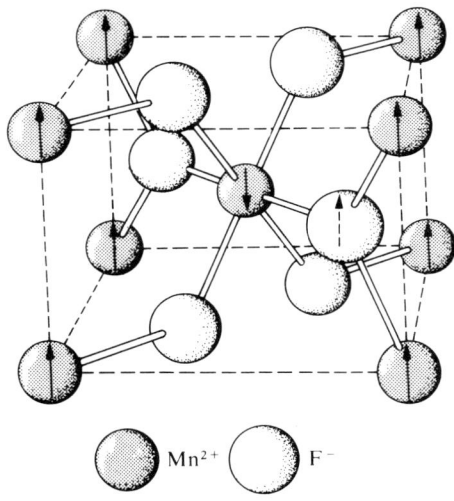

FIG. 12.6. Chemical and magnetic structure of MnF$_2$. The arrows indicate the direction and arrangement of the magnetic moments assigned to the manganese ions. The broken arrow on one fluorine ion indicates the direction of the net spin transferred from the adjacent Mn$^{2+}$ ions. (After Nathans *et al.* 1963).

In magnetic scattering experiments with unpolarized neutrons, the cross term in eqn (12.20), $2\mathbf{q}\cdot\boldsymbol{\lambda}\,(N'M'' + N''M')$, averages to zero and the scattered intensity is proportional to $N^2 + q^2M^2$. It can be seen, then, that small values of $M$ give measurable changes in the polarization ratio but make only a very small contribution to the total scattering in the unpolarized experiment (e.g., if $M/N = 1/100$ then $R = 1\cdot04$ but $M^2/N^2 = 1/10^4$). The polarized beam method enables measurements of very small magnetic cross sections to be made, thus making it the most powerful experimental technique for the detection of the small perturbations in the spin density distribution due to covalency. Three experiments which give detailed information for ionic systems have been reported and are considered separately below.

# NEUTRON DIFFRACTION AND COVALENCY

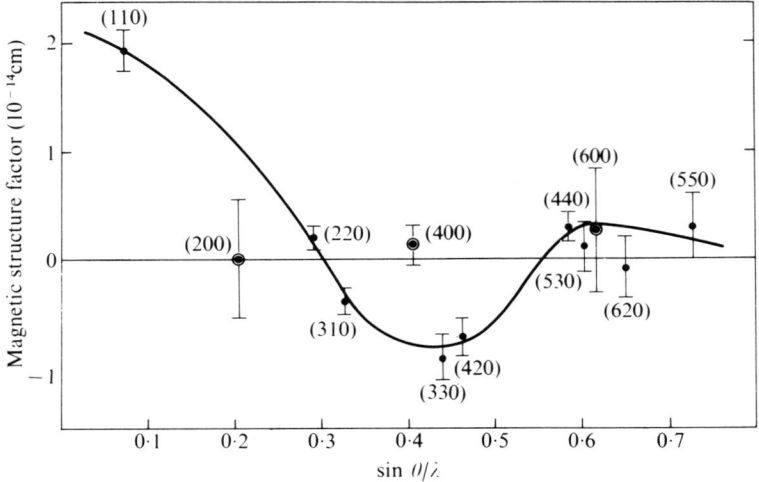

FIG. 12.7. Plot of the experimentally observed magnetic structure factors in MnF$_2$. The points with the dotted circles denote the reflections for which the magnetic scattering is expected to be identically zero. (After Nathans *et al.* 1963).

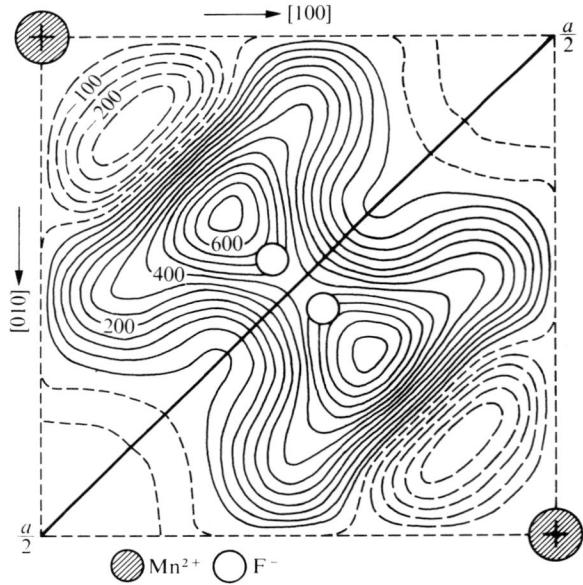

FIG. 12.8. Fourier synthesis of the covalent spin density in MnF$_2$ projected on the (001) plane, as measured from the 'forbidden' magnetic structure factors. The solid line normal to the diagonal indicates a mirror plane imposed on the spin density by use of a partial set of structure factors. Zero contour dotted; contour intervals numbered. (After Nathans *et al.* 1963.)

*12.3.1.1. MnF₂ (Nathans et al. 1963)*. Manganese difluoride has the rutile structure with a magnetic unit cell identical to the chemical unit cell (Fig. 12.6). The spins lie in ferromagnetic sheets, alternately parallel and antiparallel along the [001] direction; that is the $Mn^{2+}$ ion at (000) has its spin in the opposite direction to the $Mn^{2+}$ at $(\tfrac{1}{2}\tfrac{1}{2}\tfrac{1}{2})$. With this magnetic structure, all the magnetic scattering should appear in the reflections with $h + k + l = 2n + 1$ since the structure factor is, from eqn (12.21):

$$F_{hkl}^{mag} = p(1 - e^{\pi i(h+k+l)}) e^{-W(\mathbf{Q})}.$$

Covalency effects, however, lead to a transfer of spin density onto the ligands (see Fig. 12.6) and can give rise to small magnetic contributions to the nuclear reflections which have $h + k + l = 2n$. Nathans et al. measured the magnetic structure factors for ten of these forbidden reflections and obtained the form factor shown in Fig. 12.7. The magnitude of this form factor depends directly on the covalency parameters. The results were interpreted by Marshall (unpublished, but quoted by Nathans, Will, and Cox 1964) to give $A_\sigma^2 + 2A_\pi^2 + A_s^2 = 3 \cdot 3\%$. It is worth noting that the experiment is a very difficult one; many of the polarization ratios were less than 1 per cent and very long counting times were necessary. Nathans et al. also carried out a partial Fourier analysis of their data which shows the spin density to be concentrated along the Mn-F bond but significantly displaced towards the fluoride ion (Fig. 12.8).

*12.3.1.2. MnCO₃ (Brown and Forsyth 1967)* The rhombohedral structure of MnCO₃ (space group $R\bar{3}c$) is shown in Fig. 12.9. Below the magnetic ordering temperature, a spontaneous canting of the antiferromagnetic spin arrangement gives rise to a weak ferromagnetism. (See Fig. 12.10: the moments on the two $Mn^{2+}$ ions are nearly parallel but are canted by a small angle $\theta$ about the Mn-Mn axis.) The spins lie in planes perpendicular to the trigonal axis such that the $Mn^{2+}$ at (000) has its spin nearly antiparallel to that on the $Mn^{2+}$ at $(\tfrac{1}{2}\tfrac{1}{2}\tfrac{1}{2})$. The canting also takes place in this plane and the magnetic unit cell is identical to the chemical unit cell.

The major part of the antiferromagnetic scattering is found in reflections with $h + k + l$ odd (c.f. MnF₂) but the reflections with $h + k + l$ even have small contributions from both the antiferromagnetic and ferromagnetic components of the magnetic moment. However, the ferromagnetic component alone can give rise to magnetic $(hhl)$ reflections with $l$ even.

Brown and Forsyth measured fourteen $(hhl)$ polarization ratios at two different applied field strengths and deduced the form factors shown in Fig. 12.11. Again, the experiment was difficult since the spontaneous ferromagnetic moment is only $0 \cdot 036 \ \mu_B$.† The experimental curves lie well below the free-ion form factor for $Mn^{2+}$ (Watson and Freeman 1961), indicating that not all the moment lies near the manganese ion. The results at two field strengths enabled the magnetic structure

† $\mu_B$ is the *Bohr magneton*, defined as $e\hbar/2mc$.

factors for the field induced and spontaneous components of the magnetization to be found separately, and the Fourier projection of the ferromagnetic spin density corresponding to the field-induced moment is shown in Fig. 12.12. A significant elongation of the manganese moment and a positive spin density at the oxygen

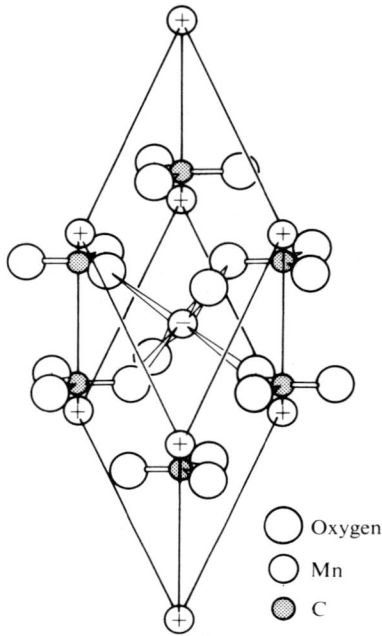

FIG. 12.9. The structure of $MnCO_3$. The spins on the atoms at the corners of the unit cell are approximately antiparallel to the spin on the atom at the centre. (After Brown and Forsyth 1967)

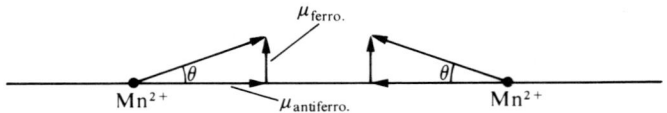

FIG. 12.10. Schematic diagram of relative spin orientations in a canted antiferromagnet.

positions can clearly be seen. Brown and Forsyth were able to obtain good agreement between this spin density map and that predicted using the covalency parameter $A_\sigma^2$ equal to 7·1 per cent. A further analysis of their data by Lingard and Marshall (1969) gave $\frac{11}{18}A_\sigma^2 + \frac{4}{9}A_\pi^2 = 6·6$ per cent.

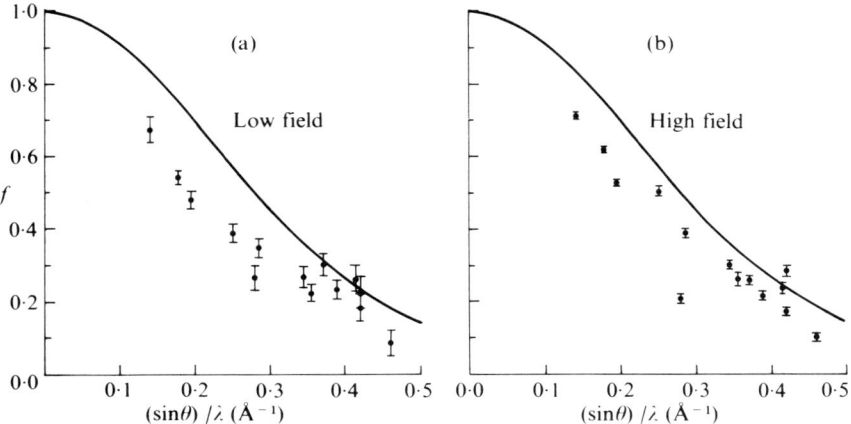

FIG. 12.11. Magnetic form factors for $MnCO_3$ derived from measurements at (a) 1·6 kOe and (b) 7·0 kOe. The full curves correspond to the theoretical calculation for $Mn^{2+}$ $3d^5$ of Watson and Freeman (1961). (After Brown and Forsyth 1967).

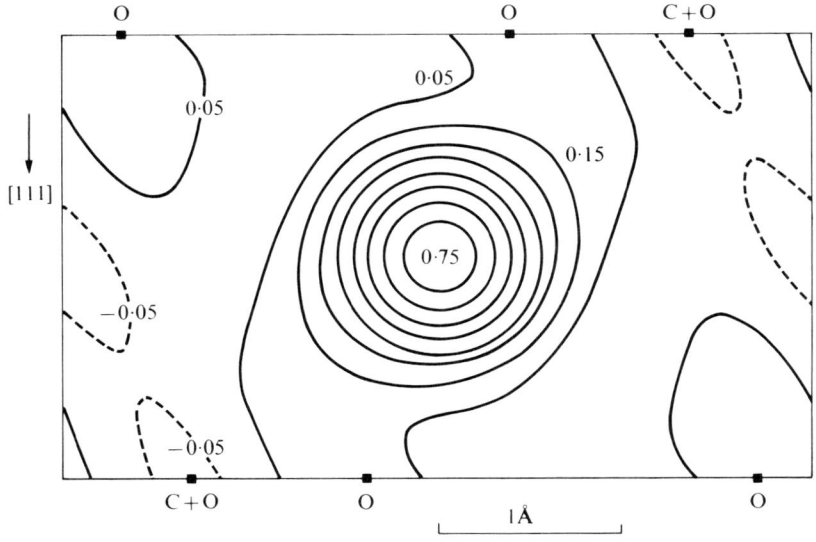

FIG. 12.12. Projection along [1 1 0] of the ferromagnetic spin density corresponding to the field-induced moment in $MnCO_3$. (After Brown and Forsyth 1967).

*12.3.1.3. $K_2NaCrF_6$ (Unpublished work by Wedgwood 1970).* $K_2NaCrF_6$ has an ordered doubled-perovskite structure (Fig. 12.13) in which the $CrF_6^{3-}$ octahedra are effectively magnetically isolated from each other (superexchange paths must go Cr-F-Na-F-Cr). As a result, the compound remains paramagnetic down to 4·2 K. However, at low temperatures and with high magnetic fields, it

# NEUTRON DIFFRACTION AND COVALENCY

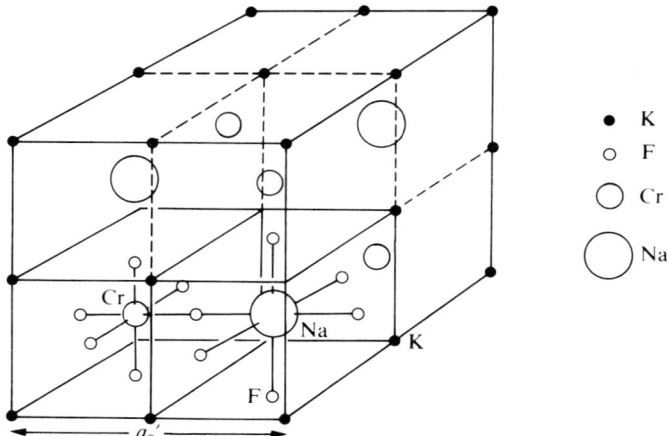

FIG. 12.13. Unit cell of cubic elpasolite, $K_2NaCrF_6$. Alternate perovskite-type sub-cells (of edge $\frac{1}{2}a_0$) are centred on Na or Cr. $a_0$ is 8·23 Å.

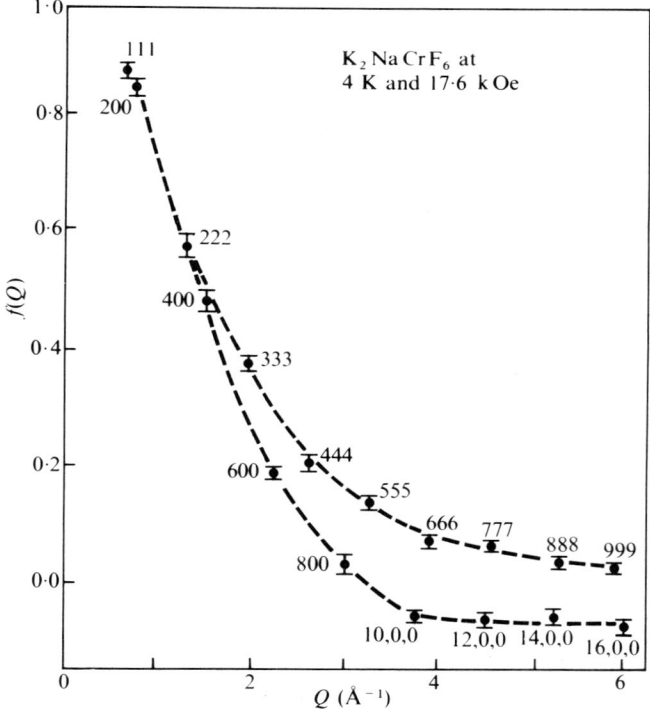

FIG. 12.14. Form factor measurements for the $(CrF_6)^{3-}$ cluster in $K_2NaCrF_6$. The difference in the form factors for *hhh* and *h00* reflections is due to the $t_{2g}$ electrons which cause a large deviation from spherical symmetry.

is possible to align the magnetic moments and obtain a large polarization dependence. Wedgwood obtained 70 per cent alignment in $K_2NaCrF_6$ at 4·2 K and with a field of 17·6 kOe and was able to measure the form factor out to high values of **Q** where large deviations from spherical symmetry occur due to the three $t_{2g}$ electrons. The $t_{2g}$ nature of the spin-density distribution is clearly seen in Fig. 12.14, which shows different form factors for the ⟨100⟩ and ⟨111⟩ directions in reciprocal space. The negative form factor for the high-**Q** $h00$ reflexions is believed to be due to the anti-bonding character of the orbitals involved in the spin transfer.

*12.3.2. Unpolarized neutron experiments*

Although the polarized neutron method must be used to determine small magnetic cross-sections, accurate values for the form factor may be obtained

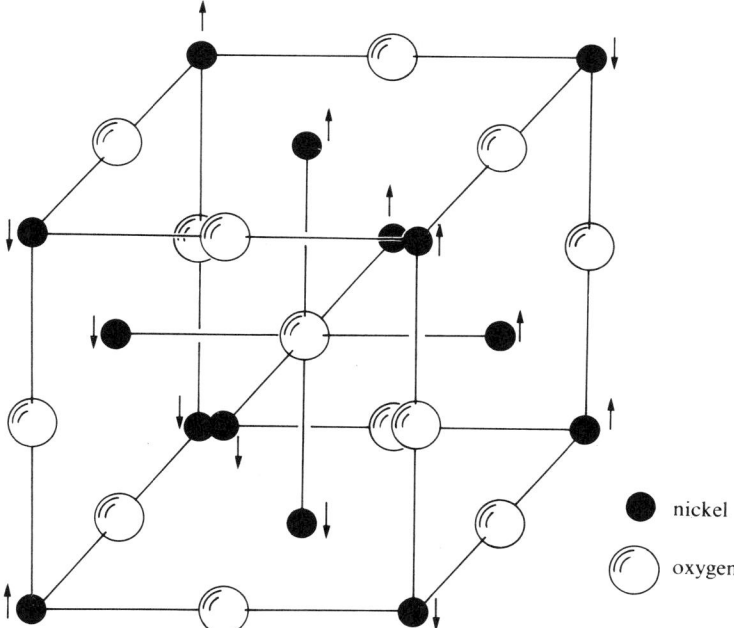

FIG. 12.15. One eighth of the magnetic unit cell of NiO showing transferred spin cancellation at the oxygens.

with unpolarized neutron beams, when the magnetic intensities are relatively large. In antiferromagnets of high symmetry, where the ligand is surrounded by an equal number of metal ions with parallel and antiparallel spins, (e.g. NiO, Fig. 12.15) the ligand moments cancel and the form factor measures only the metal and overlap contributions. Extrapolation of the form factor to the forward direction gives the moment reduction and the covalency parameters directly. This type of experiment has been carried out by Alperin (1962) for

NiO using a single crystal. The extrapolated form factor (Fig. 12.16) gave a magnetic moment for the $Ni^{2+}$ ion of $1.81 \pm 0.20 \mu_B$ compared with the theoretical value of $2.23 \mu_B$ for the free ion.

It should be noted that, in order to place the magnetic intensities on an absolute basis in this type of experiment, the distribution of magnetic domains must first be determined. These domains correspond to a change in spin orientation within a ferromagnetic sheet (e.g. in $MnF_2$) or to twinning between parts of the crystal with different contraction axes, where magnetic ordering causes a contraction normal to the ferromagnetic planes (e.g. in NiO). Alperin found an

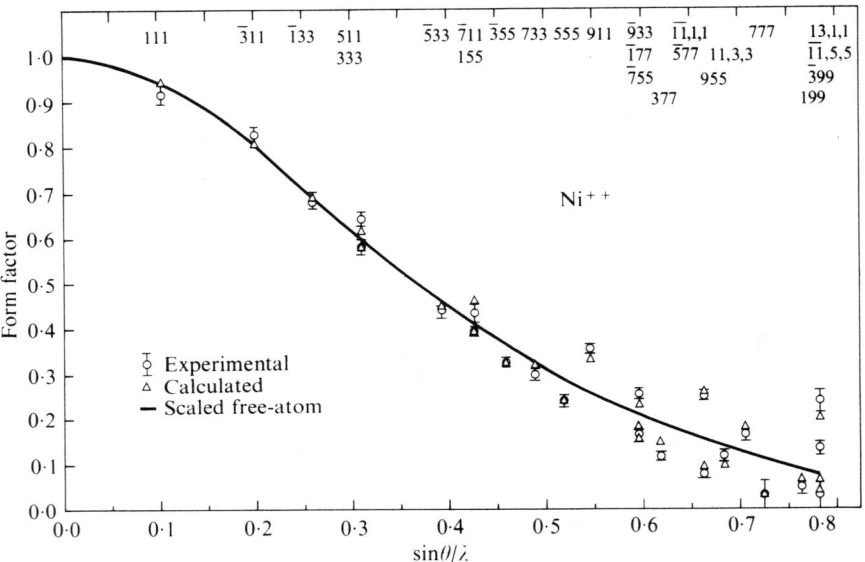

FIG. 12.16. Magnetic form factor of $Ni^{2+}$ in NiO. The indices at the top of the figure correspond (in the order shown) to the points directly below. (After Alperin 1962).

isotropic distribution of domains in NiO, but in $MnF_2$ 80 per cent of one type predominated.

A simpler method involves measurement of only one magnetic peak intensity at $Q \approx 1.3$ Å$^{-1}$. At this low value of $Q, f(Q) \simeq 0.9$, and it can be assumed to a good approximation that the effect of any deviation from ionic behaviour on the *shape* of the form factor is small and that the free-ion value may be used to calculate the magnetic moment. The attraction of this method is that experiments can be carried out using powder samples, which are relatively easy to prepare and for which systematic errors in intensity measurements (e.g. absorption and extinction) are much less important than for single crystals (see § 10.3).

The absolute magnetic intensities can be found by comparison with a nuclear peak intensity if the nuclear structure and scattering lengths are known. For example, in NiO (Fig. 12.17) the absolute intensity of the magnetic 111

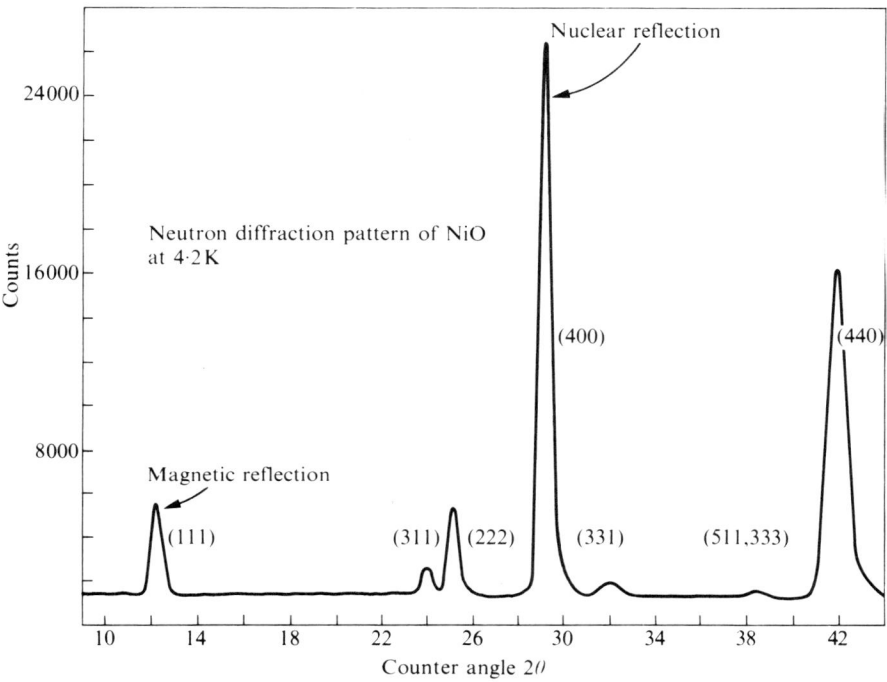

FIG. 12.17. Neutron diffraction pattern of NiO at 4·2 K.

reflection was found by comparison with the nuclear 400 reflection (Fender et al. 1969). In cases where the ligand nuclear positions are not precisely known, an external calibrant is used. Tofield and Fender (1970) determined the moment reduction in $YFeO_3$ by measuring the ratio of the magnetic intensities of the 011 and 101 reflections of $YFeO_3$ and the intensity of the nuclear reflection 111 of germanium (see Fig. 12.18).

In the interpretation of the measured magnetic moment reduction it is important to take account of two effects other than covalency which may influence $Sf(\mathbf{Q})$:

(a) The nature of the antiferromagnetic ground state gives rise to a zero-point spin deviation, so that S is effectively reduced by a few per cent. Spin-wave theory (Anderson 1952) predicts that this correction is comparable in magnitude to the covalent moment reduction; in MnO, for instance, the zero-point spin deviation is 3·1 per cent compared with a moment reduction due to covalency of 4·1 per cent.

(b) Orbital contributions can arise for some 'spin only' d-electron configurations, due to admixtures of higher states of non-zero angular momentum into the ground state by the spin-orbit interaction. (See, e.g. Blume's (1961) calculation of the form factor for $Ni^{2+}$ in NiO).

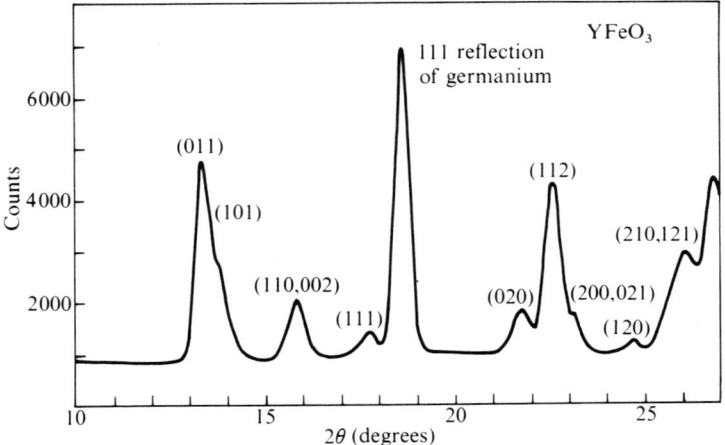

FIG. 12.18. Neutron diffraction pattern of YFeO₃. (After Tofield 1970).

### 12.3.3. Measurement of paramagnetic scattering

For a paramagnetic ion in a diamagnetic host lattice, the ligand forward peak can, in principle, be measured. This can be performed most easily using long wavelength neutrons (4–7 Å) where Bragg scattering effects are eliminated. The diffuse differential scattering cross-section for an isolated paramagnet, per magnetic atom, is

$$\left(\frac{d\sigma}{d\Omega}\right)^{\text{para}}_{\text{diffuse}} = \frac{2}{3}\left(\frac{e^2\gamma}{mc^2}\right)^2 S(S+1)\,|f(\mathbf{Q})|^2. \qquad (12.22)$$

$(d\sigma/d\Omega)^{\text{para}}_{\text{diffuse}}$ can be obtained by measuring the total differential cross-section for the sample and subtracting the contributions from other diffuse scattering processes. A rather better method is to eliminate all the non-magnetic terms, by measuring the difference in differential cross-section with a magnetic field applied along the scattering vector and with no magnetic field. The difference is

$$\left(\frac{d\sigma}{d\Omega}\right)^{\text{para}}_{\text{diffuse}} = \left(\frac{e^2\gamma}{mc^2}\right)^2 \left[\frac{S(S+1)}{3} - \langle S_z^2 \rangle\right] |f(\mathbf{Q})|^2 \qquad (12.23)$$

where $z$ is the direction of the applied magnetic field. Unfortunately, for systems sufficiently dilute so that interactions between dopant ions are negligible, this difference cross-section is too small to measure. An experiment of the first type has been carried out by Tofield and Fender in 1971, using a ruby single crystal containing 1·3 atomic per cent of chromium. Alternate measurements were made on this crystal and on a similar undoped crystal to minimize errors in the subtraction. The experiment provides the first observation of the predicted forward peak in an ionic system. The results are shown in Fig. 12.19 together with two free-ion curves for $Cr^{3+}$ normalized to 0·9 and 1·0 at $\mathbf{Q} = 0$. The errors

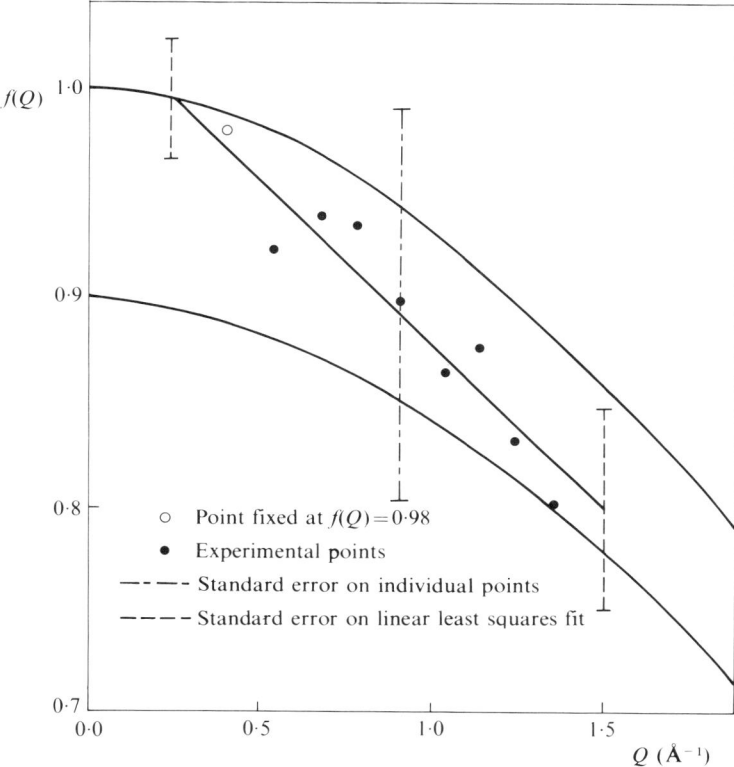

FIG. 12.19. Experimental form factor for $Cr^{3+}$ doped in $Al_2O_3$. (After Tofield and Fender 1971).

were necessarily large, because of the very small cross-sections involved (2-3 mb), and only semi-quantitative information could be obtained. In general, without higher neutron fluxes, the study of doped systems does not appear very promising at the present time; but the application of magnetic difference-scattering methods to compounds containing relatively large concentrations of magnetically isolated ions (e.g. $K_2NaCrF_6$) may prove to be a practicable way of obtaining quantitative results.

## 12.4. Survey of the experimental results

A considerable amount of data has appeared in the literature since Rimmer's (1970) survey and, although difficulties in interpretation still exist, some general conclusions may now be drawn. In §12.4.1 the neutron results are summarized and the trends examined, and in §12.4.2 a comparison is made of some neutron and resonance data.

## 12.4.1. Neutron measurements

The results in Table 12.1 with the exception of $MnF_2$, $MnCO_3$ and Alperin's NiO measurement, have all been obtained using powder samples. Although few very accurate form-factor measurements have been made for transition-metal ions in ionic systems, the assumption that the shape of the form factor can be based on the free-ion value seems to be justified. For $Ni^{2+}$ in NiO, the predicted expansion of the form factor (Hubbard and Marshall 1965), compared with the free ion, was observed by Alperin (1962) and this has been used to interpret the other $Ni^{2+}$ results. However, for $Mn^{2+}$ in $MnCO_3$ (Brown and Forsyth 1967; Lingard and Marshall 1969) the form factor was found to closely follow the free-ion curve (Watson and Freeman 1961) although early measurements indicated a contraction of about 10 per cent (Hastings et al. 1959). The results of Hastings

TABLE 12.1†

*Covalency Parameters from Neutron Diffraction Measurements*

| Ion | Compound | $A_\sigma^2 + 2A_\pi^2 + A_s^2$ | Structure-type | Reference |
|---|---|---|---|---|
| $Mn^{2+}$ | $MnF_2$ | 3·3% | rutile | 1 |
| | MnO | 3·5 ± 0·6% | NaCl | 2 |
| | $\alpha$-MnS | 7·0 ± 0·3% | NaCl | 2 |
| | $\alpha$-MnSe | 7·5 ± 0·3% | NaCl | 3 |
| | $MnSe_2$ | 7·8 ± 1·1% | pyrites | 3 |
| | MnTe | 9·8 ± 0·5% | NiAs | 4 |
| | | $\frac{11}{18}A_\sigma^2 + \frac{4}{9}A_\pi^2$ | | |
| | $MnCO_3$ | 6·7% | $NaNO_3$ | 5 |
| | | $A_\sigma^2 + A_s^2$ | | |
| $Ni^{2+}$ | $KNiF_3$‡ | 2·6 ± 1·8% | perovskite | 6 |
| | NiO | 3·8 ± 0·2% | NaCl | 2 |
| | NiO | 4·1% | | 7 |
| | | $A_\sigma^2 + 2A_\pi^2 + A_s^2$ | | |
| $Fe^{3+}$ | $LaFeO_3$ | 8·1% | perovskite | 8 |
| | $LaFeO_3$ | 10·0 ± 0·5% | | 9 |
| | $YFeO_3$ | 11·0 ± 1·0% | perovskite | 9 |
| | | $A_\pi^2$ | | |
| $Cr^{3+}$ | $LaCrO_3$ | 3·7% | perovskite | 8 |
| | $LaCrO_3$ | 1·6 ± 0·6% | | 9 |
| $Mn^{4+}$ | $CaMnO_3$ | 3·3 ± 0·8% | perovskite | 9 |

*Table references*
1. Nathans et al. 1963; Nathans, Will and Cox 1964.
2. Fender et al. 1968.
3. Jacobson and Fender 1970.
4. Fender and Coffin 1971 (unpublished).
5. Brown and Forsyth 1967; Lingard and Marshall 1969.
6. Hutchings and Guggenheim 1970.
7. Alperin 1962.
8. Nathans et al. 1964.
9. Tofield and Fender 1970.

† The errors quoted in Tables 1 to 4 include only experimental uncertainties.
‡ Corrected for zero-point spin deviation (see Anderson 1952).

*et al.* were obtained before the effects of covalency were realised and if the experimental data were normalized to $S = \frac{5}{2}$ then only an *apparent* contraction of the $Mn^{2+}$ form factor is expected.

The results for MnO and NiO are similar (using Anderson's zero-point spin-deviation correction in both cases), which is to be expected if the predominant effect is reduction of charge on the cation with both cations having similar electron affinities. It has, however, been suggested (Rimmer 1970) that the individual covalency parameters for $Mn^{2+}$ are, in fact, anomalously low. An explanation in terms of 4s orbital participation has been proposed (Hubbard *et al.* 1966): exchange interactions can lead to a small amount of spin in the low lying 4s orbital, which is polarized parallel to the existing spin in the $t_{2g}$ and $e_g$ orbitals and gives an apparent reduction in the spin transferred to the ligands through covalency.

No significant difference is found between oxide and fluoride covalency for $Mn^{2+}$ and $Ni^{2+}$, but for the more readily donating $S^{2-}$ ion there is a marked increase; in general, the variations in the covalency parameter for the Group VI ligands with $Mn^{2+}$ follow the changes in electro-negativity. For the compound $MnSe_2$ with the pyrites structure, the effect of the Se–Se bond is to reduce the number of orbitals available for $\pi$ bonding with the metal. The total covalency, however, remains closely similar to that found in MnSe. $MnCO_3$ and MnO are related in a similar way. With $CO_3^{2-}$ the number of ligand orbitals available for $\pi$ bonding (and to a lesser extent for $\sigma$ bonding) is reduced, but the covalency parameters are larger than those expected from the MnO study.

There is some discrepancy between the two sets of results for $LaCrO_3$ and $LaFeO_3$. In the more recent measurements of Tofield and Fender (1970) the sources of error were carefully investigated; particular attention was paid to the stoichiometry of the samples, so that these measurements are probably more reliable than the earlier ones of Nathans *et al.* (1964). In any case, the covalency parameters for the $d^5$ ion, $Fe^{3+}$, are considerably larger than for $Mn^{2+}$ in MnO, reflecting the increased electron affinity of the 3+ oxidation state as compared with the 2+ state. A similar change with increase of oxidation state is found for the $d^3$ ions, $Cr^{3+}$ and $Mn^{4+}$.

The magnitude of the $d^3$ values is discussed below.

### 12.4.2. Comparison with resonance data

In comparing results from the diffraction and resonance techniques, it is important to note that, although both methods give $A_\sigma^2 + A_s^2$ for $d^8$ ions and $A_\pi^2$ for $d^3$ ions, for $d^5$ ions the resonance studies measure $A_\sigma^2 - A_\pi^2$ and the diffraction methods $A_\sigma^2 + 2A_\pi^2 + A_s^2$: see Table 12.4. It is to be expected then, that for $d^3$ and $d^8$ ions the results should agree. This is indeed the case for the $d^8$ ion: (see Table 12.2). The neutron and resonance results for the $d^3$ ions are given in Table 12.3. At first sight, covalency in the fluorides appears to be greater than in the corresponding oxides, and this is contrary to expectations and to the results for the divalent oxides and fluorides. In addition, some

## NEUTRON DIFFRACTION AND COVALENCY

unpublished neutron results for $CrF_3$ and $FeF_3$ were found to be similar to those for $LaCrO_3$ and $LaFeO_3$ respectively, by Jacobson and McBride, in unpublished results (1971).

TABLE 12.2
*Resonance and Neutron Covalency Parameters for $Ni^{2+}$*

| Compound | $A_\sigma^2 + A_s^2$ | Method | Reference |
|---|---|---|---|
| $KNiF_3$ | 2·6 ± 1·8% | Neutrons | 1 |
| $KNiF_3$ | 4·3 ± 0·25% | NMR | 2 |
| $Ni^{2+}/KMgF_3$ | 3·6 ± 0·2% | EPR | 3 |
| NiO | 3·8 ± 0·2% | Neutrons | 4 |
| NiO | 4·1% | Neutrons | 5 |

*Table references*
1. Hutchings and Guggenheim 1970.
2. Shulman and Sugano 1963.
3. Hall et al. 1963.
4. Fender et al. 1968.
5. Alperin 1962.

Tofield and Fender (1970) have suggested that the difference between two sets of data arises from spin polarization of the $e_g$ orbitals, a similar effect to that suggested for the 4s orbitals of $Mn^{2+}$ (Hubbard et al. 1966). Exchange interactions can lead to small amounts of unpaired spin in the $e_g$ orbitals which are polarized parallel to the existing spin in the $t_{2g}$ orbitals. Spin in the opposite

TABLE 12.3
*Covalency Parameters for $d^3$ ions*

| Ion | Method | Compound | $A_\pi^2$ (%) | Reference |
|---|---|---|---|---|
| $Cr^{3+}$ | Neutrons | $LaCrO_3$ | 1·6 ± 0·6 | 1 |
| | EPR | $KMgF_3$ | 4·9 ± 0·15 | 2 |
| | NMR | $K_2NaCrF_6$ | 4·9 ± 0·8 | 3 |
| $Mn^{4+}$ | Neutrons | $CaMnO_3$ | 3·3 ± 0·8 | 1 |
| | EPR | $Cs_2GeF_6$ | 9·2 | 4 |

*Table references*
1. Tofield and Fender, 1970.
2. Hall et al. 1963.
3. Shulman and Knox 1960.
4. Helmholtz et al. 1961.

sense is left on the ligand. The NMR and EPR methods which detect unpaired spin on the ligand thus indicate increased covalency, whereas the neutron method gives less than the normal moment reduction. Combination of both techniques in this case appears to provide information concerning the limitations of the bonding model used.

If the assumption is made that covalency in oxides and fluorides is closely similar, then the $Mn^{2+}$ and $Fe^{3+}$ data may be combined to give the individual $\sigma$ and $\pi$ covalency parameters:

$Mn^{2+}$    $A_s^2 = 0.5\%, A_\sigma^2 = 1.2\%, A_\pi^2 = 0.9\%$
$Fe^{3+}$    $A_s^2 = 0.8\%, A_\sigma^2 = 5.5\%, A_\pi^2 = 2.1\%$.

The increase in oxidation state from +2 to +3 clearly favours $\sigma$ bonding relative to $\pi$: $A_\sigma^2$ increases twice as rapidly as $A_\pi^2$. The negative value of $A_\sigma^2 - A_\pi^2$ for $Cr^+$, when compared with $A_\sigma^2 - A_\pi^2 = 0.35$ per cent for $Mn^{2+}$, again suggests that $\sigma$ bonding is enhanced relative to $\pi$ bonding by an increase in oxidation state.

TABLE 12.4

*Covalency Parameters for some $d^5$ ions*

| Ion | Compound | Covalency (%) | Method | Reference |
|---|---|---|---|---|
| $Cr^+$ | NaF | $A_\sigma^2 - A_\pi^2 = -0.6 \pm 0.6$ | EPR | 1 |
| $Mn^{2+}$ | $KMnF_3$ | $A_s^2 = 0.52 \pm 0.02$ | | 2 |
| | | $A_\sigma^2 - A_\pi^2 = 0.35 \pm 0.05$ | | |
| | MnO | $A_\sigma^2 + 2A_\pi^2 + A_s^2 = 3.5 \pm 0.5$ | Neutrons | 3 |
| $Fe^{3+}$ | $KMgF_3$ | $A_s^2 = 0.8 \pm 0.2$ | EPR | 1 |
| | | $A_\sigma^2 - A_\pi^2 = 3.4 \pm 0.6$ | | |
| | $LaFeO_3$ | $A_\sigma^2 + 2A_\pi^2 + A_s^2 = 10.0 \pm 0.5$ | Neutrons | 4 |
| | $YFeO_3$ | $A_\sigma^2 + 2A_\pi^2 + A_s^2 = 11.0 \pm 1.0$ | Neutrons | 4 |

*Table references*
1. Hall *et al.* 1963.
2. Egashira and Hirakawa 1967.
3. Fender *et al.* 1968.
4. Tofield and Fender 1970.

The application of both neutron and resonance methods is clearly a profitable approach. One limitation of the resonance technique is the requirement of a ligand nuclear spin, and only a few experiments have been carried out on oxides and sulphides, where isotopic enrichment is necessary (e.g. MnO: Jones 1966; MnS: Lee 1968). The existence of a number of covalency parameters from the neutron method will perhaps encourage further resonance work of this kind. Correspondingly, apart from fluorides there is little neutron information for halide ligands, although NMR and EPR data are available for $Mn^{2+}$, $Ni^{2+}$ and $Cu^{2+}$ in chloride lattices (Rinneberg *et al.* 1969, Tsay and Helmholtz 1969 and Rinneberg and Hartmann 1970). The main restriction in the neutron case is one of finding magnetically suitable host lattices, although the development of low-temperature paramagnetic scattering experiments on dilute systems, using polarized neutron beams, allows a much greater degree of flexibility in this respect.

It may be possible to extend measurements to more complex ligands, particularly using neutron diffraction where covalency parameters can be obtained from the effective cation moment. Some chemically interesting systems for study would be the $\sigma$ donors $H_2O$ and $NH_3$ and the $\pi$ acceptor $CN^-$. The resonance techniques are hampered here since a knowledge of the ligand wavefunction is required to obtain the spin-transfer coefficients from the experimental data. It would also be of interest to try to detect small covalency effects in the ionic compounds of the lanthanides, especially for the IV oxidation state.

## 12.5. References

ALPERIN, H. A. (1962). *J. phys. Soc. Japan, Suppl. BIII.* **17**, 12.
ANDERSON, P. W. (1952). *Phys. Rev.* **86**, 694.
BLUME, M. (1961). *Phys. Rev.* **124**, 96.
BROWN, P. J. and FORSYTH, J. B. (1967). *Proc. Phys. Soc.* **92**, 125.
EGASHIRA, K. and HIRAKAWA, K. (1967). *J. phys. Soc. Japan.* **22**, 344.
ELLIS, D. E., FREEMAN, A. J. and ROS, P. (1968). *Phys. Rev.* **176**, 688.
FENDER, B. E. F., JACOBSON, A. J. and WEDGWOOD, F. A. (1968). *J. Chem. Phys.* **48**, 990.
HALL, T. P. P., HAYES, W., STEVENSON, R. W. H. and WILKENS, J. (1963). *J. chem. Phys.* **38**, 1977; **39**, 35.
HASTING, J. M., ELLIOTT, N. and CORLISS, L. M. (1959). *Phys. Rev.* **115**, 13.
HELMHOLZ, L., GUZZO, A. V. and SANDERS, R. N. (1961). *J. chem. Phys.* **35**, 1349.
HUBBARD, J. and MARSHALL, W. (1965). *Proc. Phys. Soc.* **86**, 561.
HUBBARD, J., RIMMER, D. E. and HOPGOOD, F. R. A. (1966). *Proc. Phys. Soc.* **88**, 13.
HUTCHINGS, M. T. and GUGGENHEIM, H. J. (1970). *J. Phys. C.* **3**, 1303.
JACOBSON, A. J. (1969). D.Phil. Thesis, University of Oxford.
JACOBSON, A. J. and FENDER, B. E. F. (1970). *J. chem. Phys.* **52**, 4563.
JONES, E. D. (1966). *Phys. Rev.* **151**, 315.
LEE, K. (1968). *Phys. Rev.* **172**, 284.
LINGARD, P. A. and MARSHALL, W. (1969). *J. Phys. C.* **2**, 276.
NATHANS, R., ALPERIN, H. A., PICKART, S. J. and BROWN, P. J. (1963). *J. appl. Phys.* **34**, 1182.
NATHANS, R., WILL, G. and COX, D. E. (1964). *Proceedings International Conference on Magnetism, Nottingham.* p. 327.
OWEN, J. and STEVENS, K. W. H. (1953). *Nature.* **171**, 836.
RIMMER, D. E. (1970). *Thermal neutron diffraction* (Ed. B. T. M. Willis). Oxford University Press. p. 211.
RINNEBERG, H. and HARTMANN, H. (1970). *J. chem. Phys.* **52**, 5814.

RINNEBERG, H., HAAS, H. *and* HARTMANN, H. (1969). *J. chem. Phys.* **50**, 3064.
SHULMAN, R. G. *and* KNOX, K. (1960). *Phys. Rev. Lett.* **4**, 603.
SHULMAN, R. G. *and* SUGANO, S. (1963). *Phys. Rev.* **130**, 506.
TOFIELD, B. C. (1970). D.Phil. Thesis. University of Oxford.
TOFIELD, B. C. *and* FENDER, B. E. F. (1970). *J. Physics Chem. Solids.* **31**, 2741.
TOFIELD, B. C. *and* FENDER, B. E. F. (1971). *J. Phys. C.* **4**, 1279.
TSAY, F. D. *and* HELMHOLTZ, L. (1969). *J. chem. Phys.* **50**, 2642.
WATSON, R. E. *and* FREEMAN, A. J. (1961). *Acta Crystallogr.* **14**, 27.
WATSON, R. E. *and* FREEMAN, A. J. (1964). *Phys. Rev.* (A) **134**, 1526.

*General*

An excellent general reference to the study of covalency by neutron scattering methods is

MARSHALL, W. C. *and* LOVESEY, S. W. (1971). *Theory of thermal neutron scattering.* Clarendon Press, Oxford.

A review of covalency and resonance techniques is given by:

OWEN, J. *and* THORNLEY, J. H. M. (1966). *Rep. Prog. Phys.* 675.

The following books provide a general introduction to bonding in transition metal compounds:

BALLHAUSEN, C. J. (1962). *Introduction to ligand field theory.* McGraw-Hill, N. York.
BALLHAUSEN, C. J. *and* GRAY, H. B. (1965). *Molecular orbital theory.* Interscience.
COTTON, F. A. (1971). *Chemical applications of group theory (Second Ed.)* Interscience.
COULSON, C. A. (1961). *Valence.* Clarendon Press, Oxford.
FIGGIS, B. N. (1966). *Introduction to ligand fields.* Interscience.

# Appendix I

## Neutron Scattering Data for Elements and Isotopes

| Atomic number $Z$ | Element or Isotope (natural abundance per cent) | $b_{coh}$ (fermis) ($= 10^{-15}$ m) | $\sigma_{coh}$ (barns) ($= 10^{-28}$ m$^2$) | $\sigma_{incoh}$ (barns) |
|---|---|---|---|---|
| 1 | H(99·98) | −3·74 | 1·8 | 79·7 |
|   | D(0·02) | 6·67 | 5·6 | 2·0 |
|   | $^3$H | 4·7 | 2·8 | |
| 2 | $^4$He | 3·0 | 1·1 | 0·0 |
| 3 | Li | −2·14 | 0·58 | 0·7 |
|   | $^6$Li (7·4) | 3·0 + 0·3i | 1·1 | |
|   | $^7$Li (92·6) | −2·33 | 0·68 | 0·8 |
| 4 | Be | 7·74 | 7·53 | 0·0 |
| 5 | B | 5·4 + 0·2i | 3·7 | 0·7 |
|   | $^{11}$B (80·4) | 6·0 | 4·5 | |
| 6 | C | 6·65 | 5·6 | 0·0 |
|   | $^{13}$C (1·1) | 6·0 | 5·5 | 1·0 |
| 7 | N | 9·40 | 11·1 | 0·3 |
|   | $^{15}$N (0·4) | 6·5 | 5·3 | |
| 8 | O | 5·80 | 4·23 | 0·0 |
|   | $^{17}$O (0·04) | 5·78 | 4·20 | |
|   | $^{18}$O (0·2) | 6·00 | 4·52 | |
| 9 | F | 5·65 | 3·98 | 0·0 |
| 10 | Ne | 4·6 | 2·7 | 0·2 |
| 11 | Na | 3·62 | 1·65 | 1·7 |
| 12 | Mg | 5·3 | 3·5 | 0·2 |
| 13 | Al | 3·45 | 1·50 | 0·0 |
| 14 | Si | 4·150 | 2·16 | 0·0 |
| 15 | P | 5·1 | 3·3 | 0·3 |
| 16 | S | 2·847 | 1·02 | 0·2 |
| 17 | Cl | 9·58 | 11·5 | 3·5 |
|   | $^{35}$Cl (75·5) | 11·8 | 17·5 | |
|   | $^{37}$Cl (24·5) | 2·6 | 0·8 | |
| 18 | A | 2·0 | 0·5 | 0·4 |
|   | $^{36}$A (0·4) | 24·3 | 74 | |
| 19 | K | 3·70 | 1·72 | 0·5 |
|   | $^{39}$K (93·1) | 3·7 | 1·7 | |
| 20 | Ca | 4·66 | 2·73 | 0·5 |
|   | $^{40}$Ca (97·0) | 4·9 | 3·0 | 0·1 |
|   | $^{44}$Ca (2·1) | 1·8 | 0·4 | |
| 21 | Sc | 11·8 | 17·5 | 6·0 |
| 22 | Ti | −3·35 | 1·4 | 3·0 |
|   | $^{46}$Ti (7·9) | 4·8 | 2·9 | |
|   | $^{47}$Ti (7·3) | 3·3 | 1·4 | |
|   | $^{48}$Ti (73·9) | −5·8 | 4·2 | |
|   | $^{49}$Ti (5·5) | 0·8 | 0·1 | |
|   | $^{50}$Ti (5·3) | 5·5 | 3·8 | |
| 23 | V | −0·52 | 0·03 | 5·1 |
| 24 | Cr | 3·53 | 1·57 | 2·5 |
|   | $^{52}$Cr (83·8) | 4·90 | 3·02 | |

# APPENDIX I

| Atomic number $Z$ | Element or Isotope (natural abundance per cent) | $b_{coh}$ (fermis) (= $10^{-15}$ m) | $\sigma_{coh}$ (barns) (= $10^{-28}$ m$^2$) | $\sigma_{incoh}$ (barns) |
|---|---|---|---|---|
| 25 | Mn | −3·87 | 1·9 | 0·4 |
| 26 | Fe | 9·51 | 11·4 | 0·4 |
|  | $^{54}$Fe (5·8) | 4·2 | 2·2 |  |
|  | $^{56}$Fe (91·7) | 10·1 | 12·8 |  |
|  | $^{57}$Fe (2·2) | 2·3 | 0·7 |  |
| 27 | Co | 2·50 | 0·79 | 5·2 |
| 28 | Ni | 10·3 | 13·3 | 4·7 |
|  | $^{58}$Ni (67·9) | 14·4 | 26·1 |  |
|  | $^{60}$Ni (26·2) | 2·82 | 1·00 |  |
|  | $^{61}$Ni (1·2) | 7·6 | 7·3 |  |
|  | $^{62}$Ni (3·7) | −8·7 | 9·5 |  |
|  | $^{64}$Ni (1·1) | −0·37 | 0·02 |  |
| 29 | Cu | 7·6 | 7·3 | 1·2 |
|  | $^{63}$Cu (69·1) | 6·7 | 5·6 |  |
|  | $^{65}$Cu (30·9) | 11·1 | 15·5 |  |
| 30 | Zn | 5·7 | 4·1 | 0·1 |
|  | $^{64}$Zn (48·9) | 5·6 | 3·9 |  |
|  | $^{66}$Zn (27·8) | 6·3 | 5·0 |  |
|  | $^{68}$Zn (18·6) | 6·7 | 5·6 |  |
| 31 | Ga | 7·2 | 6·5 | 1·0 |
| 32 | Ge | 8·186 | 8·42 | 1·0 |
| 33 | As | 6·4 | 5·1 | 2·9 |
| 34 | Se | 7·97 | 7·98 |  |
| 35 | Br | 6·79 | 5·79 | 0·3 |
| 36 | Kr | 7·4 | 6·8 |  |
| 37 | Rb | 7·08 | 6·3 | 0·0 |
|  | $^{85}$Rb (72·2) | 8·3 | 8·7 |  |
| 38 | Sr | 6·92 | 6·02 | 4·0 |
| 39 | Y | 7·93 | 7·90 |  |
| 40 | Zr | 7·14 | 6·41 | 0·3 |
| 41 | Nb | 7·11 | 6·35 | 0·2 |
| 42 | Mo | 6·88 | 5·95 | 0·6 |
| 43 | $^{99}$Tc | 6·8 | 5·8 |  |
| 44 | Ru | 7·3 | 6·7 | 0·1 |
| 45 | Rh | 5·9 | 4·4 | 1·2 |
| 46 | Pd | 6·0 | 4·5 | 0·3 |
| 47 | Ag | 5·97 | 4·48 | 1·8 |
|  | $^{107}$Ag (51·8) | 8·3 | 8·7 | 1·3 |
|  | $^{109}$Ag (48·2) | 4·3 | 2·3 | 3·7 |
| 48 | Cd | 3·7 + 1·6i | 2·0 |  |
|  | $^{113}$Cd (12·3) | −15 + 12i | 46 |  |
| 49 | In | 3·9 | 1·9 |  |
| 50 | Sn | 6·1 | 4·7 | 0·2 |
|  | $^{116}$Sn (14·3) | 5·8 | 4·2 |  |
|  | $^{117}$Sn (7·6) | 6·4 | 5·1 |  |
|  | $^{118}$Sn (24·0) | 5·8 | 4·2 |  |
|  | $^{119}$Sn (8·6) | 6·0 | 4·5 |  |
|  | $^{120}$Sn (32·9) | 6·4 | 5·1 |  |
|  | $^{122}$Sn (4·9) | 5·5 | 3·8 |  |
|  | $^{124}$Sn (5·9) | 5·9 | 4·4 |  |
| 51 | Sb | 5·64 | 4·00 | 0·2 |
| 52 | Te | 5·43 | 3·71 | 0·8 |
|  | $^{120}$Te (0·1) | 5·2 | 3·4 |  |

# APPENDIX I

| Atomic number Z | Element or Isotope (natural abundance per cent) | $b_{coh}$ (fermis) ($= 10^{-15}$ m) | $\sigma_{coh}$ (barns) ($= 10^{-28}$ m$^2$) | $\sigma_{incoh}$ (barns) |
|---|---|---|---|---|
| | $^{123}$Te (0·9) | 5·7 | 4·1 | |
| | $^{124}$Te (4·6) | 5·5 | 3·8 | |
| | $^{125}$Te (7·0) | 5·6 | 3·9 | |
| 53 | I | 5·28 | 3·50 | 0·4 |
| 54 | Xe | 4·8 | 2·9 | |
| 55 | Cs | 5·42 | 3·69 | 3·2 |
| 56 | Ba | 5·25 | 3·46 | 2·5 |
| 57 | La | 8·3 | 8·7 | 0·6 |
| 58 | Ce | 4·82 | 2·92 | 0·0 |
| | $^{140}$Ce (88·5) | 4·7 | 2·8 | |
| | $^{142}$Ce (11·1) | 4·5 | 2·6 | |
| 59 | Pr | 4·4 | 2·4 | 1·6 |
| 60 | Nd | 7·2 | 6·5 | 9·5 |
| | $^{142}$Nd (27·1) | 7·7 | 7·5 | |
| | $^{144}$Nd (23·9) | 2·8 | 1·0 | |
| | $^{146}$Nd (17·2) | 8·7 | 9·5 | |
| 62 | $^{149}$Sm (13·8) | $-19 + 45i$ | 300 | |
| | $^{152}$Sm (76·7) | $-5$ | 3 | |
| | $^{154}$Sm (22·7) | 8 | 8 | |
| 63 | Eu | 5·5 | 3·8 | |
| 64 | Gd | 15 | 28 | |
| | $^{157}$Gd (15·7) | $43 + 40i$ | 433 | |
| 65 | Tb | 7·6 | 7·2 | |
| 66 | Dy | 16·9 | 36 | |
| | $^{160}$Dy (2·3) | 6·7 | 5·6 | |
| | $^{161}$Dy (18·9) | 10·3 | 13·3 | |
| | $^{162}$Dy (25·5) | $-1·4$ | 0·2 | |
| | $^{163}$Dy (25·0) | 5·0 | 3·1 | |
| | $^{164}$Dy (28·2) | 49·4 | 307 | |
| 67 | Ho | 8·5 | 9·1 | 4 |
| 68 | Er | 7·9 | 7·8 | 7 |
| 69 | Tm | 7·2 | 6·5 | |
| 70 | Yb | 12·8 | 20·6 | |
| 71 | Lu | 7·3 | 6·7 | |
| 72 | Hf | 7·8 | 7·6 | |
| 73 | Ta | 6·91 | 6·00 | 0·0 |
| 74 | W | 4·77 | 2·86 | 2·8 |
| | $^{182}$W (26·4) | 8·3 | 8·7 | |
| | $^{183}$W (14·4) | 4·3 | 2·3 | |
| | $^{184}$W (30·6) | 7·6 | 7·3 | |
| | $^{186}$W (28·4) | $-1·2$ | 0·2 | |
| 75 | Re | 9·2 | 10·6 | |
| 76 | Os | 10·7 | 14·4 | 0·5 |
| | $^{188}$Os (13·3) | 7·8 | 7·6 | |
| | $^{189}$Os (16·1) | 11·0 | 15·2 | |
| | $^{190}$Os (26·4) | 11·4 | 16·3 | |
| | $^{192}$Os (41·0) | 11·9 | 17·8 | |
| 77 | Ir | 10·6 | 14·1 | |
| 78 | Pt | 9·5 | 11·3 | 0·7 |
| 79 | Au | 7·6 | 7·3 | 2 |
| 80 | Hg | 12·66 | 20·14 | 6 |
| 81 | Tl | 8·89 | 9·93 | 0·1 |
| 82 | Pb | 9·40 | 11·1 | 0·3 |

APPENDIX I

| Atomic number Z | Element or Isotope (natural abundance per cent) | $b_{coh}$ (fermis) (= $10^{-15}$ m) | $\sigma_{coh}$ (barns) (= $10^{-28}$ m$^2$) | $\sigma_{incoh}$ (barns) |
|---|---|---|---|---|
| 83 | Bi | 8·52 | 9·12 | 0·25 |
| 90 | Th | 10·3 | 13·3 | 0·0 |
| 92 | U | 8·53 | 9·14 | |
|  | $^{235}$U (0·7) | 9·8 | 12·1 | |
|  | $^{238}$U (99·3) | 8·4 | 8·9 | |
| 93 | Np | 10·6 | 14·1 | |
| 94 | $^{239}$Pu | 7·5 | 7·1 | |
|  | $^{240}$Pu | 3·5 | 1·5 | |
|  | $^{242}$Pu | 8·1 | 8·2 | |

*Notes:* (a) The coherent amplitudes $b_{coh}$ are derived largely from a compilation by Professor C. G. Shull; the values for H, D, C, O, and the halogens were provided by Dr. L. Koester and for $^{15}$N by Dr. F. A. Wedgwood (all in private communications to the editor, 1972). The remaining values of $b_{coh}$ have been published in the literature.

(b) The complex amplitudes for Li, Cd, Sm, and Gd correspond to a neutron wavelength of 1 Å.

(c) The absorption cross-sections $\sigma_{abs}$ are from Hughes and Schwartz (1958)—see reference in Chapter 1—and correspond to a neutron wavelength of 1·8 Å.

# Appendix II

## Properties of the δ-Function

The Dirac $\delta$-function, $\delta(x)$, can be defined by

$$\delta(x) = 0 \quad \text{if } x \neq 0$$
$$= +\infty \quad \text{if } x = 0$$

and

$$\int_{-\infty}^{\infty} \delta(x)\,dx = 1.$$

It can be visualized as a function that is zero everywhere except in a very small domain around the point $x = 0$, where it is so large that its integral over the entire domain is unity. The exact shape of the function inside this domain does not matter.

The most important property of $\delta(x)$ is denoted by the equation

$$\int f(x)\,\delta(x - x_0)\,dx = f(x_0), \tag{A.1}$$

where $f(x)$ is any function which is well-defined at the point $x = x_0$.

The $\delta$-function also satisfies the following equations, which are used in several chapters,

$$\delta(ax) = \frac{1}{|a|}\,\delta(x), \tag{A.2}$$

and

$$\delta(x) = \frac{1}{2\pi} \int_{-\infty}^{\infty} e^{ixt}\,dt. \tag{A.3}$$

(see §1.4.2.).

# Appendix III

## Fourier Transforms

Let r denote position in space and $t$ the time variable. The Fourier transform of any function $G(\mathbf{r}, t)$ of these two variables is

$$S(\mathbf{Q}, \omega) = \frac{1}{2\pi} \iint G(\mathbf{r}, t)\, e^{i(\mathbf{Q}\cdot\mathbf{r} - \omega t)}\, d\mathbf{r}\, dt. \tag{A.4}$$

$\mathbf{Q}$ and $\omega$ are 'Fourier variables'; $\mathbf{Q}$ corresponds to $\mathbf{r}$ and has dimensions of reciprocal distance, and $\omega$ corresponds to $t$ and has dimensions of frequency.

The inversion theorem for Fourier transforms allows us to express $G(\mathbf{r}, t)$ in eqn (A.4) in the inverse form:

$$G(\mathbf{r}, t) = \frac{1}{2\pi} \iint S(\mathbf{Q}, \omega)\, e^{-i(\mathbf{Q}\cdot\mathbf{r} - \omega t)}\, d\mathbf{Q}\, d\omega. \tag{A.5}$$

Note the change in sign in the exponential term in going from eqn (A.4) to eqn (A.5). dr in eqn (A.4) represents an element of volume in real space; similarly $d\mathbf{Q}$ in eqn (A.5) is an element of volume in reciprocal space.

Fourier transforms are mathematical functions which appear in any discussion of the scattering of radiation by matter. They relate, for instance, the structure in space of condensed matter with the change in momentum ('momentum transfer') of the radiation which it scatters. As such, they are well known in X-ray crystallography in the form

$$A(\mathbf{s}) = \int \rho(\mathbf{r})\, e^{2\pi i \mathbf{s}\cdot\mathbf{r}}\, d\mathbf{r}, \tag{A.6}$$

with the inverse relation

$$\rho(\mathbf{r}) = \int A(\mathbf{s})\, e^{-2\pi i \mathbf{s}\cdot\mathbf{r}}\, d\mathbf{s}. \tag{A.7}$$

Here $A(\mathbf{s})$ is the total amplitude of X-rays scattered from the distribution of electron density $\rho(\mathbf{r})$. (By including the factor $2\pi$ in the exponential of eqn (A6) and eqn (A7), so that the vector in reciprocal space is $2\pi\mathbf{s}(\equiv\mathbf{Q})$, there is no constant factor outside the integral signs.) In neutron scattering problems, in contrast with the X-ray case, the quantum of the incident radiation is comparable with the energy of the excited states of the solid or liquid under investigation. For this reason, there is both a change in momentum and a change in energy of the scattered radiation, and the appropriate Fourier transforms involve two variables, as in eqn (A4) and eqn (A5). Thus the structure in both space ($\mathbf{r}$) and time ($t$), often called the *structure and dynamics* of the scattering system, are connected, *via* Fourier transform relations discussed in Chapter 1, with the momentum transfer ($\hbar\mathbf{Q}$) and the energy transfer ($\hbar\omega$) of the scattered radiation.

The derivation of an analytical expression for $S(\mathbf{Q}, \omega)$, by inserting an appropriate expression for $G(\mathbf{r}, t)$ in eqn (A4) and performing the integration, is (as a rule) a straightforward but tedious exercise. The resultant formulae, discussed,

# APPENDIX III

for example, in Chapter 6, are quoted and not derived in the main text of the book. We shall sketch here the steps involved in deriving one such formula, viz. the Fourier transform relation (eqn (6.13))

$$S(\mathbf{Q}, \omega) = \frac{1}{\pi} \left( \frac{DQ^2}{\omega^2 + (DQ^2)^2} \right), \tag{A.8}$$

obtained by putting the solution of Fick's diffusion equation

$$G(\mathbf{r}, t) = (4\pi D |t|)^{-\frac{3}{2}} \exp(-r^2/4D|t|) \tag{A.9}$$

into eqn (A.4).

We must evaluate the expression

$$S(\mathbf{Q}, \omega) = \frac{1}{2\pi} \iint (4\pi D|t|)^{-\frac{3}{2}} \exp(-r^2/4D|t|) e^{i(\mathbf{Q} \cdot \mathbf{r} - \omega t)} \, d\mathbf{r} \, dt. \tag{A.10}$$

$G(\mathbf{r}, t)$ in eqn (A.9) is a function which is radially symmetric in $\mathbf{r}$ space (i.e. $G(\mathbf{r}, t) = G(r, t)$), and so we can perform the integration over $\mathbf{r}$ by transforming eqn (A.10) to spherical polar co-ordinates $r, \theta, \phi$:

$$S(\mathbf{Q}, \omega) = \frac{1}{2\pi} \int_{-\infty}^{\infty} (4\pi D|t|)^{-\frac{3}{2}} e^{-i\omega t} \, dt \int_0^\infty \int_0^\pi \int_0^{2\pi} e^{-r^2/(4D|t|)}$$
$$\times e^{iQr\cos\theta} r^2 \sin\theta \, dr \, d\theta \, d\phi$$

$$= \frac{1}{2\pi} \int_{-\infty}^{\infty} (4\pi D|t|)^{-\frac{3}{2}} e^{-i\omega t} \, dt \cdot 2\pi \int_0^\infty$$
$$\times e^{-r^2/(4D|t|)} \left[ \frac{e^{iQr\cos\theta}}{-iQr} \right]_0^\pi r^2 \, dr$$

$$= \int_{-\infty}^{\infty} (4\pi D|t|)^{-\frac{3}{2}} e^{-i\omega t} \, dt \cdot \frac{2}{Q} \int_0^\infty e^{-r^2/(4D|t|)} r \sin Qr \, dr. \tag{A.11}$$

The integration over $r$ in eqn (A.11) is obtained from the standard expression,

$$\int_0^\infty \exp(-\alpha r^2) r \sin \beta r \, dr = \frac{\beta \pi^{\frac{1}{2}}}{4\alpha^{3/2}} \exp(-\beta^2/4\alpha).$$

Putting $\alpha = 1/(4D|t|)$ and $\beta = Q$, eqn (A.11) becomes

$$S(\mathbf{Q}, \omega) = \frac{1}{2\pi} \int_{-\infty}^{\infty} e^{-i\omega t} e^{-Q^2 D|t|} \, dt$$

$$= \frac{1}{\pi} \int_0^\infty \cos \omega t \, e^{-Q^2 Dt} \, dt,$$

and the integration over $t$ leads directly to the final expression (A.8).

# Author Index

Abrahams, S. C., 3, 29, 210, 211, 224
Agraval, A. K., 142, 144
Aldred, B. K., 72, 76, 143, 144
Aleksandrov, V. B., 242, 247
Allen, G., 97, 110, 112, 117
Alperin, H. A., 285, 286, 290, 292, 294, 296
Anderson, J. S., 228, 229, 244, 247, 249
Anderson, O. L., 147, 171
Anderson, P. W., 287, 290, 294
Antal, J. J., 251, 268
Arndt, U. W., 34, 48
Ascarelli, P., 184, 198
Ashcroft, N. W., 185, 199
Averbach, B. L., 152, 171, 180, 199
Axmann, A., 166, 171

Bacon, G. E., 48, 216, 217, 224
Baenziger, N. C., 288, 247, 248
Baker, J. M., 239, 247
Ballhausen, C. J., 270, 271, 295
Barker, W., 247
Bedford, L. A. W., 38, 48
Belbeoch, B., 237, 247
Bell, R. J., 154, 157, 171
Berger, M., 117
Bevan, D. J. M., 228, 236, 247, 248
Birchenall, C. E., 248
Bird, N. F., 171
Bird, R. B., 199
Bleaney, B., 237, 248
Blech, I. A., 180, 199
Blume, M., 287, 294
Borie, B., 256, 268
Boutin, H., 117
Brand, J. C. D., 224
Brier, P. N., 110, 112, 117
Broadley, J. S., 204, 224
Brockhouse, B. N., 145
Brown, A. N., 59, 76
Brown, P. J., 281-3, 290, 294
Buhler, H. F., 187, 199
Bunce, L. J., 63, 76
Bursill, L. A., 229, 248

Caglioti, G., 181, 198
Carpenter, J. M., 179, 200
Carter, R. E., 236, 248

Castro, G., 72, 76
Chandler, W. T., 246, 249
Cheetham, A. K., 225, 232, 239-41, 244, 247, 248, 266, 268
Chernoplekov, M. A., 249
Chertkov, A. A., 249
Childs, P. E., 266, 267, 268, 269
Chudley, G. T., 126, 144
Chumbley, L. C., 48
Clark, C. D., 251, 258, 264, 265, 268
Clayton, G. T., 194, 199
Cochran, W., 93, 96, 268
Cocking, S. J., 131, 144, 154, 162, 164, 166, 171, 180, 199
Coffin, P., 290
Cohen, J. B., 243, 245, 248, 268
Collins, M. F., 259, 268
Cooper, M. J., 233, 235, 248
Corliss, L. M., 294
Costa, P., 248, 268
Cotton, F. A., 271, 295
Coulson, C. A., 205, 224, 295
Cowley, J. M., 251, 253, 268
Cox, D. E., 281, 290, 294
Cross, P. C., 50, 77
Cruickshank, D. W. J., 224
Cunliffe, A. V., 112, 117
Curry, N., 216, 217, 224
Curtiss, C. F., 199
Cyvin, S. J., 22, 26, 29, 84, 85, 95

Dahlborg, U., 136, 144, 164, 171
Danford, M. D., 187, 199
Danner, H. R., 105, 117
Dasannacharya, B. A., 133, 144, 200
Davies, E. R., 247
Davies, J. A., 248
Dean, P., 154, 157, 171
Decius, J. C., 50, 77
Desai, R. C., 142, 144
Dienes, G. J., 147, 171, 268
Donovan, J. L., 177
Dorner, B., 171
Douglas, R. W., 172
Dows, D. A., 72, 76
Duffill, C., 199
Dyer, R. F., 38, 48

303

# AUTHOR INDEX

Eden, R. C., 72, 77, 143, 144
Egashira, K., 293, 294
Egelstaff, P. A., 48, 54, 56, 76, 127, 129, 132, 137, 143-5, 162, 171, 173, 176, 181, 182, 185, 194-7, 199
Eick, H. A., 247
Elcombe, M. M., 170, 171
Ellinger, F. H., 248
Elliott, N., 294
Elliott, R. J., 126, 144
Ellis, B., 172
Ellis, D. E., 270, 294
Enderby, J. E., 145, 149, 171, 181, 187, 188, 193, 199
Eyring, L., 228, 247, 248

Feldkamp, L. A., 107, 108, 117
Fender, B. E. F., 243, 244, 246, 247, 248, 250, 253, 257, 259, 261, 263, 266, 268, 269, 287-94
Ferguson, C. A., 153, 171
Figgins, R., 139, 141, 144
Figgis, B. N., 270, 271, 295
Flubacher, P., 159, 171
Foote, F., 242, 248
Forsyth, J. B., 281, 282, 283, 290, 294
Fréchette, V. D., 172
Freeman, A. J., 281, 283, 290, 294, 295
Froidevaux, C., 246, 248
Fumi, F. G., 199

Garashina, L. S., 242, 247, 248
Gaskell, P. H., 147, 171
Gissler, W., 171
Gordon, R. G., 142, 144
Gray, H B., 270, 295,
Gray, P., 200
Greenwood, N. N., 244, 248, 249
Gruebel, R. W., 194, 199
Guggenheim, H. J., 290, 292
Guzzo, A. V., 294

Haas, M., 153, 171
Halder, N. C., 200
Harada, I., 87, 96
Hall, J. W., 38, 48
Hall, T. P. P., 241, 248, 293, 294, 297
Hamilton, W. C., 216, 218, 220, 223, 224
Haraldsen, H., 227, 248
Harris, D. H. C., 63, 76
Hartmann, H., 293-5
Hastings, J. M., 290, 294
Hayes, W., 294
Haywood, B. C., 54, 76
Heard, C. R. T., 180, 199
Helmholz, L., 292, 295

Henfrey, A. W., 246, 248, 259, 261, 263, 268
Henrich, V. E., 233, 248
Henshaw, G. D., 186, 199
Herzberg, G., 69, 77
Himmel, L., 243, 248
Hickman, B. S., 269
Hirakawa, K., 293-4
Hirschfelder, J. O., 184, 199
Hochstrasser, R. M., 72, 76
Holley, C. E., 247, 248
Holmryd, S., 140, 141, 144
Hopgood, F. R. A., 294
Howe, A. T., 244, 248
Van Hove, L., 2, 19, 30, 73, 77, 135, 145
Hrizo, J., 249
Huang, K., 232, 248, 256, 268
Hubbard, J., 270, 271, 279, 290-2, 294
Hughes, D. J., 29
Hurrell, J. P., 247
Hutchings, M. T., 290, 292, 294
Hutchinson, P., 199
Hvoslef, J., 212, 213, 244
Hyde, B. G., 227, 229, 248

Ibers, J. A., 216, 218, 220, 222, 223, 224
Ippolitov, E. G., 248
Isherwood, S. P., 187, 199

Jacobson, A. J., 276, 290, 292, 294
Jellinek, F., 225, 248
Jette, E. R., 242, 248
Johansson, N. G. E., 248
Johnson, M. D., 184, 199
Johnson, R. H., 145
Jones, D. A., 248
Jones, E. D., 293, 294
Jones, G. O., 146, 172
Jones, S., 59, 76
Jovic, D., 136, 144
Jude, R. J., 216

Kahn, D., 185, 249
Kaplow, R., 152, 171
Khan, A., 186, 199
Kihlborg, L., 228, 248
King, J. S., 105, 107, 108, 117
Kitaigorodskii, A. I., 93, 96
Kitigawa, T., 106, 117
Kittel, C., 102, 117
Kjems, J., 43, 48, 73, 76
Knox, K., 292
Koch, F., 243, 245, 248, 268
Koehler, W. C., 248
Koester, L., 299
Kofstad, P., 259, 268
Kollmar, A., 171
Kordis, J., 228, 247

# AUTHOR INDEX

Kost, M. E., 249

Von Laue, M., 252, 268
LaPlaca, S. J., 224
Larsson, K. E., 125, 126, 131, 132, 136, 138, 144, 145, 164, 171
Leadbetter, A. J., 146 et seq., 151-3, 156, 158, 166, 171
Lee, K., 293, 294
Leggeatt, A., 248
Lergh, R. S., 2, 29
Llewellyn, P. M., 248
Levy, H. A., 187, 199, 220, 224
Lingard, P. A., 282, 290, 294
Litchinsky, D., 156, 166, 171
Lomer, W. M., 30
Longster, G. F., 49, 55-7, 72, 76
Lorch, E. A., 151, 153, 171, 184, 199
Lorenzelli, R., 248, 268
Lovesey, S. W., 1, 30, 278, 295
Low, G. G., 30, 259, 268
Lye, R. G., 249

McClellan, A. L., 216, 224
McCrum, N. G., 117
Macdonald, A. L., 215, 224
McGraw, B. L., 222-4
Mackay, K. M., 246, 248
Mackenzie, J. D., 146, 172
McQuillan, A. D., 247, 248, 259, 268
Maklaskov, A. G., 248
Maradudin, A. A., 155, 171
March, N. H., 145, 199
Margoshes, M., 218, 224
Marshall, W. C., 1, 20, 271, 278, 279, 281, 282, 290, 294, 295
Martin, D. G., 251, 254, 261, 268, 269
Martin, R. L., 247
Matzke, M., 239, 248
Mehl, R. F., 248
Mika, K., 187, 189-92, 199
Mitchell, E. W. J., 251, 256, 258, 264, 268, 269
Miyazawa, T., 103, 106, 117
Morrison, J. A., 171
Morrison, J. D., 224
Mössbauer, R. L., 244
Moss, S. C., 251, 269
Mott, N. F., 172
Mulford, R. N. R., 248
Myers, W. R., 105, 117

Nakamoto, K., 218, 224
Narten, A. H., 194, 197-9
Nathans, R., 279-81, 290, 291, 294
Nelin, G., 140, 141, 144
de Novion, C. H., 245, 248, 263, 268
North, D. M., 181, 184, 187, 199, 200

Orton, B. R., 187, 199
Owen, J., 270, 271, 294, 295

Paalman, N. H., 179, 199
Page, D. I., 143, 144, 173, 182, 184, 186, 187, 189, 190-2, 197-9
Parks, T. C., 247
Paskin, A., 144
Pauling, L., 219
Pawley, G. S., 14, 22, 25, 26, 29, 73-6, 78, 84, 85, 93-6, 211, 213, 224
Pério, P., 247
Peterson, S. W., 220, 224
Petrunin, V. F., 249
Pickart, S. J., 294
Piekarski, C., 247
Pings, C. J., 179, 199
Pimentel, G. C., 216, 224
Placzek, G., 199
Plesser, T., 171
Pope, N. K., 145
Powell, M. J. D., 183, 200
Powles, J. G., 188, 135, 139, 141, 143, 144, 197-9
Prewitt, C. T., 237, 249
Prins, J. A., 172
Pryor, A. W., 269

Raccah, P. M., 233, 248
Rahman, A., 122, 128, 129, 133, 144, 171, 186
Rainey, V., 199
Randolph, P. D., 124, 131, 144, 154, 164, 166, 171
Rao, K. R., 133, 144, 180, 194, 200
Rawson, H., 146, 172
Read, B. E., 117
Reynolds, P. A., 51, 63-8, 73-6
Rice, S. A., 174, 200
Riley, F. D., 243, 244, 248
Rimmer, D. E., 289, 294
Rinaldi, R. P., 78, 96
Rinneberg, H., 293, 295
Roberts, L. E. J., 239, 248, 249
Robertson, J. M., 3, 39, 203, 210, 214-7, 224
Rooksby, H. P., 243, 249
Ros, P., 294
Rossier, D., 246, 248
Roth, W. L., 236, 243-5, 248, 249
Rouse, K. D., 235, 248
Rundle, R. E., 218, 224
Russell, M. C. B., 38, 48

Sabine, T. M., 269
Safford, G. H., 117
Sahni, V. C., 93, 96, 100
Sakamoto, M., 125, 126, 127, 145

305

# AUTHOR INDEX

Sanders, R. N., 294
Schaufele, R. F., 108, 117
Schiff, L. I., 4, 30
Schlemper, E. O., 224
Schottky, W., 225, 249
Schuldt, M. S., 247
Schwartz, R. B., 29
Sears, V. F., 55, 76
Shannon, R. D., 237, 249
Shearer, H. M. M., 224
Sherrer, J. R., 51, 63, 76
Shil'stein, S. S., 249
Shimanouchi, T., 87, 96, 103, 108, 117
Shulman, R. G., 270, 292, 295
Singwi, K. S., 154, 163, 164, 166, 171
Slaggie, E. L., 190, 199
Somenkov, V. A., 246, 247, 249
Speakman, J. C., 201, 215, 224
Springer, T., 171
Steeb, S., 187, 199
Steele, D., 236, 248, 249, 253, 257, 263, 266, 268, 269
Stevens, K. W. H., 294
Stevenson, R. W. H., 294
Stewart, R. J., 251, 256, 258, 264, 265, 268, 269
Stiller, H., 171
Stirling, G. C., 31, 63, 76
Stoicheff, B. P., 171
Strang, S. L., 171
Striffler, C. D., 179, 200
Stringfellow, M. W., 158, 171
Subbarao, E. C., 236, 249
Sugano, S., 270, 295
Summerfield, G. C., 105, 117
Sutter, P. H., 249
Szigeti, B., 29

Tasumi, M., 103, 117
Taylor, R. I., 248
Thornley, J. H. M., 270, 271, 295
Tien, T. V., 236, 249
Tiwari, 29
Tofield, B. C., 187-93, 295
Tosi, M. P., 199
Treloar, L. R. G., 117
Trevino, S., 117

Tsay, F. D., 293, 295
Twidell, J. W., 248

Venables, J. D., 246, 249
Venkataraman, G., 93, 96, 100, 107, 108, 117
Vineyard, G. H., 128, 138, 145, 180, 200

Wadsley, A. D., 249
Waeber, W. B., 27, 30
Wagner, C. N. J., 187, 199, 200, 225, 249
Wallace, W. E., 259, 268
Wallbank, A. D., 259, 268
Wallis, D. E., 48
Walters, R. J., 246, 247
Warren, B. E., 249, 254, 256, 269
Watson, R. E., 270, 281-3, 290, 295
Webb, F. J., 54, 76
Wedepohl, P. T., 251, 269
Wedgwood, F. A., 283, 294, 299
Weiss, R. J., 268
de Wette, F. W., 163, 171
White, J. G., 3, 29, 210, 224
White, J. W., 49, 51, 55-7, 60, 61, 63-8, 70-7, 143, 144
Wilkens, J., 294
Wilkinson, G. R., 72, 77
Will, G., 281, 290, 294
Williams, G., 177
Willis, B. T. M., 34, 48, 237, 243, 248, 249, 268
Wilson, E. B., 50, 77
Windsor, C. G., 1, 78, 96
Wingfield, B. R., 199
Woodward, I., 214, 224
Woolfson, M. M., 214, 224
Wright, A. C., 151-3, 166, 171
Wright, C. J., 60, 61, 77
Wycherley, K. E., 159, 171

Yeats, E., 14, 29, 211, 213, 224
Yip, S., 142, 144, 200

Zachariasen, W. H., 194, 248
Zbinden, R., 117
Zemlyanov, M. G., 249

# Subject Index

absolute magnetic intensities, 286
absorption corrections, 178, 179
accuracy in measuring Bragg intensity, 233
acoustic branch, 155
    modes, 26, 103
    phonons, 52
activation energy, diffusion of polymers, 116
adamantane, 96
admixture parameters, 271
alkaline-earth fluorides, 237, 239
alumina, 289
aluminium, 36
amorphous materials, 98
analyser crystal, 43
angular-velocity correlation function, 141
anharmonic effects, 86
anion-deficient fluorite compounds, 235
anion-excess fluorite compounds, 237
anomalous solid solutions, 225
antibonding orbital, 271
antisymmetric mode, 84
Apollo model, 196, 197
argon, 2, 127, 181, 182, 184, 185
atomic motions in glasses, 154
    liquids, 118$ff$
*Aufbau* principle, 274
'average unit cell', 229, 236

barium fluoride, 233
barn, 8
bending vibrations, 74
benzene, 139-41
beryllium, 36
    filter, 45, 46, 158, 258
    filter-detector spectrometer, 28, 29, 44, 47, 59
    fluoride, 146
    oxide, 35, 251
binary alloys, 186
    liquids, 186
bismuth filter, 258
black-and-white space groups, 88
Bloch's theorem, 24
Bloch wave functions, 27
body-centred cubic structure, 246, 260
Bohr magneton, 281

bonding orbital, 270
bonds (very short), 223
Born approximation, 5
boron, 40
Bose-Einstein population factors, 26
bound-atom scattering length, 7
Bragg diffraction, 107
    scattering, 183, 225
    in liquids, 136
Bragg's law, 174
Brillouin lines, 176
    scattering, 159
    zone, 24, 50, 82, 91, 102, 108
de Broglie wavelength, 122
bromobenzene, 139
Brownian motion, 131
Buckingham potential, 94

cadmium, 40
carbon tetrachloride, 195-7
carbonyl bending vibrations, 61
catalysts, 98
cell model of liquids, 118
cerium, 235
    hydride, 230, 247
chalcogenide glasses, 147
characteristic times and distances for atomic motion in liquids, 127, 128
curved-slot chopper, 45
chopper monochromator, 38
chromium sulphide, 227
close-neighbour size effect, 256
cluster, 254
    2:2:2, 237, 266
    2:4:2, 240, 266
clustering, 250
cobalt carbonyl hydride, 62
cobalt-iron alloy, 41
coherent and incoherent scattering, 7, 20
    cross-sections, 19, 177
    elastic scattering cross-section, 230
    inelastic neutron scattering, 107
    scattering, 1
        cross-section, 8
        from phonons, 26
    structure factor, 133
    translational motion in liquids, 133

# SUBJECT INDEX

cold neutrons, 37
communal entropy problem, 118
Compton modified scattering, 256
computer argon, 121, 128, 133
configurational entropy, 250
constant-Q method, 165
continuous flux reactors, 35
convolution approximation, 163
co-operative modes, 176
    motions, 135
co-ordination shells, 194
copper-gold, 254
copper-tin alloy, 187
corrections to neutron diffraction data, 178
correlation functions, 19
    of molecular orientation, 196
covalency, 3, 270
    in magnetic salts, 41
    parameters, 271, 282, 291
covalent form factor, 277
crystal-field theory, 270
C-type rare-earth oxide, 228
    $M_2O_3$ structure, 264
cuprous chloride, 120, 151, 187, 191

Debye-Scherrer cone, 12
Debye-Waller factor, 7, 24, 26, 52, 53, 69, 148, 130, 150, 231, 234, 278
defect cluster, 266
    complex, 237
    solid, 250
delta function, 11, 15, 27, 29, 300
delocalization of d-electron, 270
density of phonon states, 29
    of states spectrum, 29, 56, 73
    of liquids, 143
depolarized Rayleigh scattering, 143
detailed balance condition, 20, 23
detector efficiency, 177
detectors, 39
deuteration, 8, 107
deuterium substitution, 72
diamond, 207
DIDO reactor, 31, 33, 38, 46
dielectrical energy loss, 100
    measurements, 143
    relaxation, 109
difference synthesis, 205
differential scattering apparatus, 259
    cross-section, 5
diffractometer, 41
diffuse scattering, 246, 250
diffusion of molecules, 55
    equation for liquids, 128
    in liquids, 129, 138
    in polymer chains, 99
    in polymers, 116
di-interstitial complex, 265

direct unit cell, 10
dispersion curves, 82, 107
    of polymers, 103
    of polyethylene, 108
    of glasses, 160
distinct correlation function, 2, 21
divacancy complex, 265
double differential scattering cross-section, 123
    perovskite structure, 283
dynamic structure factor, 27, 162

eigen-frequencies of vibration, 49
eigen-state formation, 15
eigen-value, 80, 93
eigen-vector, 80, 93
eigen-vectors of lattice modes, 73
Einstein solid, 130
elastic cross-section, 2
    differential cross-section, 148
    diffuse scattering, 42
    scattering, 4, 174
electron affinity, 291
    nuclear double resonance, 239
    spin resonance, 237
electronic structure, 268
electrostatic model, 270
elpasolite, 284
energy analysis, 45
    selection, 35
    transfer, 32
enthalpy of defect formation, 250
    of mixing, 264
equipartition relation, 121
experimental techniques, 31
external dispersive modes, 25
    mode of vibration, 85
extinction, 13, 42, 234, 235

fermi, 209
ferromagnetic spin density, 282
ferrous oxide, 225, 267
Fick's diffusion equation, 302
    law, 120
fluorite structure, 228, 253
flux distribution, 35
force constant, 93
form factor, 234, 286
Fourier transorm, 6, 20, 174, 183, 301
free-atom scattering length, 7

gas ionization detector, 39
gas-like behaviour of liquids, 141
gaussian approximation, 69, 128
germania, 251
germanium, 36, 146, 251, 264, 265
    tetrabromide, 196, 197
glass transformation temperature, 146

308

glasses, structure and atomic motion, 146, 179
glass-to-rubber transition, 99
glycerol, 155
graphite, 35, 251, 261
group theoretical assignment of modes, 27, 88

hard-sphere potential, 119, 183, 184
harmonic approximation, 51, 154
  crystal, 122
  oscillator, 26
  solid, 130
heats of formation, 270
HERALD reactor, 31, 59, 258
hexachlorobenzene, 75
hexamethylene tetramine, 93, 96
hindered methyl rotation, 111
Huang diffuse scattering, 156
hydrogen, 127
  amplitudes in molecular crystals, 63, 69
  bonding, theory of, 61, 201, 204, 207, 223
hydrogenous compounds, 48
  material for shielding, 40

incoherent approximation, 155, 156, 165
  cross-section, 113
  inelastic neutron scattering spectroscopy, 50, 109
  scattering, 1, 27, 175
    amplitudes, 53
    cross-sections, 7, 8
    law, 113, 123, 125
    for glasses, 163
inelastic scattering, 31
  cross-section, 2
infra-red scattering, 22
  spectra, 56
  spectroscopy, 1, 2, 20, 49, 69, 100, 104, 143
instruments for elastic scattering, 41
  for inelastic scattering, 43
integral representation of delta function, 17
integrated intensity, 12
interatomic force constants, 49
internal non-dispersive modes, 25
  vibrational modes, 2
interstitial atoms, 225, 250, 252, 253, 259
ionized molten salts, 186
irreducible representation, 271
isothermal compressibility, 183
isotopic and spin-incoherent scattering, 256
  substitution, 143, 187, 194, 198
  substitution method, 69

jump diffusion, 139
  models for liquids, 126, 127, 129

Koch-Cohen cluster, 244, 267
Kronecker delta function, 186

Langevin equation for liquids, 121
lattice defects, 42, 225
  energetics, 267
  modes of polythene, 105, 108
  of polymers, 107
Laue monotonic scattering, 253
Lennard-Jones potential, 119, 121, 122, 129, 185, 186
librational modes, 80, 86, 90
lifetime of longitudinal phonons in liquids, 138
ligand form factor, 276
  forward peak, 288
light scattering, 135, 136
linac, 35
liquid aluminium, 136
  argon, 132, 138
  crystals, 3
  dynamics, 45
  hydrogen, 54
  hydrogen-deuterium moderator, 258
  lead, 137, 166
  methane, 142
  methanol, 71
  sodium, 124, 131, 132
localized vibrations in glasses, 155
longitudinal sound waves in glasses, 159
long-wavelength neutrons, 288

macroscopic diffusion, 140
magnesium oxide, 251
magnetically ordered solids, 278
magnetic form factor, 271, 275, 278
  interaction vector, 278
  scattering, 3, 267
  space group, 88
  structure factor, 278
magnetite, 244
magnetized mirrors, 41
manganese carbonate, 281
  difluoride, 279, 281
  oxide, 293
  sulphide, 293
Martin approximation, 266
Maxwellian distribution of neutrons, 34, 35
mechanical relaxation in polymers, 109
  wavelength selector, 258
mechanical velocity selectors, 35, 37
metal and ligand orbitals, 271
metal-ligand interactions, 270
metallic hydrides, 259
methane, 139, 201

## SUBJECT INDEX

methyl alcohol, 143, 144
  group re-orientation, 143
  rotation, 111
  torsional vibrations, 49, 109
Miller indices, 11
models for calculating phonons, 78
modified Buckingham potential, 185
  diffusion model, 138
molecular association, 201
  dynamics, 78
    calculations, 133, 186
    of polymers, 109
  energy levels, 45
  liquids, 50, 139, 193
  motion in liquids, 118
  orbital bonding, 271
    energy levels, 274
  point group, 271
  re-orientation in liquids, 139
  spectroscopy, 49
  translations in liquids, 139
  vibration frequencies, 49
molten cuprous chloride, 189
molybdenum oxide, 228
momentum transfer, 32, 52
monatomic liquids, 181
monochromation, 35, 46
monochromatic-beam method of structure analysis, 39
monochromators, 35
monotonic scattering, 252, 253, 260
Monte-Carlo calculations, 192
multi-channel analyser, 43
multi-cluster, 256
multi-phonon cross-section, 130
  scattering, 69
multiple-angle crystal spectrometer, 43, 45
multiple Bragg scattering, 234, 256
  scattering corrections, 180

naphthalene, 3, 12-14, 22, 25, 26, 28, 29, 41, 73 74, 78, 95, 107, 210, 211
Néel temperature, 267
neutron coherent scattering amplitude, 299
  covalency parameter, 292
  diffractometer, 14
  elastic scattering, 31
  energy, 4
  energy-gain spectrometer, 47
  flux, 34
  flux spectrum, 34
  polarization, 279
  rest mass, xiii
  scattering amplitude, 210
  scattering data, 296
  speed, 4, 32
nickel oxide, 285
niobium-deuterium system, 259

niobium hydride, 246, 247
non-stoichiometric compounds, 225, 256
  hydrides, 246
normal-beam equational geometry, 42
normal coordinates, 51
  mode of vibration, 80, 154
normalized auto-correlation function, 128
normalizing correction, 179
nuclear Bragg scattering, 9
  magnetic relaxation, 109
  resonance, 100, 135, 139, 141, 143, 246
  spin-lattice relaxation, 111
  structure factor, 278
null-matrix alloy, 179
nylon, 100

octahedral complex, 271, 273
one-eletron molecular-orbital description, 270
one-phonon coherent scattering cross-section, 24, 136
  incoherent scattering cross-section, 24
optical activity, 72
optic modes, 26
  of polymers, 103
  phonons, 52
optical selection rules, 49
  spectroscopy, 50, 53, 107
order contamination, 35
orientational correlation, 193
oriented polymers, 107
orthogonality condition, 271
overlap term, 275, 276

pair distribution function, 134, 150, 174
  potential, 173
para dichloro benzene, 63, 65, 74
  di-iodo benzene, 64, 66, 74
paramagnetic scattering, 256, 267, 288
partial structure factors, 187, 188, 191
pentane, 127
perfect-gas equation, 125
  model, 130
perovskite structure, 290
phonon annihilation, 26
  creation, 26
  density of states, 28, 52, 64, 72, 130
  dispersion, 2
  curves, 76, 50
    in glasses, 170
    in liquids, 135, 137
  frequencies, 79
  frequency distribution, 154, 155
  lifetime in glasses, 164
  modes, character of, 154
  spectrum of liquids, 131

# SUBJECT INDEX

phonons in glasses, 164
  in molecular crystals, 78
  in polymers, 107,
photomultiplier tube, 40
Placzek correction, 179-81, 197
  expansion, 181
plastic crystal, 3, 96
PLUTO reactor, 31, 33, 43
polarization analysis, 107
  efficiency, 41
  factor, 23
  ratio, 279
  vector, 2, 26
polarized beam method, 279
  neutron experiments, 278
  neutrons, 41
polycrystalline aluminium, 136
  filters, 36
polydimethyl siloxanes, 112, 114, 115, 116
polyethylene, 97, 98, 104, 107, 108
polymers, 72, 97, 155
polymethylene oxide, 97, 98
polymethyl methacrylate (Perspex), 98, 111
polypropylene, 98, 99, 109
  oxide, 97, 110
  oxide-$d_3$, 110
position-sensitive detector, 43
potential function, 107
powder diffractometer, 41
  scattering, 12
praseodymium, 235
  oxide, 228
preferred orientation, 235
proportional counters, 39, 40
proton acceptor, 202
  channelling, 239
  donor, 202
pulsed fast reactors, 35
  neutron sources, 39
pyrites structure, 291
pyrolytic graphite, 36

quartz glass, 162, 170
quasi-crystalline model of glass, 151
quasi-elastic scattering, 55, 100, 112, 113, 115, 124, 125, 140, 141, 143, 148

radial distribution function, 183, 192, 256
radiation-damaged solids, 250, 251, 268
Raman scattering, 51, 86
  spectroscopy, 26, 31, 49, 69, 100, 104, 143
randomly packed spheres, 185
random point defects, 250
  solid solutions, 252
  vacancies, 252, 253
rare-earth fluorides, 237

reciprocal cell, 10
re-orientation correlation time, 141
re-orientational diffusion, 141, 142
resonance covalency parameter, 292
  method, 270, 292
  modes in glasses, 170
    in vitreous germania, 158
Reststrahl frequency, 59
rigid-body model, 15
rock-salt structure, 207, 261
rotating-crystal spectrometer, 47
rotational angular momentum, 54
  diffusional constant, 140
  energy, 54
  motion in polymers, 112, 116
rubber, 99
rubidium, 184, 185
rutile structure, 290

scattered wave vector, 32
scattering amplitude, 209
  from single nucleus, 6
  from many nuclei, 7
  function, 174
  law, 180
    for liquids, 162
  length, 6
  potential, 6
  theory for elastic scattering, 4
    for inelastic scattering, 15
  vector, 6, 251, 275
segmental diffusion in polymers, 112
selection rules in infra-red spectroscopy, 55
  in Raman spectroscopy, 55
self correlation function, 2, 12
  diffusion coefficient, 128, 143
  structure factor, 133
shear plane, 229
  structure, 228
shielding, 40
short-range ordering, 250, 253, 263
Shubnikov space group, 88
side-group motion in polymers, 109
silica, 146, 251
silicon, 146, 251
simple diffusion, 124
  liquids, 181, 185
single-crystal diffractometer, 41
  scattering, 14
single-particle modes, 176
sodium, 127, 132, 184
  chloride structure, 242, 270
  fluoride, 223
  silicate glasses, 147
soft modes in glasses, 166
solid hydrogen, 53
sound velocity, 176
  in glasses, 169

311

## SUBJECT INDEX

sound-wave modes, 175
space-time correlation function, 2, 154
    representation, 18
spherically symmetric potential, 183
spin density distribution, 279
    echo n.m.r., 116
    polarization, 292
spin-transfer coefficient, 274
spin-wave mode, 3
    theory, 287
static approximation, 149, 177, 178
steric effect, 93
stress relaxation time, 146
stretching mode of polythene, 104
structure factor, 11, 175, 190, 195, 196
    of glasses, 169
    of glasses, 148
    of liquids, 173
    of non-stoichiometric compounds, 225
substitutional defect, 225
$\beta$-succinic acid, 203, 204
sulphur, 78
symmetric mode of vibration, 82

terbium, 235
tetrachloro benzene, 63, 67
tetra-methyl ammonium bromide, 59
thermal diffuse scattering, 235, 256
    displacement, 232
    flux, 33
time-correlation function, 128
time-dependent formulation, 16
time-of-flight, 35, 37, 51, 166
    diffraction, 38
    powder, 13
    spectrometer, 38, 46, 63, 113
    spectrum, 56
    techniques, 39, 157
torsional energy levels, 111
    modes of polythene, 104
    of polymers, 112
    oscillations in liquids, 141
    quantum number, 111
    vibrations, 59, 74, 75
total scattering, 149
    cross-section, 8, 255
tracer methods for diffusion, 116
transition elements, 259, 270
transition-metal ions, 267
    oxides, 270
translation of polymer chains, 99
translational diffusion, 142
    modes, 88, 130
    motion in liquids, 121, 126, 127
    in polymers, 116

transverse phonons in liquids, 138
    in glasses, 159
triple-axis spectrometer, 36, 43-6, 105, 150, 166
twin-chopper spectrometer, 43, 45

units, xiii, 31
unpaired spin-density distributions, 275
unpolarized neutron experiments, 285
uranium oxide, 230, 237
Urey-Bradley force field, 51, 63

vacancy, 225
vacancy-interstitial pair, 255
vacant sites, 250
vanadium, 179, 181
    carbide, 245, 262
velocity auto-correlation function, 128, 131
    selector, 46
vibrational frequencies of polymers, 101
    modes of liquids, 131
    of polymers, 109
    motion of liquids, 129
    spectroscopy, 55
    states of molecular crystals, 21
virial coefficients, 119
viscosity, 146
    of glasses, 146
vitreous germania, 153
van der Waals radius, 203
van Hove correlation functions, 20, 174
    singularity, 73

water, 3, 125, 126, 127, 141, 197, 201
wavelength flux, 34
    resolution, 43
    selection, 35
white-beam method of structure analysis, 39
wustite, 242

X-ray atomic form factor, 152
    beam, 34
    diffraction, 31
    diffuse scattering, 251
    scattering, 20
X-rays, 2

Young's modulus of polymer, 108

zero-order approximation, 118, 119
zero-point energy, 223
    spin deviation, 287, 290, 291
zinc, 182
zinc-blende structure, 270

QC
793.5
T428
C46

OCT 8 1974